Design of Earthquake Resistant Structures

CONTRIBUTORS

E. Rosenblueth,	*Professor of Engineering, National University of Mexico*	(Chapter 1)
A. K. Chopra, and	*Professor of Civil Engineering, University of California, Berkley*	
N. M. Newmark,	*Professor of Civil Engineering and in Center for Advanced Study, Emeritus, University of Illinois at Urbana-Champaign*	(Chapter 2)
L. Esteva,	*Professor, Graduate School of Engineering and Research Professor, Institute of Engineering, National University of Mexico*	(Chapter 3)
O. de Buen,	*Professor and Head, Division of Civil Engineering, School of Engineering, National University of Mexico*	(Chapter 4)
R. Park, and	*Professor and Head, Department of Civil Engineering, University of Canterbury, New Zealand*	
T. Paulay,	*Professor of Civil Engineering, University of Canterbury, New Zealand*	(Chapter 5)
M. J. N. Priestley,	*Reader in Civil Engineering, University of Canterbury, New Zealand*	(Chapter 6)
R. V. Whitman, and	*Professor of Civil Engineering, Massachusetts Institute of Technology*	
J. Bielak,	*Associate Professor of Civil Engineering, Carnegie-Mellon University, Pennsylvania*	(Chapter 7)
A. Goldberg, and	*Consulting Engineer, Codes and Standards, President, Goldberg Research & Development Assoc. Corp.*	
E. A. Rukos,	*Research Professor, Instituto de Ingeniería, National University of Mexico*	(Chapter 8)

DESIGN OF EARTHQUAKE RESISTANT STRUCTURES

Edited by

Emilio Rosenblueth,
Professor of Engineering, National University of Mexico

PENTECH PRESS
London : Plymouth

First published 1980
by Pentech Press Limited
Estover Road, Plymouth,
Devon.

ISBN 0 7273 0402 X

British Library Cataloguing in Publication Data
Design of earthquake resistant structures.
1. Earthquakes and building
I. Rosenblueth, Emilio
624'.176 TA658.44
ISBN 0-7273-0402-X

Set by Mid-County Press
Merivale Road, London SW15
Printed and bound in Great Britain by
Redwood Burn Limited, Trowbridge & Esher

Contents

1 CHARACTERISTICS OF EARTHQUAKES 1
- 1.1 Causes of earthquakes 1
- 1.2 Tectonic generating mechanisms 2
- 1.3 Seismic waves 2
- 1.4 Accelerograms 2
- 1.5 Magnitude, moment and intensity 4
- 1.6 Properties of linear, single-degree systems 5
- 1.7 Kinds of spectra 6
- 1.8 Spectra on firm ground 7
- 1.9 Local effects 13
- 1.10 Ground-motion components 16
- 1.11 Earthquake simulation 17
- 1.12 Nonvibratory seismic effects 19
- References 23

2 ANALYSIS 27
- 2.1 Scope 27
- 2.2 Analysis procedures: overview 27
- 2.3 Design response spectra 29
- 2.4 Building properties and allowable ductility for analysis 33
- 2.5 Modal analysis procedure 36
- 2.6 Equivalent lateral force procedure 41
- 2.7 Special factors 45
- 2.8 Limitations of equivalent lateral force and modal analysis procedures 49
- 2.9 Choice between equivalent lateral force and modal analysis procedures 50
- References 52

3 DESIGN: GENERAL 54
- 3.1 Nature and objectives of earthquake resistant design 54
- 3.2 Structural response and control variables 55
- 3.3 Design principles 70
- 3.4 Safety criteria 80
- References 97

4 STEEL STRUCTURES 100
- 4.1 Introduction 100
- 4.2 Behaviour of steel structures under strong earthquakes 100
- 4.3 Moment resistant frames 101
- 4.4 Behaviour and design of structural members 102
- 4.5 Design of fully rigid frames 130
- 4.6 Drift control 131
- 4.7 Bracing 132
- 4.8 Diaphragms 134
- 4.9 Other types of structure 136
- 4.10 Brittle fracture and lamellar tearing 136
- References 138

5 CONCRETE STRUCTURES 142
5.1 Introduction 142
5.2 Strength and ductility provisions 142
5.3 Stress–strain behaviour of concrete and steel 144
5.4 Reinforced concrete members 148
5.5 Reinforced concrete building frames 164
5.6 Reinforced concrete shear walls in buildings 173
5.7 Reinforced concrete bridge piers 183
5.8 Prestressed concrete building frames 184
References 193

6 MASONRY 195
6.1 Introduction 195
6.2 Methods of construction 197
6.3 Behaviour of constituent materials 198
6.4 Behaviour and design of masonry shear walls 203
6.5 Masonry infilled frames 215
6.6 Masonry columns 221
References 221

7 FOUNDATIONS
7.1 General considerations 223
7.2 Formulation of the problem of dynamic soil-structure interaction 224
7.3 Kinematic interaction 229
7.4 Inertial interaction 235
7.5 Pile foundations 241
7.6 Comparison of techniques for analysis 243
7.7 Site stability 247
7.8 Design considerations 255
References 257

8 NONSTRUCTURAL ELEMENTS 261
8.1 Types of nonstructural elements 261
8.2 Distinction of element importance 262
8.3 Rigid nonstructural elements 265
8.4 Flexible nonstructural elements 267
8.5 General design criteria 272
8.6 Element interaction 280
8.7 Connections, anchorages and details 285
References 288

Index 290

Preface

Design of Earthquake Resistant Structures is written primarily for structural engineers, present and future. It is intended to fill a gap: there are excellent books covering the observational, experimental, and theoretical backgrounds or giving the specialist the tools for deciding on design criteria applicable when he deals with a major job or participates in code writing; these books do not tell how to design. At the other end of the spectrum there are building codes that specify criteria to be met; they too do not tell how to design. The gap is that of a guide to designers providing sufficient background so that bases for recommendations can be understood and so that the engineer will even be in the position of knowing when to ignore the recommendations and why.

Recent trends in earthquake resistant design reflect a better understanding of the phenomena at play and the widespreading availability of computers and computer programs. As a consequence there is growing use of dynamic analysis coupled with approximate methods for realistically modelling the disturbances and structural behavior based on reliable field and laboratory tests. The most sophisticated analyses now include consideration of inelastic behavior, continuous media treated by variational methods, often combined with finite-element techniques, and simultaneous consideration of several ground-motion components. Results of accurate analyses give rise to increasingly accurate methods of equivalent static analysis that devote explicit consideration to the most significant parameters and variables, in particular ductile behavior. In turn these methods have served to develop extremely simplified procedures applicable to the vast majority of low-cost dwellings.

In a rapidly evolving discipline every book is born obsolete. It is less obsolete if written by a team, with one or two authors per chapter, so that it can be produced in a short time, and if it emphasizes the less controversial, more durable issues. The present work was done in this way. And although not pretending to furnish a guide to the design of every conceivable structure, it attempts to be sufficiently general to be of wide interest. As in designing a curriculum: one cannot pretend to be universal but should descend to the level of minute details in some specialized fields so that the alert student will be able to extrapolate to problems not covered in school.

Bibliographic material is listed in the references of each chapter so as to cover the more specialized background, both empirical and theoretical, without encumbering the text. Those interested in delving into a subject are encouraged to consult these works.

The level at which the text has been written assumes ordinary undergraduate familiarity with statics, dynamics, structural behavior, matrix methods of analysis, and design. In addition, some familiarity with elementary concepts of theory of probabilities, such as expectation, standard deviation, and conditional probability, is taken for granted in

Chapters 1 and 3, while Chapter 1 and the one on foundations presuppose superficial knowledge of soil mechanics.

The book begins by giving an overall view of seismic phenomena and seismicity, including a description of focal mechanisms, earth waves, the various components of ground motion, correlations among the components, and simulation. Attention is also given to nonvibratory effects of earthquakes. The book turns into structural analysis for dynamic disturbances going directly into simplifications applicable when the disturbances are seismic. The rest of the book is devoted to design proper. The first chapter on design embraces matters common to all types of materials and structural solutions. Then specific chapters refer to steel, concrete, masonry, foundations, and nonstructural components. The Appendix contains an abridged version of the Modified Mercalli intensity scale.

It is the authors' intent that this work be used both as textbook in advanced undergraduate or elementary graduate courses and as reference for the practicing structural engineer. In both functions, Chapters 2 and 3 are crucial for they concern analysis and design of common structures in general. Having mastered these one can turn to the specific problems of interest in the last five chapters.

Emilio Rosenblueth

1 Characteristics of earthquakes

Emilio Rosenblueth

1.1 CAUSES OF EARTHQUAKES

Many phenomena make the earth tremble: natural and man-made explosions, meteoritic impacts, cave-roof collapses, volcanic activity, tectonic movements. The latter by far have the greatest interest in engineering. Meteoritic impacts contribute insignificantly to world seismicity in the usual range of return periods. At return periods of 10^4 yr their contribution is, however, of the same order as that of tectonic earthquakes[1.1]. Other earthquakes have local importance but worldwide they are hardly worth considering.

1.2 TECTONIC GENERATING MECHANISMS

Most if not all tectonic earthquakes originate in relative displacements at geologic faults. The main faults are plate boundaries, for the crust consists of several plates that move relative to each other. Some intraplate faults also exhibit relative motion that causes significant tremors.

Forces that move the plates are the subject of speculation. One plausible explanation[1.2] theorizes that our planet was formed by aggregation of meteorites. Increase in mass has increased radioactivity. The planet is consequently getting warmer and its nucleus grows at the expense of mantle melting. Molten materials have diverse compositions, and hence densities, because of their different origins. The heavier materials fall toward the earth's centre and form the solid, inner nucleus. The lighter ones migrate and reach the surface as lava. This pushes the plates in the same directions as the currents set up by differences in density of the various materials as they melt. Temperature differentials in a homogeneous earth would be insufficient to explain the phenomena observed.

The theory that earthquakes occur when friction is overcome at faults has been amply confirmed (even though we cannot rule out other mechanisms as generators of some tectonic earthquakes[1.3]). It began to be formalized in Reid's theory[1.4] of elastic rebound. The crust is viewed as subjected to stresses associated with shearing strains. When the strength at a fault is overcome, the crust tends to recover its undeformed shape, and this rebound gives rise to an earthquake. The phenomenon implies a stick-slip process: the static frictional resistance exceeds that under dynamic conditions. Earthquakes are therefore associated with a stress drop. Owing to heterogeneities in stress and resistance, motion must begin at a point and propagate. This takes place at a speed comparable to that of shear waves (see below). When static frictional resistance does not exceed the dynamic value, creep, rather than earthquakes, is the result.

1.3 SEISMIC WAVES

Focal disturbances emit body waves. When these reach the ground surface they originate surface waves, which travel parallel to the surface and whose amplitude tends to zero as the depth increases. Body waves are of two kinds: P (also called primary, longitudinal, compressional, or dilatational) and S (secondary, transverse, shear, or equivoluminal). There are many kinds of surface waves. Of greatest interest in earthquake engineering are L (Love) and R (Rayleigh) waves. L waves occur only in stratified media. Particle motion in L waves is parallel to the free surface and perpendicular to the direction of wave travel. There is therefore correlation between the rotational component of motion about a vertical axis and translational horizontal components. In R waves of given frequency, particle motion is elliptical; it takes place in planes perpendicular to the free surface. There is, thus, correlation between the vertical and the horizontal components.

Velocities of P and S waves in elastic materials are frequency independent. They are, respectively

$$v_p = \left[\frac{(1 - \nu E)}{(1 - \nu - 2\nu^2)\rho} \right]^{1/2} \tag{1.1}$$

$$v_s = \left(\frac{G}{\rho} \right)^{1/2} \tag{1.2}$$

where E = modulus of elasticity, $G = E/2(1+\nu)$ = modulus of rigidity, ρ = mass density, and ν = Poisson's ratio. In all materials $v_s < v_p$. The velocity v_r of R waves is slightly lower than that of S waves. The ratio v_r/v_s depends on ν, it is frequency dependent except in an elastic halfspace[1.5,1.6]. If $\nu = 0.25$, then $v_p/v_s = \sqrt{3}$ and, in an elastic halfspace, $v_r/v_s = 0.92$. Velocities v_l of L waves are always frequency dependent. In a stratum overlying a bedrock halfspace, v_l varies monotonically between v_s for rock when the wave frequency is zero and v_s for the soil as the frequency tends to infinity.

All wave velocities are frequency dependent in viscoelastic media.

1.4 ACCELEROGRAMS

From the outset, wave trains are irregular owing to complexities in focal mechanisms. As waves traverse geologic formations they become more irregular because of multiple reflections and refractions at interfaces, and

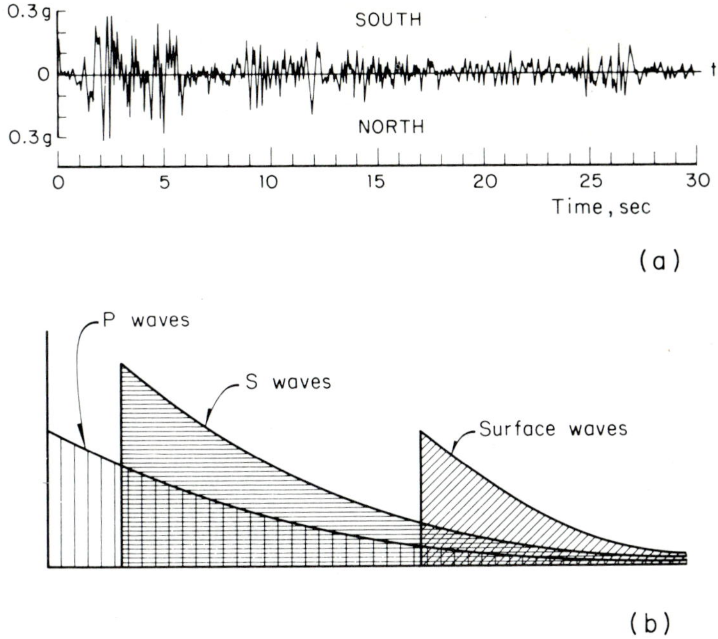

Fig. 1.1 *Typical earthquake accelerogram. (a) NS component of* 1940 *earthquake recorded at El Centro, CA. (b) Types of waves in record*

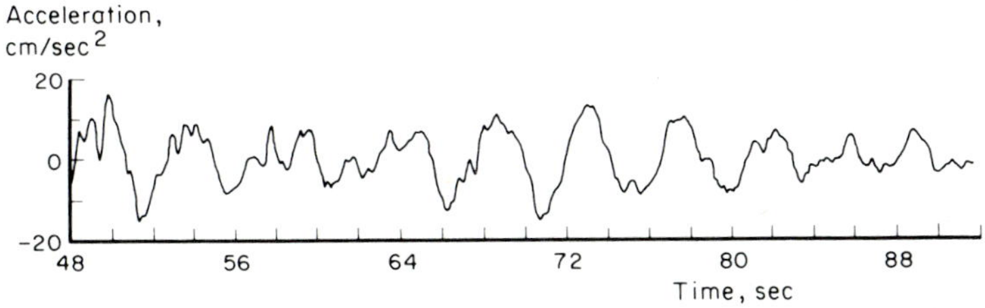

Fig. 1.2 *Segment of accelerogram of NE component of* 28 *August* 1973 *earthquake near Sports Palace, Mexico City (after E. Faccioli)*

they are further complicated by dispersions due to frequency dependence of wave speeds. As a result, earthquake accelerograms on firm ground are extremely irregular (Figure 1.1a). At the surface of soft strata, filtering of waves and their multiple reflections at the rock-soil interface and at the free surface make the records resemble harmonic motion (Figure 1.2).

P waves are first to arrive, followed by S waves and finally surface waves. A record may be viewed as having three segments (Figure 1.1b). The first contains P waves only; the second contains P and S; and the third has some P and mostly S, L, and R waves. Prevailing periods increase systematically along an accelerogram.

Optical and mechanical recording accelerographs require a starting device lest they consume inordinate amounts of power and paper. These instruments lose the initial portion of every earthquake. Interpretation of their records is beset with errors of many other sources. Computed structural responses for very high and very low frequencies can be in error by more than 100% and the same is true of ground displacements inferred from accelerograms. In the intermediate range of periods, say between 0.5 and 3 s, the mean spectral ordinates are more trustworthy, with coefficients of variation of perhaps 0.2 due to these sources of error.

Errors of this nature are practically eliminated by recording on magnetic tape using frequency modulation and resorting to a memory device so as to record the beginning of every earthquake.

1.5 MAGNITUDE, MOMENT, AND INTENSITY

Richter magnitude M (or M_s) is computed from the surface-wave portion of seismograms. It is an indirect measure of the energy released[1.7]

$$\log_{10} W \simeq 11.4 + 1.5M \tag{1.3}$$

where the energy W is in ergs. The expression is not exact since energies carried by body and surface waves are not in a constant ratio. The magnitude m (or M_b) inferred from the body-wave portion of seismograms is less sensitive than M to focal depth. On the average the two measures are related through

$$M = 1.59m - 3.97 \tag{1.4}$$

Seismic moment[1.8] also measures the size of an earthquake but is directly related to slip, stress drop, and area of fault slip. There is not a one-to-one relation between seismic moment and magnitude (Figure 1.3).

Intensity measures the local potential destructiveness of an earthquake. Subjective scales have been necessary wherever no strong-motion instruments have been operating. This means all sites for all earthquakes before the 1930s, and most sites for most earthquakes at present. The most commonly used subjective scale in the West is the modified Mercalli (MM) (see Appendix 1.1). The MSK scale, used in socialist countries, is similar but includes description of geomorphologic effects[1.10].

Instrumental scales are intended to overcome wide individual variations in assessment of earthquake destructiveness. The best known instrumental scales include *Housner's* and *Arias'*. The former[1.11] defines intensity as the area under the pseudovelocity spectrum in ft/s between periods of 0.1 and 2.5 s, often for a damping ratio of 0.2, and averaged in two orthogonal

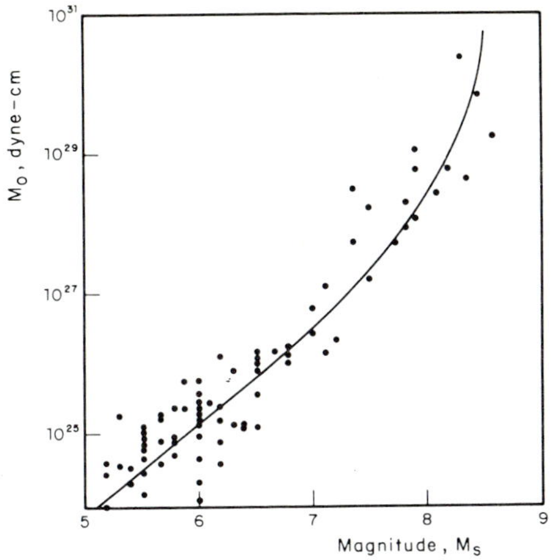

Fig. 1.3 *Relation between magnitude and seismic moment*[1.9]

horizontal directions. The latter scale[1.12] is $\pi/2g$ times the integral of the squared ground acceleration, where g = acceleration of gravity. Usefulness of instrumental scales is limited since instrumental intensity can be computed only when one has ground-motion records and can infer effects on any given structure.

Measures of earthquake parameters are less precise than is normally recognized. Until recently, focal coordinates could only be estimated within several tens of kilometers. Most maps containing information on these coordinates are rough. Recomputation has shown strong correlation with tectonic features which was not apparent in the original charts.

Following a strong earthquake, different stations report magnitudes often differing more than one degree from each other. Within 48 hr these values are unified to within 0.2 degrees, giving the impression of great precision. The spread in intensities assessed by different observers is even greater.

1.6 PROPERTIES OF LINEAR, SINGLE-DEGREE SYSTEMS

Figure 1.4 depicts a structure that behaves linearly and has a single degree of freedom. According to D'Alembert's principle

$$M\ddot{x} + C\dot{y} + Ky = -P \tag{1.5}$$

in which M = mass, C = dashpot constant, K = stiffness (spring constant), $P = P(t)$ = applied force, t = time, x = absolute displacement, $y = x - x_0$ = deformation (relative displacement), x_0 = ground displacement, and dots denote time derivatives. A structure like this, in which damping force $-C\dot{y}$ is

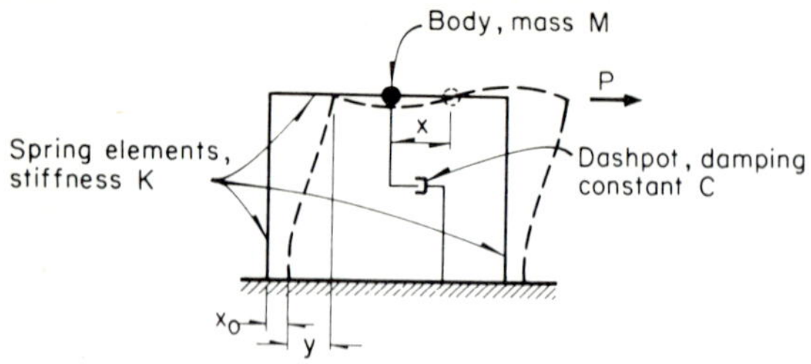

Fig. 1.4 *Single-degree system*

proportional to deformation velocity, is said to have *linear* or *viscous damping.*

In free motion, P and x_0 equal zero, so $y=x$. Substituting into Equation 1.5 and dividing through by M we get

$$\ddot{y}+2\zeta\omega y+\omega^2 y=0 \tag{1.6}$$

where $\omega=(K/M)^{1/2}$ = undamped natural circular frequency (rad/s), $\zeta=C/C_{cr}$ = damping ratio, and $C_{cr}=2\sqrt{KM}$ = critical damping.

When $\zeta<1$ the structure is said to be *underdamped*. The general solution of Equation 1.6 is then

$$y=y_1 e^{-\zeta\omega(t-t_1)} \sin \omega'(t-t_1) \tag{1.7}$$

where y_1 and t_1 are constant and $\omega'=\omega\sqrt{1-\zeta^2}$ = *(damped) natural circular frequency*. The structure describes damped harmonic oscillations of frequency $f'=2\pi\omega'$ = *(damped) natural frequency* (cps or Hertz) and period $T'=f'^{-1}$ = *(damped) natural period*. The undamped natural frequency and period, corresponding to the simple harmonic oscillations that solve Equation 1.7 when $\zeta=0$, are respectively $f=2\pi\omega$ and $T=f^{-1}$.

When $\zeta\geq 1$, ω' is imaginary and the structure does not oscillate but tends asympotically to its undeformed state. If $\zeta>1$ it is said to be *supercritically damped*. In the borderline case $\zeta=1$ it is said to have *critical damping.*

1.7 KINDS OF SPECTRA

A *spectrum* is a representation of a function in the frequency domain. Types of spectra common in earthquake engineering are:

(1) Fourier spectrum: $F(\omega)=\int_0^s e^{i\omega t} x_0(t)\mathrm{d}t$; x_0 = ground motion, ω = circular frequency, $i^2=-1$, t = time, s = duration of motion; $F(\omega)$ is a complex function that contains information about amplitude and phase.

(2) Fourier amplitude spectrum: $|F(\omega)|=\{[\int_0^s x_0(t) \sin \omega t\ \mathrm{d}t]^2+[\int_0^s x_0(t) \cos \omega t\ \mathrm{d}t]^2\}^{1/2}$.

(3) Response spectra: $R=\max_t|r(t)|$; $r=r(t)$ = structural response. Usually

the structure in which r is a response is a linear single-degree system with natural circular frequency ω and damping ratio ζ.

The response can be computed from

$$r(t)=\int_0^t \ddot{x}_0(\tau)\psi(t-\tau)\mathrm{d}t \tag{1.8}$$

where ψ is the *impulse response function* for the type of response in question. If r stands for deformation, $\psi(t)=-\omega'^{-1}\exp(-\zeta\omega t)\sin\omega' t$.

Usually R is plotted in terms of the system's natural frequency or period, one curve for every value of ζ chosen. Common types of response spectra are those for (absolute) acceleration, in which $r=\ddot{x}$; pseudoacceleration: $r=\omega^2 y$; pseudovelocity (sometimes called velocity spectrum): $r=\omega y$; and deformation (sometimes called displacement spectrum): $r=y$.

Often the logarithm of pseudovelocity spectral ordinates is plotted versus the logarithm of period or of frequency. Ascending and descending 45° lines correspond then to the logarithm of pseudoacceleration and deformation spectral ordinates. The three kinds of spectrum are thus represented on a single plot.

Since response spectra directly furnish the maximum numerical values of responses of linear, single-degree systems they are the bases for analyzing and designing structures idealized as systems of this nature. Each natural mode of vibration of a linear multidegree system behaves as a single-degree system. Hence response spectra also provide the modal responses of such structures. By applying certain rules to combine modal responses it is possible to estimate the overall structural responses, and even to get an idea of the responses of nonlinear systems (Chapter 2).

1.8 SPECTRA ON FIRM GROUND

1.8.1. General features

Typical response spectra for horizontal motion of firm ground are shown in Figure 1.5. Also shown are maximum ground acceleration, velocity, and displacement. As the period tends to zero, spectral acceleration and pseudoacceleration tend to the maximum ground acceleration, for a structure having zero natural period is infinitely rigid or has no mass and hence it undergoes the same motion as the ground. As the period tends to infinity, spectral deformation tends to the maximum ground displacement, since the structure approaches the condition of a mass standing in mid-air. In the middle portion the expected spectral pseudovelocity, for given maximum ground velocity, is roughly independent of period.

1.8.2 Ground-motion parameters

Parameters of ground motion can be estimated from the following relations, which furnish expected values at a station in terms of focal characteristics and

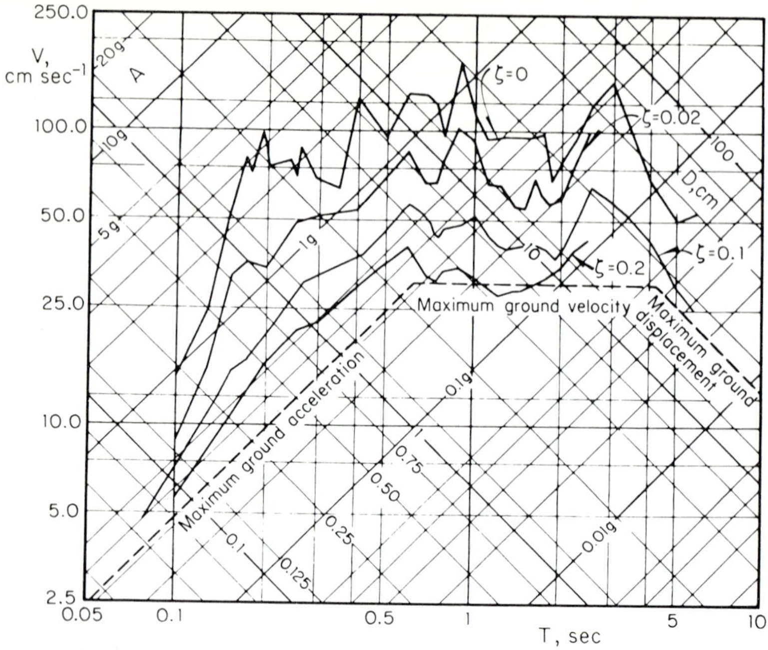

Fig. 1.5 Spectra of NS component of 1940 earthquake recorded at El Centro, CA[1.13]

distance. The relations have been derived for western United States and can be expected to yield reasonable values in geologically similar areas. Attenuation with distance is appreciably less pronounced near the Peruvian coast[1.14], in eastern United States, and in continental shields generally.

In all cases probability distributions for given magnitude and focal distance are approximately lognormal.

When only magnitude and focus coordinates are known, the maximum ground acceleration a, velocity v, and displacement d in an arbitrary horizontal direction are often computed from expressions of the form

$$E(\ln Y) = \ln b_1 + b_2 M - b_3 \ln L' \tag{1.9}$$

in which $E(\cdot)$ = expectation of variable $(\cdot)$; Y stands for a, v, or d; $b_{1,2,3}$ are constants; and L' equals focal distance L plus some quantity. Pertinent values proposed by different authors appear in Table 1.1 in units such that a, v, and d turn out in cm/s^2, cm/s, and cm respectively when L is in km.

Values reported by McGuire are based on 68 records obtained in the western coast of the United States. Those by Esteva and Villaverde consider also some records in Mexico. And the ones by Donovan are based on 678 records having $a > 1$ cm/s^2. Special care was taken in McGuire's study to eliminate data that would reflect correlation between L and site conditions.

TABLE 1.1 Parameters in Equation 1.9

Y	b_1	b_2	b_3	$L' - L$	$COV^{(1)}$	*Source reference*
a	467	0.640	1.301	25	0.548	1.15
a	1 059	0.500	1.320	25	0.805	1.16
a	5 829	0.800	2.000	40	0.711	1.17
v	5.64	0.923	1.202	25	0.696	1.15
v	36	1.000	1.700	25	0.838	1.17
d	0.393	1.000	0.885	25	0.883	1.15
$\bar{A}^{(2)}$	69 600	0.800	2.000	70	0.869	1.17
$V_1^{(3)}$	0.155	0.880	0.587	25	0.699	1.15
$\bar{V}^{(4)}$	265	1.000	1.700	60	0.708	1.17

(1) Coefficient of variation of Y
(2) Maximum smoothed-spectrum acceleration for zero damping
(3) Smoothed-spectrum pseudovelocity at 1 Hz and 2% damping ratio
(4) Maximum smoothed-spectrum pseudovelocity for zero damping

Results obtained by the different criteria differ widely, especially for small and large L(Figure 1.6).

Better correlations are possible in terms of magnitude and the coordinates of the 'centre of energy release' (centre of area of fault slip), or, if the focal depth is roughly constant, in terms of magnitude and the coordinates of the intensity epicentre. Application of these criteria ordinarily requires, however, knowledge about the length and direction of the area of fault slip and therefore permits application of a much more accurate simulation procedure based on a physical conceptual model of the earthquake[1.18]

1.8.3. Expected spectral ordinates

Given the maximum ground accelerations a, velocity v, and displacement d, the following rules[1.19] permit estimating the expected spectral ordinates as functions of period and damping ratio. (In these very simplified rules d is taken proportional to a. For $a=g$, d is taken as 36 in = 91 cm.) Pseudovelocities at the frequencies identified as those of control points in Figure 1.7 are found by multiplying the corresponding ordinates of the a, v, d lines by the factors in Table 1.2.

Let $\bar{V}(\zeta)$ = expected spectral pseudovelocity for a damping ratio ζ. According to Esteva and Villaverde and, more noticeably so, according to McGuire the ratio $\bar{V}(\zeta)/v$ or V_1/v increases with L for small ζ (say $\zeta<0.25$). (See parameters in the last three lines of Table 1.1.) The conclusion is consistent with the observation that ground motions approach harmonic oscillations as $L\rightarrow\infty$ because of the faster damping of short-period waves

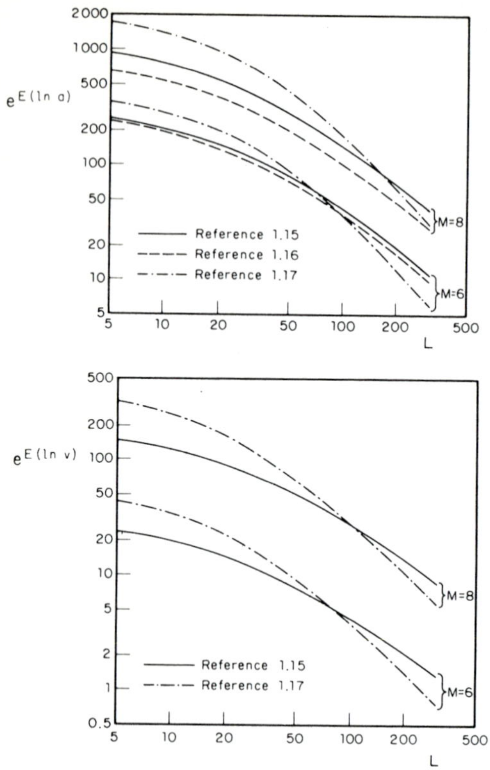

Fig. 1.6 Maximum ground accelerations and velocities according to various criteria

with distance[1.6]. According Newmark[1.20] and to McGuire, $\bar{V}(\zeta)$ may be taken independent of frequency in the intermediate-frequency range (see Figure 1.7). According to still another criterion[1.6] the a, v, d lines are taken as the spectrum for $\zeta = 0.25$. Spectra for other values of ζ are computed using Equation 1.11.

A combination of the various criteria gives some guidance for drawing expected spectra given the magnitude and focal coordinates.

1.8.4 Probability distributions of normalized spectral ordinates

Idealizing earthquakes as segments of stationary white noise one may compute the probability distribution of $R/\bar{R}_0$, where R is a spectral response corresponding to some frequency and damping ratio while $\bar{R}_0$ is the expected value of the response associated with the same frequency but zero damping[1.21, 1.22]. For sufficiently high $R/\bar{R}_0$ these probability distributions approach extreme type 2 distributions. The coefficient of variation in the intermediate frequency range varies as

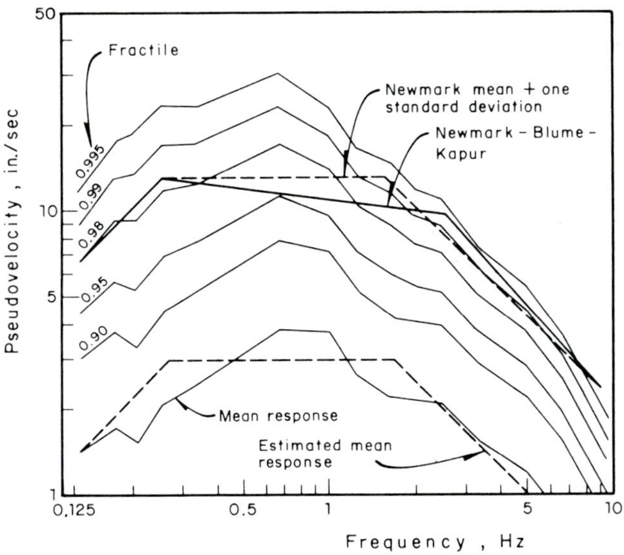

Fig. 1.7 Design spectra according to McGuire[1.15], Newmark[1.20], and Newmark, Blume, and Kapur[1.19]

$$\mathrm{COV}\left(\frac{R}{R_0}\right) \simeq 0.32 \left(\frac{\bar{R}}{\bar{R}_0}\right)^{0.82} \tag{1.10}$$

where

$$\frac{\bar{R}}{\bar{R}_0} \simeq (1 + 0.785\zeta\omega s)^{-0.41} \tag{1.11}$$

s = duration of white-noise segment[1.23]. This duration may be estimated from [1.6]

$$\bar{s} = 0.002\mathrm{e}^{-0.74}M + 0.3L \tag{1.12}$$

where s is in seconds and L in km. The resulting distribution must be combined with that of $\bar{R}_0$, which is practically lognormal, with the parameters given in the previous section.

In the ranges of extremely long or extremely short periods, effects of damping are less pronounced than for intermediate periods, for, when a and d are given, spectral ordinates tend to deterministic values independent of ζ.

1.8.5 Description of ground motions in random-vibration theory

For many purposes it is convenient to use random-vibration theory in

TABLE 1.2 Newmark–Blume–Kapur amplification factors[1.19]

	Amplification factors at control points			
ζ, %	*A* (33 Hz)	*B* (9 Hz)	*C* (2.5 Hz)	*D* (0.25 Hz)
0.5	1.0	4.96	5.95	3.20
2.0	1.0	3.54	4.25	2.50
5.0	1.0	2.61	3.13	2.05
7.0	1.0	2.27	2.72	1.88
10.0	1.0	1.90	2.28	1.70

describing earthquake motions. This is best done by characterizing a stochastic process by the first few moments of its power spectral density[1.24]. (The power spectral density or spectral density function is a function of circular frequency ω, proportional to the energy associated with ω. it is defined as $G(\omega)=\lim_{s\to\infty} E|F_s(\omega)|/s$, where $F_s(\omega)$ is the Fourier transform of the segment transform of the segment $-s/2 \leq t \leq s/2$ of a member $x(t)$ of the stochastic process of interest[1.25, 1.26].) Iit is often assumed that earthquakes on firm ground have a Kanai-Tajimi power spectral density[1.27], which is of the form

$$G(\omega)=\frac{(1+4\zeta_g^2\omega^2/\omega_g^2)G_0}{(1-\omega^2/\omega_g^2)^2+4\zeta_g^2\omega^2/\omega_g^2} \tag{1.13}$$

or the analytically more tractable

$$G(\omega)=\frac{G_0(\omega/\omega_g)^4}{\omega_g}\,e^{-4\omega/\omega_g} \tag{1.14}$$

in which G_0, ζ_g, and ω_g are constants and the latter is the prevailing circular frequency[1.28].

The ith moment of $G(\omega)$ is defined as

$$\lambda_i=\int_0^\infty \omega_i G(\omega)\,d\omega \tag{1.15}$$

These moments serve to estimate several pertinent properties of families of ground motions. For example, the ground-acceleration variance equals λ_0 (area under the $G(\omega)$ curve); the statistics of the maximum ground acceleration can be estimated from λ_0, λ_1, and λ_2; and the statistics of structural response can be estimated from the first few moments of the response power spectral density, which in turn is obtained by convoluting $G(\omega)$ and the *transfer function* (*Fourier transform of the impulse response function*). The approach serves as basis for an approximate but effective

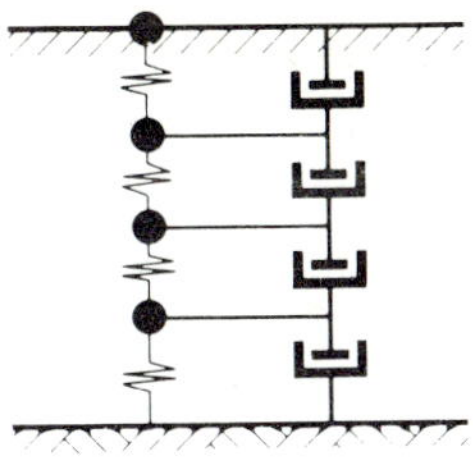

Fig. 1.8 Lumped-mass idealization of soil layer

treatment of nonlinear multidegree systems[1.29]. Power spectral density is also useful in the generation of simulated earthquakes having a specified spectrum[1.30]

1.9 LOCAL EFFECTS

Seismic waves are modified as they travel. Of particular significance are local effects of topography and geology.

1.9.1 Effects of topography

There is amplification of ground motions in some period ranges at the tops of ridges and mountains, with reduction near the ridge and mountain bases. Ground depressions have the opposite effects. Some observations[1.31] and two- and three-dimensional finite-difference and finite-element studies have begun throwing light on the subject[1.32–1.35]. Three-dimensional studies have been limited by computer capacity, especially since the underlying formations cannot be assumed interrupted very near the free surface because of spurious reflections at the fictitious boundaries[1.36].

Records obtained during aftershocks of the 1975 Haicheng earthquake[1.37] indicate a magnification factor of 1.8 in maximum ground acceleration at the top of even a small hill.

1.9.2 Linear one-dimensional filtering

Major contrasts in soil properties are usually found near the ground surface, especially under valleys, on which most cities lie. When the contrast between rock and soil is pronounced and stratification is practically horizontal it is sufficiently accurate in the intermediate and far fields of earthquakes to treat bedrock as infinitely rigid and replace the soil with lumped masses simply coupled through massless springs and dashpots[1.38] (Figure 1.8). Bedrock motion is then taken identical with that at outcrops.

The assumption of infinite rigidity of bedrock is unnecessary. In reality some energy is lost through radiation damping, that is, partial refraction of down-travelling waves into bedrock which modifies the motion of the rock-

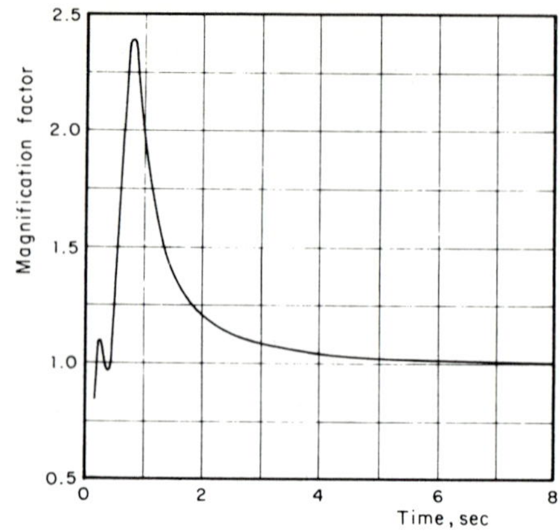

Fig. 1.9 Magnification factor for single layer[1.6]

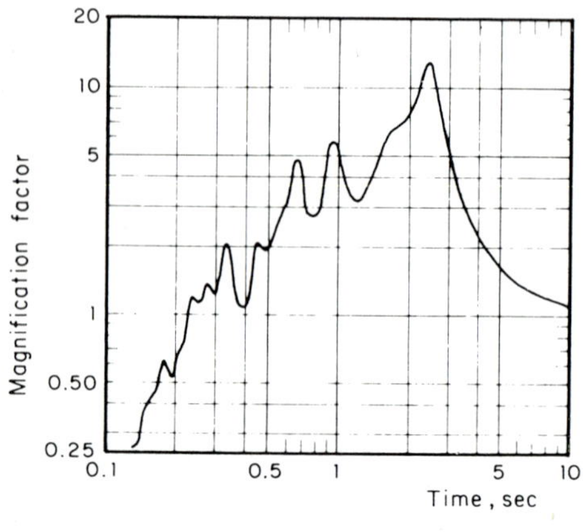

Fig. 1.10 Computed magnification factor for one site in the Valley of Mexico[1.40]

soil interface. The phenomenon can be treated exactly by introducing a dashpot at the interface[1.39].

One can infer the shape of spectra at the surface of soft ground either using this numerical model in conjunction with earthquake simulation or, in a stochastic-process treatment when the soil is idealized as of linear behaviour, resorting to an exact analytical procedure applicable to Fourier spectra[1.40]. Under appropriate circumstances the latter approach has been satisfactory[1.41]. Figures 1.9 and 1.10 illustrate the magnification factors computed.

1.9.3 More realistic models

(This first paragraph refers to greater realism in terms of material behaviour preserving one-dimensional approximation.) The intensity

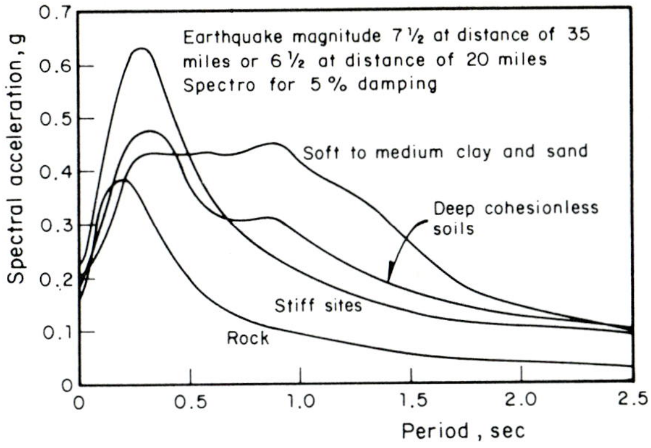

Fig. 1.11 *Average spectra on different kinds of soil*[1.49]

threshold beyond which one should recognize nonlinear behaviour of soil is often quite low. For simulation studies one may use nonlinear analysis or retain the assumption of linearity but using a reduced soil rigidity and increased internal damping. It has been found[1.42] that the secant modulus and viscous damping equivalent to hysteretic damping of soil strained cyclically at 2/3 the actual peak strain leads to satisfactory results. This linearization technique requires use of iterative analysis. A linearization technique for stochastic-process treatment of nonlinear soil has been proposed[1.43].

Ambraseys[1.44] has shown that in one-dimensional elastoplastic problems the presence of a weak layer limits the accelerations that can be transmitted to the surface and, for strong motions, introduces prevailing ground periods absent at low earthquake intensities.

To take into account conditions not adequately represented by horizontal stratification one may use the method of characteristics[1.45], two- or three-dimensional finite elements, or lumped masses interconnected by springs and dashpots, and either linearization techniques[1.46], the boundary-integral approach[1.47], or direct nonlinear analysis[1.48]. Again there are limitations owing to computer capacity and boundary conditions in bedrock and again effects of surface waves have not received adequate attention.

The main objection to most studies on local amplification done to date is that incident disturbances are ordinarily oversimplified. Despite availability of powerful numerical tools, rarely are surface waves taken into consideration.

1.9.4 Typical spectra

Seed *et al.*[1.48] have averaged the spectra of several earthquakes, grouping them according to site conditions and focal distance (Figure 1.11). Particularly for spectra at the surface of soft soil the sample is small and extremely heterogeneous. (See also Reference 1.50.) Since every site has its own geologic and seismic features, curves in Figure 1.11 should not be taken at face value although they serve as guides of what can be expected. For design purposes, moreover, spectra derived from records must be modified to take into account various forms of soil-structure interaction (Chapter 7), structural behaviour (Chapters 2 and 3), and optimization in design (Chapter 3).

1.9.5 Soil compaction and liquefaction

Earthquake damage to structures is often not due to direct vibrational effects but to large, permanent deformations of the soil. In all types of soils these deformations are mostly due to shear failure. In noncohesive soils they can be associated with compaction. In saturated noncohesive soils under certain drainage conditions there is also the possibility of loss of strength through cyclic mobility or liquefaction. These matters as well as shear failure of soils are covered in Chapter 7.

1.10 GROUND-MOTION COMPONENTS

Given a structure, a ground motion has as many potentially relevant components as there are degrees of freedom in the structure's base. Six-component earthquakes act on every rigid base. The corresponding generalized coordinates may be chosen arbitrarily provided they are linearly independent; for example, three orthogonal translational and three orthogonal rotational components. For a bridge on several supports there can be many more components. In individual cases we may anticipate, though, that only some components will have significant effects.

Consider a structure for which only translational components of base motion are significant. Let $X_i = 1, 2, 3$ denote three orthogonal directions and x_i the corresponding base displacements. Consider covariances

$$\mathrm{Cov}(x_i, x_j; t) = \int_0^t \ddot{x}_i(\tau)\ddot{x}_j(\tau)\, d\tau;\ i, j = 1, 2, 3 \qquad (1.16)$$

where dots mean time derivatives. There is always a set of cartesian coordinates X_i, known as *principal directions*, such that

$$\mathrm{Cov}(x_i, x_j; t) = 0;\ i \neq j;\ i, j = 1, 2, 3 \qquad (1.17)$$

After the first few seconds of motion, these directions change little with time[1.51]. For practically every t these accelerograms are uncorrelated.

It has been found that one principal direction, say X_3, is roughly vertical while there is debate about the general orientation of the two roughly horizontal directions[1.51, 1.52].

For earthquakes sufficiently intense to be of practical interest the intensities in all horizontal directions are comparable. It follows that there is little error in assuming that ground motions are uncorrelated in the vertical and any two orthogonal horizontal directions[1.53].

If each of these accelerograms is idealized as a zero-mean Gaussian process, responses of every linear system will be of this same nature. For any given exceedance probability, design structural responses are proportional to their standard deviation. Lack of correlation of the principal accelerograms insures that responses thereto are also uncorrelated and hence that the corresponding variances are additive. Consequently the envelope of structural responses describes the correlation ellipsoid defined by adding the variances of responses to the principal components in state space for the structure of interest[1.53]. This is the basis of the design procedure for multicomponent earthquakes in Chapter 2.

For some types of waves of given frequency there is complete correlation among some components of ground motion (Section 1.3). The situation can be dealt with by redefining the generalized coordinates of base motion so as to do away with the correlation or, when the natural modes excited by the different components have different frequencies, by making use of the reduction or elimination of correlation among the responses despite correlations among components.

1.11 EARTHQUAKE SIMULATION

The most accurate means for analyzing a structure subjected to a given ground motion is the step-by-step computation of its responses. Whatever the structural properties the most convenient methods include Newmark's β method[1.48], the central-difference predictor scheme, and Park's stiffly stable method[1.54]. The approach is expensive. A single run on a nonlinear multidegree system consumes considerable computer time. Moreover, the high dispersion of responses to a family of earthquakes appropriate for a given site makes it mandatory to analyze a structure for a large number of disturbances and to process the results statistically. It is justified to do so only in special instances: when the structure is particularly important or when the number of nominally identical structures is high. On the other hand, the number of available strong-motion records associated with given foci and magnitudes is scarce for all sites. Hence the need to simulate earthquakes in a way consistent with pertinent tectonic features.

Which model we choose for earthquake simulation depends on the properties of the structure we wish to analyze. Each among the following models is best under a given set of circumstances.

(1) The simplest model that preserves the high irregularity of earthquakes is a segment of stationary white noise produced in an analog or digital computer. (White noise is a stochastic process having equipartition of energy among all frequencies.) This is adequate for single-degree linear systems if the expected spectral ordinates of the earthquake family being simulated vary smoothly in the neighbourhood of the structure's natural period[1.55] and this

period is not too short (for earthquakes on firm ground, not much shorter than $v/2\pi a$, where a, v = maximum ground acceleration and velocity respectively). It is also adequate for single-degree, mildly nonlinear structures if, in addition, the expected acceleration spectrum is practically flat considerably beyond the structure's natural period.

(2) Next in complexity comes a segment of stationary Gaussian process[1.6,1.24]. (A stochastic process is Gaussian when the probability distribution of the stochastic variable is Gaussian at every instant.) With it we can generate a family of simulated motions whose expected spectral ordinates follow a specified law (see below). Such motions are adequate for the analysis of linear multi-degree systems. One limitation, however, of this and most of the following processes is that, if bedrock motion is Gaussian, it cannot remain Gaussian after filtering through a soil of nonlinear behaviour. Apparently implications of this have not been studied.

(3) Perhaps the most commonly used model is the product of a stationary Gaussian process and a deterministic time function. The latter ordinarily reproduces the energy build-up, approximately stationary, and dying-out phases of real earthquakes[1.56, 1.57]. The model is more accurate than the previous one for linear systems and may also be used for nonlinear ones provided their hysteretic energy absorption is high enough so that the time dependence of frequency content has no significant effect.

(4) The response of a damped harmonic oscillator to a segment of white or Gaussian noise changes in frequency content with time, much in the manner of real seisms. Its amplitude modulation also resembles that of true seisms. It may be adequate for analyzing all sorts of structures, even elastic nonlinear ones, whose responses are quite sensitive to the time dependence of frequency content. Since, however, this feature is not controlled in the simulation, the model leads generally to less accurate results than the next model.

(5) By superimposing three amplitude-modulated segments of stationary Gaussian processes (see Figure 1.1b), each starting at a different instant and each with a different power spectral density function we simulate the typical three phases of earthquakes (predominantly P, S, and surface waves respectively) and can closely approximate amplitude and frequency variation[1.6].

(6) The foregoing models simulate one earthquake component at a time. Simultaneous action of the translational components can be coped with by generating uncorrelated processes along three principal directions (see Section 1.10). Consideration of simultaneous action of several components is important in nonlinear systems, for which the superposition principle does not hold.

(7) More trustworthy than the foregoing models is the generation of simulated motions according to a physically based model. In one version[1.18] we assume that P, S, and surface waves are generated at a series of closely spaced foci along a straight line, thus simulating the propagation of fault slip in nature. Each wave generated is then repeated many times with random phase shift and amplitude again simulating the natural process. The model

can reproduce directional effects, including the Doppler effect, and simultaneously reproduce as many ground-motion components as desired — translational and rotational included — provided the real soil behaves as modelled.

Through use of finite-element and other techniques it is also possible to simulate effects of local topography and of various geologic features, as well as nonlinear soil behaviour.

Techniques have been developed for generating simulated earthquakes whose expected spectral ordinates conform to a specified shape[1.29]. In essence, the power spectral density function is computed from the specified spectral shape. This determines how much energy should be assigned to each of a large number of frequencies. Phases are then randomized. Available methods are still not capable, though, of incorporating the wide spread of spectral ordinates about the mean, which is characteristic of true earthquake families[1.29].

A variation of this method permits constructing simulated ground motions having smooth response spectra[1.58]. These spectra have ordinates equal to or very slightly greater than those of a specified design spectrum. Although the spectra look strikingly different from those of true earthquakes, inspection of the accelerograms does not disclose the abnormality. Use of these computer-generated motions has been advocated on the basis that they are supposedly conservative but not too much so. For structures of linear behaviour, though, they are certainly not always conservative, as two natural modes sometimes combine not additively but subtractively to give the maximum overall response; overestimate of a spectral ordinate can therefore err on the unsafe side.

An appealing technique for generating a family of accelerograms from a single specimen consists in solving members of a family of stochastic differential equations whose parameters are governed by the given specimen[1.59]. It is not possible at this stage to advocate this method for, although the accelerograms generated look convincingly like real earthquakes, there is no basis for assuming that they are consistent with known tectonic features.

The cost of simulating ground motions in a computer grows sharply with degree of sophistication. It is therefore well to devote thought in every instance to the type of model that will be adopted and, more important, whether simulation will be used at all. Often it is preferable to use approximate methods, such as modifications of modal analysis, to incorporate in a probabilistic sense the effects of possible future earthquakes than to accurately compute responses to a few past or simulated motions which, it is known, will never occur in the future.

1.12 NONVIBRATORY SEISMIC EFFECTS

Soil compaction and loss of strength are caused by seismic vibrations but their effects on structures are nonvibratory. The same holds for other forms of large permanent deformations and cracking of soils, which are often manifest in damage to underground pipes. Slope failures triggered by earthquakes have had catastrophic effects. There are direct nonvibratory effects on structures. These effects are caused directly by fault motion, either during an earthquake, immediately following one, or through fault creep not associated with earthquakes. These matters are covered in Chapter 3.

ACKNOWLEDGMENTS

The writer is grateful to Jacobo Bielak, Luis Esteva, and Ezio Faccioli for their valuable suggestions and criticisms.

NOTATION

A = spectral acceleration or pseudoacceleration
a = maximum ground acceleration
b_i = parameters in Equation 1.9
C = dashpot constant
COV(·) = coefficient of variation of variable (·)
Cov(·,·) = covariance of variables (·,·)
D = spectral deformation
d = maximum ground displacement
E = Young's modulus
$E(\cdot)$ = expectation of variable (·)
$F(\omega)$ = Fourier spectrum
$|F(\omega)|$ = Fourier amplitude spectrum
$F_s(\omega)$ = Fourier transform of segment $-s/2 \leq t \leq s/2$ of a member of a stochastic process
f = undamped natural frequency
f' = damped natural frequency
G = modulus of rigidity
G_0 = constant in equations 1.13 and 1.14
$G(\omega)$ = power spectral density (spectral density function)
g = acceleration of gravity
K = stiffness
L = focal distance
L' = L plus some quantity
M = mass; M_s
M_b = m
M_0 = seismic moment
M_s = Richter magnitude (surface-wave magnitude)
m = body-wave magnitude
P = force applied to a single-degree system
R = $\max_t|r(t)|$

R_0 = value of R in undamped system
r = structural response
s = duration of segment of stationary process equivalent to family of earthquakes
T = undamped natural period
T' = damped natural period
t = time
V = spectral velocity or pseudovelocity
v = maximum ground velocity
v_l = velocity of Love waves
v_p = velocity of P waves
v_r = velocity of Rayleigh waves
v_s = velocity of S waves
X = cartesian coordinate
x = displacement
x_0 = ground displacement
Y = ground-motion parameter
y = $x - x_0$
ζ = damping ratio
ζ_g = constant in Equation 1.13
λ_i = ith moment of $G(\omega)$
ν = Poisson's ratio
ρ = mass density
τ = dummy time variable
ψ = impulse response function
ω = circular frequency; undamped natural circular frequency
ω' = damped natural circular frequency
ω_g = prevailing circular frequency of ground motion

APPENDIX 1.1 THE MODIFIED MERCALLI SCALE

Mercalli's (1902) improved intensity scale served as a basis for the scale advanced by Wood and Neumann (1931), known as the modified Mercalli scale and commonly abbreviated MM. The modified version is described below with some improvements by Richter (1958). The following remarks are taken almost verbatim from *Elementary Seismology*, Charles F. Richter (W. H. Freeman and Company, San Francisco, copyright © 1958).

To eliminate many verbal repetitions in the original scale, the following convention has been adopted. Each effect is named at that level of intensity at which it first appears frequently and characteristically. Each effect may be found less strongly, or in fewer instances, at the next lower grade of intensity; more strongly or more often at the next higher grade. A few effects are named at two successive levels to indicate a more gradual increase.

Masonry A, B, C, D. To avoid ambiguity of language, the quality of masonry, brick or otherwise, is specified by the following lettering (which has no connection with the conventional Class *A*, *B*, *C* construction).

Masonry A. Good workmanship, mortar, and design; reinforced, especially laterally, and bound together by using steel, concrete, etc.; designed to resist lateral forces.

Masonry B. Good workmanship and mortar; reinforced, but not designed in detail to resist lateral forces.

Masonry C. Ordinary workmanship and mortar; no extreme weaknesses like failing to tie in at corners, but neither reinforced nor designed against horizontal forces.

Masonry D. Weak materials, such as adobe; poor mortar; low standards of workmanship; weak horizontally.

Modified Mercalli Intensity Scale of 1931 (*Abridged and rewritten by C. F. Richter*)*

1. Not felt. Marginal and long-period of large earthquakes.
2. Felt by persons at rest, on upper floors, or favourably placed.
3. Felt indoors. Hanging objects swing. Vibration like passing of light trucks. Duration estimated. May not be recognized as an earthquake.
4. Hanging objects swing. Vibration like passing of heavy trucks; or sensation of a jolt like a heavy ball striking the walls. Standing motor cars rock. Windows, dishes, doors rattle. Glasses clink. Crockery clashes. In the upper range of 4, wooden walls and frames crack.
5. Felt outdoors; direction estimated. Sleepers wakened. Liquids disturbed, some spilled. Small unstable objects displaced or upset. Doors swing, close, open. Shutters, pictures move. Pendulum clocks stop, start, change rate.
6. Felt by all. Many frightened and run outdoors. Persons walk unsteadily. Windows, dishes, glassware broken. Knickknacks, books, and so on, off shelves. Pictures off walls. Furniture moved or overturned. Weak plaster and masonry *D* cracked. Small bells ring (church, school). Trees, bushes shaken visibly, or heard to rustle.
7. Difficult to stand. Noticed by drivers of motor cars. Hanging objects quiver. Furniture broken. Damage to masonry *D* including cracks. Weak chimneys broken at roof line. Fall of plaster, loose bricks, stones, tiles, cornices, unbraced parapets, and architectural ornaments. Some cracks in masonry *C*. Waves on ponds; water turbid with mud. Small slides and caving in along sand or gravel banks. Large bells ring. Concrete irrigation ditches damaged.
8. Steering of motor cars affected. Damage to masonry *C*; partial collapse.

* The author C. F. Richter requests that, should it be necessary to specify it exactly, the reference be 'Modified Mercalli Scale, 1956 version', without attaching his name. The expression 'Richter scale' is popularly attached to the magnitude scale; it is desirable to forestall confusion between magnitude and intensity.

Some damage to masonry *B*; none to masonry *A*. Fall of stucco and some masonry walls. Twisting, fall of chimneys, factory stacks, monuments, towers, elevated tanks. Frame houses moved on foundations if not bolted down; loose panel walls thrown out. Decayed piling broken off. Branches broken from trees. Changes in flow or temperature of springs and walls. Cracks in wet ground and on steep slopes.

9. General panic. Masonry *D* destroyed; masonry *C* heavily damaged, sometimes with complete collapse; masonry *B* seriously damaged. General damage to foundations. Frame structures, if not bolted, shifted off foundations. Frames racked. Conspicuous cracks in ground. In alluviated areas sand and mud ejected, earthquake fountains, sand craters.

10. Most masonry and frame structures destroyed with their foundations. Some well-built wooden structures and bridges destroyed. Serious damage to dams, dikes, embankments, Large landslides. Water thrown on banks of canals, rivers, lakes, etc. Sand and mud shifted horizontally on beaches and flat land. Rails bent slightly.

11. Rails bent greatly. Underground pipelines completely out of service.

12. Damage nearly total. Large rock masses displaced. Lines of sight and level distorted. Objects thrown into the air.

REFERENCES

1.1 Rosenblueth, E., 'Research in Earthquake Engineering, Challenge to Imagination', *Proc. First Canadian Conference on Earthquake Engineering*, Vancouver, Canada, 7–26 (1971)

1.2 Artyushkov, E. V., 'Density Differentiation of the Earth's Matter and Processes at the Core–Mantle Interface', *J. Geophysical Res.*, **77**, 6454–6458 (1972)

1.3 Evison, F. F., 'On the Occurrence of Volume Change of the Earthquake Source', *Bull. seism. Soc. Am.*, **57**, No. 1, 9–25 (1967)

1.4 Reid, H. F., 'Mechanics of the Earthquake', in *The California Earthquake of April* 18, 1906, 2, Carnegie Institute of Washington, D.C., updated (1969)

1.5 Bullen, K. E., *An Introduction to the Theory of Seismology*, Cambridge University Press, London (1953)

1.6 Newmark, N. M. and Rosenblueth, E., *Fundamentals of Earthquake Engineering*, Prentice-Hall, Inc., Englewood Cliffs, N.J. (1971)

1.7 Richter, C. F. *Elementary Seismology*, W. H. Freeman and Co., San Francisco, Cal. (1958)

1.8 Brune, J. N., 'The Physics of Earthquake Strong Motions', in *Seismic Risk and Engineering Decisions*, C. Lomnitz and E. Rosenblueth editors, Elsevier, Amsterdam, 145–147 (1976)

1.9 Chinnery, M. A. and North, R. G., 'The Frequency of Very Large Earthquakes', *Science*, **190**, No. 4220, 1197–1198 (1975)

1.10 Medvedev, S. V. and Sponheuer, W., 'Scale of Seismic Intensity', *Proc. Fourth World Conference on Earthquake Engineering*, Santiago, Chile, 1, A-2, 143–153 (1969)

1.11 Housner, G. W., 'Spectrum Intensities of Strong-Motion Earthquakes', *Proc. Symposium on Earthquake and Blast Effects on Structures* (Edited by C. M. Duke and M. Feigen) Univ. California, Los Angeles, 21–36 (1952)

1.12 Arias, A., 'A Measure of Earthquake Intensity', in *Seismic Design for Nuclear Power Plants* (Edited by R. Hansen) M.I.T. Press, Cambridge, Mass. (1969)

1.13 Blume, J. A., Newmark, N. M. and Corning, L. H., *Design of Multistory Reinforced Concrete Buildings for Earthquake Motions*, Portland Cement Association, Chicago (1961)

1.14 Cloud, W. K. and Perez, V., 'Strong Motion Records and Acceleration', *Proc. Fourth World Conference on Earthquake Engineering*, Santiago, Chile, A-2.119-2.132 (1969)

1.15 McGuire, R. K., 'Seismic Structural Response Risk Analysis, Including Peak Response Regressions on Earthquake Magnitude and Distance', *Report* R74-51, Department of Civil Engineering, MIT, Cambridge, Mass. (1974)

1.16 Donovan, N. C., 'Site Response Procedures for Major Buildings', Paper presented at ASCE National Structural Engineering Meeting, San Francisco, Cal. (Apr. 1973)

1.17 Esteva, L. and Villaverde, R., 'Seismic Risk, Design Spectra and Structural Reliability', *Proc. Sixth World Conference on Earthquake Engineering*, New Delhi (1977)

1.18 Rascón, O. A. and Chávez, M. 'On an Earthquake Simulation Model', *Proc. Fifth World Conference on Earthquake Engineering*, Rome, Vol. 2, 2899–2907 (1973)

1.19 Newmark, N. M., Blume, J. A. and Kapur, K. K., 'Seismic Design Spectra for Nuclear Power Plants', *J. Pwr. Div. Am. Soc. civ Engrs*, **99**, No. PO2, 873–889 (1973)

1.20 Newmark, N. M., Consulting Engineering Services, 'A Study of Vertical and Horizontal Earthquake Spectra', Urbana, Ill., USAEC Contract No. AT(49-5)-2667, WASH-1255 (1973)

1.21 Rosenblueth, E. and Bustamente, J. I., 'Distribution of Structural Response to Earthquakes', *J. Engng Mech. Div. Am. Soc. civ. Engrs*, **88**, No. EM3, 75–106 (1962)

1.22 Brady, A. G., 'Studies of Response to Earthquake Ground Motion', doctoral thesis, *California Institute of Technology*, Pasadena, Cal. (1966)

1.23 Rosenblueth, E., 'Optimum Resource Allocation in Earthquake Engineering', *Proc. Fifth European Conference on Earthquake Engineering*, Istanbul, Turkey, Vol. 2 (1975)

1.24 Vanmarcke, E., 'Structural Response to Earthquakes', in *Seismic Risk and Engineering Decisions* (Edited by C. Lomnitz and E. Rosenblueth) Elsevier, Amsterdam, 287–337 (1976)

1.25 Franklin, J. N., 'Digital-Computer Simulation of a Gaussian Random Process with Given Power Spectral Density', Technical Report 120, *Computing Center California Institute of Technology*, Pasadena, Cal. (1963)

1.26 Rosenblueth, E., 'Probabilistic Design to Resist Earthquakes', *J. Engng Mech. Div. Proc. Am. Soc. civ. Engrs*, **90**, No. EM5, 189–219 (1964)

1.27 Tajimi, H., 'A Statistical Method of Determining the Maximum Response of a Building Structure During an Earthquake', *Proc. Second World Conference on Earthquake Engineering*, Tokyo and Kyoto, 2, 781–797 (1960)

1.28 Goto, H. and Kameda, H., 'Statistical Inference of the Future Ground Motion', *Proc. Fourth World Conference on Earthquake Engineering*, Santiago, Chile, 1, A-1, 39–54 (1969)

1.29 Gazetas, G., 'Random Vibration Analysis of Inelastic Multi-Degree-of-Freedom Systems Subjected to Earthquake Ground Motions', *Report* R76-39, Department of Civil Engineering, MIT, Cambridge, Mass. (1976)

1.30 Luyties, III, W. H., Anagnostopoulos, S. A. and Biggs, J. M., 'Studies on the Inelastic Dynamic Analysis and Design of Multi-Story Frames', *Report* R76-29, Department of Civil Engineering, MIT, Cambridge, Mass. (1976)

1.31 Davis, L. L. and West, L. R., 'Observed Effects of Topography on Ground Motion', *Bull. seism. Soc. Am.*, **63**, No. 1, 283–298 (1973)

1.32 Boore, D. M., 'A Note on the Effect of Simple Topography on Seismic SH Waves', *Bull. seism. Soc. Am.*, **62**, No. 1, 275–284 (1972)

1.33 Trifunac, M. D., 'Scattering of Plane SH Waves by a Semi-Cylindrical Canyon', *Earthquake Engineering and Structural Dynamics*, **1**, 267–281 (1973)

1.34 Reimer, R. B., Clough, R. W. and Raphael, J. M., 'Evaluation of the Pacoima Dam Accelerogram', *Proc. Fifth World Conference on Earthquake Engineering*, Rome, Italy, Vol. 2, 2328–2337 (1973)

1.35 Lefebvre, G., Duncan, J. M. and Wilson, E. L., 'Three-Dimensional Finite Element Analysis of Dams', *J. Soil Mech. Fdns Div. Am. Soc. civ. Engrs*, **99**, No. SM7, 495–507 (1973)

1.36 Ayala, G. A. and Aranda, G. R., 'Boundary Conditions in Soil Amplification Studies', *Preprints, Sixth World Conference on Earthquake Engineering*, New Delhi, India, 6.117-6.122 (1977) .

1.37 Liu, Hui-hsien, personal communication (1976)

1.38 Idriss, I. M. and Seed, H. B., 'Seismic Response of Horizontal Soil Layers', *J. Soil Mech. Fdns Div. Am. Soc. civ. Engrs*, **94**, No. SM4, 1003–1031 (1968)

1.39 Rosenblueth, E. and Elorduy, J., 'Characteristics of Earthquakes on Mexico City Clay' [in Spanish] in *Nabor Carrillo. El Hundimiento de la Ciudad de Mexico y Proyecto Texcoco*, contribución de Proyecto Texcoco al Séptimo Congreso Internacional de Mecánica de Suelos e Ingeniería de Cimentaciones. Secretaría de Hacienda y Crédito Publico, Fiduciaria: Nacional Financiera, S.A., Mexico, 287–328 (1969)

1.40 Herrera, I. and Rosenblueth, E., 'Response Spectra on Stratified Soil', *Proc. Third World Conference on Earthquake Engineering*, Auckland and Wellington, New Zealand, 1.44-1.60 (1965)

1.41 Herrera, I., Rosenblueth, E. and Rascón, O. A., 'Earthquake Spectrum Prediction in the Valley of Mexico', *Proc. Third World Conference on Earthquake Engineering*, Auckland and Wellington, New Zealand, 1.61-1.74 (1965)

1.42 Constantopoulos, I. V., 'Amplification Studies for a Nonlinear Hysteretic Soil Model', *Report R*73-46, Department of Civil Engineering, MIT, Cambridge, Mass. (1973)

1.43 Faccioli, E., 'A Stochastic Approach to Soil Amplification', *Bull. seism. Soc. Am.*, **66**, No. 4, 1277–1292 (1976)

1.44 Ambraseys, N. N., 'Behavior of Foundation Materials during Strong Earthquakes', Abstract, *Proc. Fourth European Symposium on Earthquake Engineering, London*, Bulgarian Academy of Sciences, Earthquake Engineering, 11–12 (1972)

1.45 Papadakis, C. N., 'Soil Transients by Characteristics Method', thesis, Department of Civil Engineering, Univ. Michigan, Ann Arbor, Mich. (1973)

1.46 Lysmer, J., Udaka, T., Seed, H. B. and Hwang, R., 'LUSH: A Computer Program for Complex Response Analysis of Soil Structure Systems', *Report EERC* 74-4, Earthquake Engineering Research Center, Univ. California, Berkeley, Cal. (1974)

1.47 Zienkiewics, O. C., 'The Finite Element Method and Boundary Solution Procedures as General Approximation Methods for Field Problems', *Proc. World Congress on Finite Element Methods in Structural Mechanics*, Bournemouth, Dorset, England, pp S1-S31 (1975)

1.48 Newmark, N. M., 'A Method of Computation for Structural Dynamics', *Trans. Am. Soc. civ. Engrs*, **127**, 1406–1435 (1962), reprinted in *Selected Papers* by N. M. Newmark, Civil Engineering Classics, American Society of Civil Engineers, New York, 601–630 (1976)

1.49 Seed, H. B., Ugas, H. and Lysmer, J., 'Site-Dependent Spectra for Earthquake-Resistant Design', *Bull. seism. Soc. Am.*, **66**, 221–243 (1946)

1.50 Mohraz, B., 'A Study of Earthquake Response Spectra for Different Geologic Conditions', *Bull. seism. Soc. Am.*, **66**, No. 3, 915–935 (1976)

1.51 Kubo, T. and Penzien, J., 'Characteristics of Three-Dimensional Ground Motions Along Principal Axes, San Fernando Earthquake', *Proc. Sixth World Conference on Earthquake Engineering*, New Delhi (1977)

1.52 Arias, A. and Sandoval, H., 'Analysis of Epicenter Identification through Eigenvectors of the Acceleration Tensor Obtained from Accelerograms' [in Spanish], internal report, Instituto de Ingenieria, Universidad Nacional Autónoma de México (1976)

1.53 Rosenblueth, E. and Contreras, H., 'Approximate Design for Multicomponent Earthquakes', *J. Engng Mech. Div. Am. Soc. civ. Engrs*, No. EM5, Vol. 103 (1977)

1.54 Adeli, H., 'Solution Techniques for Linear and Nonlinear Dynamics of Structures Modeled by Finite Elements', The John A. Blume Earthquake Engineering Center, Department of Civil Engineering Stanford University, *Report No.* 23 (1976)

1.55 Caughey, T. K. and Gray, A. H., discussion of Reference 1.21, *J. Engng Mech. Div. Am. Soc. civ. Engrs*, **89**, No. EM2, 159–168 (1962)

1.56 Jennings, P. C., Housner, G. W. and Tsai, N. C., 'Simulated Earthquake Motions for Design Purposes', *Proc. Fourth World Conference on Earthquake Engineering*, Santiago, Chile, Vol. 1, A-1, 145–160 (1969)

1.57 Amin, M. and Ang, A. H., 'A Nonstationary Stochastic Model for Strong-Motion Earthquakes', *Civil Engineering Studies, Structural Research Series No.* 306, University of Illinois, Urbana (1966)

1.58 Tsai, N.-C., 'Spectrum-Compatible Motion for Design Purpose', *J. Engng Mech. Div. Am. Soc. civ. Engrs*, **98**, No. EM2, 345–356 (1972)
1.59 Saragoni, G. R. and Hart, G. C., 'Simulation of Artificial Earthquakes', *Int. Journal of Earthquake Engineering and Structural Dynamics*, Vol. 2, 249–267 (1974)

2 Analysis

Anil K. Chopra and Nathan M. Newmark

2.1 SCOPE

Procedures described in this chapter are restricted essentially to buildings although some aspects of the topics discussed may be applied to other structures. In general the object of earthquake resistant design is to enable a structure to resist with slight or no damage the earthquake motions that might reasonably be expected to occur during the lifetime of the structure, thus avoiding expensive repairs if a minor earthquake should occur. However, a more important purpose is to provide a large measure of resistance to prevent collapse or failure that might cause major property damage or loss of life, even in the event of a major earthquake of rare probability of occurrence. Although, for the first case, a structure may be designed to remain nearly elastic with high probability, in the latter instance it is unreasonably uneconomic to design for elastic behaviour unless the structure is of such a character and of such importance that it might not be usable after the earthquake. For example, a smaller amount of inelastic deformation would be permitted at the maximum design level for a hospital compared with an apartment building.

Procedures for analysis described herein, in Section 2.2, are of two types: (1) rigorous procedures of analysis to determine the response of structures essentially in the elastic range; and (2) approximate procedures suitable for use in design. In this chapter only a brief summary is given of the more rigorous procedures and attention is focused primarily on approximate procedures for design.

Because a structure is supported on a foundation, the interaction of the structure and its foundation, including the soil or rock underneath, is an important factor in determining the structural response. This topic is considered in Chapter 7.

2.2 ANALYSIS PROCEDURES: OVERVIEW

From the most general point of view, analysis of structural response to earthquake motion should consider as many components of ground motion as the number of degrees of freedom of the base: six for rigid base, many more for flexible foundations (spread footings without foundation girders), dams, bridges, etc. However, standard strong motion instruments do not record rotational components of ground motion and, in the past, have usually not been installed in a dense enough grid to provide information on variation of ground motion over distances of the order of base dimensions of structures. In order to study their possible effects on structural response, ground rotations and spatial variations in translational motions have been estimated by idealized assumptions about the ground motion[2.1, 2.2]. In most analyses, however, no more than the three translational components (two horizontal and one vertical) of ground motion are considered.

Existing computer programs for inelastic, three-dimensional dynamic analysis of structures[2.3, 2.4] do not have extensive collection of structural elements appropriate for buildings and are therefore of limited application in building analysis. Complete three-dimensional analysis of buildings is a formidable task. Usually separate planar models in the orthogonal horizontal directions are analyzed for one component of ground motion; computer programs have been developed to implement such analyses[2.5–2.7]. However, planar analyses may not always be reasonable even for a building with nearly coincident centers of mass and resistance, because strong coupling between lateral motions in two orthogonal directions and torsional motions can arise if the lower natural frequencies are nearly coincident or due to asymmetric inelastic effects. In a model which has been employed to account for these coupling effects with a minimum of computational effort, the building is idealized as an assemblage of plane frames linked by rigid floor diagrams, with no enforcement of compatibility for vertical and rotational displacements at joints common to two or more frames[2.8, 2.9]. This model may be employed for buildings with floor diaphragms sufficiently stiff in their own plane relative to the vertical elements of the lateral force resisting structural system, and where the axial shortening of columns and lack of compatibility at joints common to two or more frames is not a significant factor in the response.

Results of nonlinear response history analysis of mathematical models of complex buildings would be reliable only if the model were representative of the building vibrating at large amplitudes of motion — large enough to cause significant yielding. Extensive experimental research on force-deformation behaviour of structural components at large deformations and shaking table studies on models of complete structures have improved our understanding of inelastic properties of structures. However, it is still difficult to construct mathematical models that lead to satisfactory results and are not complicated to the point of becoming impractical for analysis of complex structures. Furthermore, because dispersion in nonlinear responses to several design ground motions is rather large[2.10, 2.11], reliable results can be achieved only by calculating response to several representative ground motions — recorded accelerograms and simulated motions — and examining the statistics of response. Inelastic response history analyses, especially three-dimensional analyses, would therefore be an impractical requirement in the design of most buildings.

The most commonly used of the simpler methods of analysis are based on the approximation that effects of yielding can be accounted for by linear analysis of the building using the design spectrum for inelastic systems, determined from the elastic design spectrum and allowable ductility factor. Forces and displacements due to each horizontal component of ground motion are separately determined by analysis of an idealization of the building with one lateral degree of freedom per floor in the direction of the ground motion component being considered. Such analysis may be carried out by either the modal analysis procedure or a simpler method which will be referred to as the equivalent lateral force procedure. Both procedures lead directly to lateral forces in the direction of the ground motion component being considered. The main difference between the two procedures lies in the magnitude and distribution of the lateral forces over the height of the building. In the modal analysis procedure the lateral forces are based on properties of the natural vibration modes of the building, which are determined from the distribution of mass and stiffness over height; but in the equivalent lateral force procedure the magnitude of forces is based on an estimate of the fundamental period, and their distribution on simple formulas appropriate for regular buildings. Otherwise the two procedures have similar capabilities and are subject to the same limitations; the latter are discussed in Section 2.8.

The direct results from either procedure are the effects of lateral forces in the direction under consideration: story shears, floor deflections, and story drifts. The story moments also are obtained directly in the modal analysis procedure but a correction factor has to be introduced in the equivalent lateral force procedure. The following effects are all considered in the same simple, indirect manner in the two procedures: effects of the horizontal component of ground motion perpendicular to the direction under consideration in the analysis, of torsional motions of the structure, of vertical motions of the structure due to horizontal ground motions, of the vertical component of ground motion, and of gravity loads — the so-called *P*-delta effects.

A preliminary design of the building must be available before the modal analysis procedure or any of the more rigorous procedures can be implemented, because these procedures require the mass and stiffness properties of the building. The equivalent lateral force procedure is ideally suited for preliminary design, and is needed even if more refined analysis procedures will be used in the final design process.

2.3 DESIGN RESPONSE SPECTRA

The response spectrum[2.10, 2.11] is defined as a graphical relationship of maximum response of a single-degree-of-freedom elastic system with damping to dynamic motion or forces. The most usual measures of response are maximum displacement, D, which is a measure of the strain in the spring

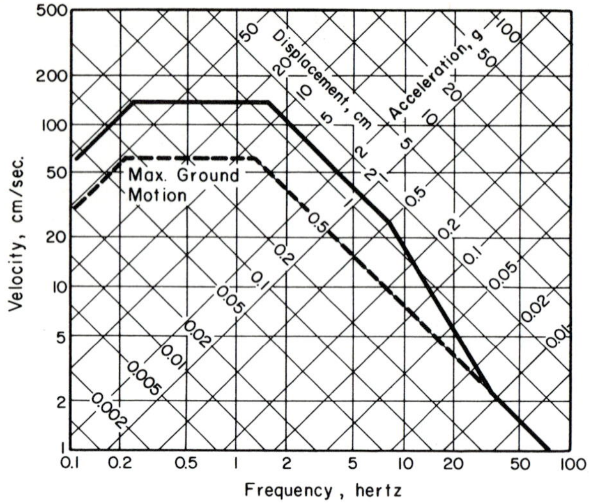

Fig. 2.1 Elastic design spectrum, horizontal motion, (0.5 *g max. acceleration,* 5% *damping, one sigma cumulative probability*)

element of the system, maximum pseudo relative velocity, V, which is a measure of the energy absorption in the spring of the system, and maximum pseudo acceleration, A, which is a measure of the maximum force in the spring of the system. Although actual response spectra for earthquake motions are quite irregular, they have the general shape of a trapezoid or tent: a simplified spectrum is shown in Figure 2.1, plotted on a logarithmic tripartite graph, and modified so that the various regions of the spectrum are smoothed to straight line portions. On the same graph are shown the maximum values of ground acceleration, velocity and displacement. The figure therefore indicates that the spectral values may be interpreted as the ground motion maxima multiplied by amplification factors depending on the frequency region — whether displacement, velocity or acceleration controlled — of the spectrum. Values of amplification factors for various amounts of damping are shown in Table 2.1 for two levels of probability considering the variation as lognormal. (See Section 1.8.)

At any specific frequency, f, the relations between the values of D_f, V_f, and A_f are defined as follows:

$$V_f = \omega D_f \tag{2.1}$$

$$A_f = \omega V_f = \omega^2 D_f \tag{2.2}$$

where ω is the circular natural frequency, $2\pi f$.

Recommended damping values for various materials and structural types are given in Table 2.2.

TABLE 2.1 Spectrum amplification factors for horizontal elastic response

Damping, % critical	One sigma (84.1%)			Median (50%)		
	A	*V*	*D*	*A*	*V*	*D*
0.5	5.10	3.84	3.04	3.68	2.59	2.01
1	4.38	3.38	2.73	3.21	2.31	1.82
2	3.66	2.92	2.42	2.74	2.03	1.63
3	3.24	2.64	2.24	2.46	1.86	1.52
5	2.71	2.30	2.01	2.12	1.65	1.39
7	2.36	2.08	1.85	1.89	1.51	1.29
10	1.99	1.84	1.69	1.64	1.37	1.20
20	1.26	1.37	1.38	1.17	1.08	1.01

Let us consider the case in which a simple mass-spring oscillator deforms inelastically with a relation of resistance to relative displacement as in Figure 2.2. It is convenient to use an elasto-plastic resistance displacement relation because one can draw response spectra for such a relation in generally the same way as spectra are drawn for elastic conditions.

In Figure 2.2, the elasto-plastic approximation to the actual curvilinear resistance-displacement curve is drawn so that the areas under both curves are the same at the 'effective' elastic limit displacement u_y and at the selected value of maximum displacement u_m. The 'ductility' factor μ is defined as

$$\mu = u_m / u_y \qquad (2.3)$$

In Figure 2.3 are shown both elastic and inelastic spectra for both acceleration and total displacement. Here the symbols D, V, A, A_0 refer to the bounds of the elastic spectrum, the symbols D', V', A', A_0 to the bounds of the elasto-plastic spectrum for acceleration, and the symbols D, V, A'', A_0'' to the bounds for the elasto-plastic spectrum for displacement. The symbol A_0 refers to the maximum ground acceleration.

In general, for small excursions into the inelastic range, when the latter is considered to be approximated by an elasto-plastic resistance curve, the acceleration response spectrum is decreased by a factor which is one over the ductility factor, $1/\mu$. Then the reduction for the two left-hand portions (D and V) of the elastic response spectrum shown in Figure 2.3 (to the left of a frequency of about 2 Hz) is by the factor $1/\mu$, and by the factor of $1/(2\mu-1)^{1/2}$ in the constant acceleration portion (A) to the right, roughly between frequencies of 2 and 8 Hz. There is no reduction beyond about 33 Hz. With this concept, one can arrive at design spectra that take account of inelastic action.

TABLE 2.2 Recommended damping values

Stress level		*Type and condition of structure*	*Percentage of critical damping*
Working stress, no more than about $\frac{1}{2}$ yield point	(a)	Vital piping or equipment	1 to 2
	(b)	Welded steel, prestressed concrete, well reinforced concrete (only slight cracking)	2 to 3
	(c)	Reinforced concrete with considerable cracking	3 to 5
	(d)	Bolted and/or riveted steel, wood structures with nailed or bolted joints	5 to 7
At or just below yield point	(a)	Vital piping or equipment	2 to 3
	(b)	Welded steel, prestressed concrete (without complete loss in prestress)	5 to 7
	(c)	Prestressed concrete with no prestress left	7 to 10
	(d)	Reinforced concrete	7 to 10
	(e)	Bolted and/or riveted steel, wood structures with bolted joints	10 to 15
	(f)	Wood structures with nailed joints	15 to 20

2.3.1 Modification of spectra for large periods or very low frequencies

The response spectra of Figures 2.1 and 2.3 have a constant velocity response, V, in the range of frequencies below about 2 Hz, with a cutoff to a constant displacement line below about 0.2 Hz. For structures with long periods, greater than about 1 s, the spectral values are not conservative enough for the lower modes of vibration in the modal analysis procedure and definitely not conservative enough for the equivalent lateral force procedure described in Section 2.2. In order to arrive at a more conservative spectrum for the design, taking account of various uncertainties involved in the combination of modal responses, and other factors, the following procedure is suggested. In the range of frequencies below 1 Hz, the velocity spectral response value should be taken to vary as the reciprocal of the frequency to the $\frac{1}{3}$ power, or as the period to the $\frac{1}{3}$ power, instead of being constant as shown in the figures. This corresponds to spectral acceleration values in this range varying as the reciprocal of the period to the $\frac{2}{3}$ power, or directly as the frequency to the $\frac{2}{3}$ power, instead of varying as the reciprocal of the period or directly as the frequency, as in the figures. In either case, it is suggested that the spectrum be considered to correspond to a constant displacement equal to the amplified ground displacement for periods longer than about 6 s, or frequencies lower than about $\frac{1}{6}$ Hz.

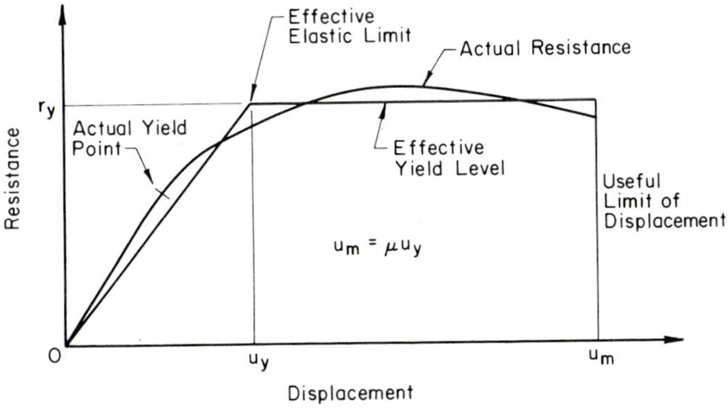

Fig. 2.2 Resistance–displacement relationship

2.4 BUILDING PROPERTIES AND ALLOWABLE DUCTILITY FOR ANALYSIS

2.4.1 Mass and stiffness

The masses to be used in the analysis of a building include all of the fixed dead and live loads, plus reasonably probable values of the variable or movable live loads. In general it need not be assumed that the design live load for which the floor is designed, for example, should be applied everywhere on each floor at the same time in determining the masses at the various levels in a building subjected to earthquake motions. It is appropriate to use the same proportion of live load that is used as a cumulative factor in the design of the columns in the building. Snow loads can be treated as live loads in the same way, with some average snow load mass taken into account in the design of the building frame.

In the computation of stiffness of steel members and framing the behaviour of the joints and connections must be taken into account by using center-to-center distances of members between joints, with consideration of the increased stiffness properties near the joints of welded construction, or possibly a decrease for riveted or bolted construction in which the bolts are not high tensile bolts. In reinforced concrete sections, the members are not likely to be completely uncracked, nor is it reasonable to assume that all sections have the reduced stiffness that a cracked section has. For this reason, it is appropriate to use an average of the moments of inertia for stiffnesses between cracked and completely uncracked sections or as between net and gross sections for reinforced concrete members, unless they clearly are stressed at such low levels or that shrinkage is so limited that cracking is not likely. Interaction of floor systems with other transverse members such as

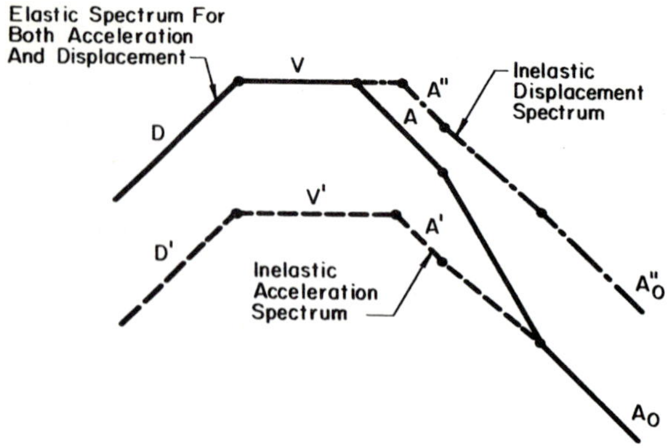

Fig. 2.3 Design spectra for earthquakes

beams and girders should be taken into account when the floor system acts as a stiffening element in flexure.

The contribution of nonstructural elements both to mass and to stiffness should be considered in the design even though these may not be used in developing the required strength of the structure. The added stiffness and mass may contribute to increase in forces which must be resisted by the load-bearing elements of the structure.

2.4.2 Damping and ductility

Damping levels are of course dependent on the level of deformation or strain in a structure. This is reflected in the recommended damping values given in Table 2.2, where values are suggested in which the percentages of critical damping are given for working stress levels or stress levels no more than $\frac{1}{2}$ the yield point, and for levels of deformation corresponding to stresses at or just below yield levels. In the table the lower values are to be used for structures in which considerable conservatism in the design is desirable, and the upper levels for ordinary structures in general.

Ductility factors for structures are used in a way which involves a general reduction in the design spectrum. Hence some reasonable assessment of the allowable ductility factor is required. For this purpose one must be aware of the differences between the various kinds of ductility factors involved in the response of structures to dynamic loading. In this respect one must make a distinction between ductility factor of a member, such as the rotational hinge capacity at a joint in a flexural member, the ductility factor of a floor or story in a building, and the overall ductility factor of the building for use in the computation of base shear from the response spectral values.

The ductility factor of a member, or of a floor, or the overall ductility factor are all governed by the development of a resistance-displacement relation, with the displacement being the longitudinal deformation in a tensile or compression member, the rotation at a joint or connection in a flexural member, or the total shearing deformation in a shear wall. The story ductility factor is essentially defined by use of a relationship in which the displacement is the relative story deflection between the floor above and the floor beneath. The overall ductility factor is some weighted average in general of the story ductility factors, and is defined best by considering a particular pattern of displacement corresponding to the preferred mode of deformation of the structure in a response condition in which inelastic energy is absorbed as generally as it is possible to develop such deformation throughout the structure.

It can be seen from the discussion that the member ductility factor may be considerably higher than the story ductility factor, which in turn may be somewhat higher than the overall ductility factor. In order to develop an overall ductility factor of 3 to 5 in a structure, the story ductility factors may have to vary between 3 to 8 or 10, and the individual member ductility factors probably will lie in the range of 5 to 15 or even more. In this regard it must be remembered that the ductility factor as defined herein is given by the ratio of the maximum permissible deformation to the deformation at the 'effective' yield displacement, rather than the ratio of the maximum deflection to the elastic limit deflection or displacement, as shown in Figure 2.2.

Ductility factors for steel are generally higher than for reinforced concrete, and for steel structures ductility factors are higher for tension than for bending and higher for bending than for compression. Ductility factors in shear are intermediate between the values in bending and compression, generally. However, the development of high ductility factors in flexure or in compression requires that the thickness of outstanding unsupported legs of members be limited in general to a value of the order of a thickness greater than $\frac{1}{6}$ the width of the outstanding leg to develop a ductility factor of the order of 6 or so in compression or on the compression side of flexural members. In any case, the yield value for a member that buckles either overall or locally is at some level just below the load where buckling or wrinkling begins, and is generally quite limited.

For reinforced concrete, the ductility is a function of the state of stress and the arrangement of reinforcement. Ductility factors of the order of 10 or more in flexure of reinforced concrete beams with equal amounts of compression and tension reinforcement are not difficult to attain. However, without the compressive reinforcement, the ductility factor is less for higher percentages of steel and is inversely proportional to the amount of steel, with a value of the order of 10 being the maximum for 1% of tensile steel reinforcement. In members in shear the ductility factor reached is also a function of the arrangement and placement of steel, and in general does not exceed values of the order of about 3, and possibly is even less for members with a high amount of compression, such as reinforced concrete columns. However, higher ductility factors can be reached if the concrete subjected to high compression is contained in some manner, such as by spiral reinforcement in which case ductility factors of the order 4 to 6 may be reached. For shear walls with diagonal as well as horizontal and vertical reinforcement, ductility factors of 4 to 6 are possible. In timber, ductility factors of the order of 2 to 4 are possible, and in masonry lower ductility factors, of the order of 1 to 3, may be reached depending upon whether the masonry is reinforced or not. A maximum of about 1.3 is the upper limit for unreinforced masonry.

The preceding paragraph provides a basis for selecting, depending on the structural material, a ductility factor for calculation of design forces by the procedures presented in Sections 2.5 and 2.6. However, if the ductility demands imposed by the design earthquake are expected to vary considerably over the various stories, ductility factors smaller than those recommended above should be selected for use in the analysis procedures. In general, for use of a ductility factor in the range of 5 to 6, the smallest ratio of story shear capacity to computed story shear should not be less than 80% of the average of these ratios for all stories. For use of a ductility factor of 4 the corresponding ratio should not be less than 67% of the average ratio.

2.5 MODAL ANALYSIS PROCEDURE

The modal method or the mode superposition method[2.10,2.11,2.12] is generally applicable to analysis of dynamic response of complex structures in their linear range of behaviour, in particular to analysis of forces and deformations in multistory buildings due to medium intensity ground shaking causing moderately large but essentially linear response of the structure. The method is based on the fact that, for certain forms of damping — which are reasonable models for many buildings — the response in each natural mode of vibration can be computed independently of the others, and the modal responses can be combined to determine the total response. Each mode responds with its own particular pattern of deformation, the mode shape, with its own frequency, the modal frequency, and with its own modal damping, and the time history of each modal response can be computed by analysis of a single-degree-of-freedom (SDOF) oscillator with properties chosen to be representative of the particular mode and the degree to which it is excited by the earthquake motion. Independent SDOF analysis of the response in each natural vibration mode is, of course, a very attractive feature of modal analysis. Even more significant is the fact that, in general, the response need be determined only in the first few modes because response to earthquakes is primarily due to the lower modes of vibration. For buildings, numerous full-scale tests and analyses of recorded motion during earthquakes have shown that the use of modal analysis, with viscously damped single-degree-of-freedom oscillators describing the response of vibration modes, is an accurate approximation for analysis of linear response.

A complete modal analysis provides the history of response — forces, displacements, and deformations — of a structure to a specified ground acceleration history. However, the complete response history is rarely needed for design; the maximum values of response over the duration of the earthquake usually suffice. Because the response in each vibration mode can be modeled by the response of a SDOF oscillator, the maximum response in the mode can be directly computed from the earthquake response spectrum, and procedures for combining the modal maxima to obtain estimates (but not the exact value) of the maximum of total response are available.

Strictly speaking, the modal method, which is applicable only to analysis of linear response, cannot be used for calculation of the design forces for buildings, because they are usually designed to deform significantly beyond the yield limit during moderate to very intense ground shaking. However, it is believed that for many buildings satisfactory approximations to the design forces and deformations can be obtained from the modal method by using he design spectrum for inelastic systems (Section 2.3) instead of the elastic response spectrum. Situations where such approximations may be in considerable error are discussed in Section 2.8. In what follows, the modal method is presented first for elastic systems and later for inelastic systems; the presentation is based on recent design recommendations[2.13].

In its most general form, the modal method for linear response analysis is applicable to arbitrary three-dimensional structural systems. However, for purposes of design of buildings it can often be simplified from the general case by restricting consideration to lateral motion in a plane. Planar models appropriate for each of two orthogonal lateral diretions are analyzed separately and results of the two analyses and the effects of torsional motions of the structures are combined in accordance with Section 2.7.

2.5.1 Structural idealization

The mass of the structure is lumped at the floor levels; only one degree of freedom — the lateral displacement in the direction for which the structure is being analyzed — per floor is required, resulting in as many degrees of freedom as the number of floors.

2.5.2 Modal periods and shapes

The periods and shapes of vibration are required for each of those natural modes of vibration which may contribute significantly to the total design quantities. These should be associated with moderately large, but essentially linear response of the structure, and the calculations should include only those building components which are effective at these amplitudes. Such periods may be longer compared to those obtained from a small-amplitude vibration test of the building when completed, or the response to small earthquake motions, because of the stiffening effects of the nonstructural and architectural components at small amplitudes. During response to strong ground shaking, however, the measured responses of buildings have shown that the periods lengthen, indicating the loss of stiffness contributed by these components.

Several methods for calculating natural periods and associated mode shapes of vibration of a structure are available[2.10,2.11,2.14]. These calculations can be carried out readily by standard computer programs.

As mentioned earlier, responses of buildings to earthquake motion are usually due mostly to the first few modes of vibration. For determining design values of forces and deformations, three modes of vibration in each lateral direction are nearly always sufficient for low- and medium-rise buildings, but more modes may be necessary in case of high-rise buildings; six modes in each direction would generally be sufficient, although more would be required for chimneys and for tall buildings in which the behaviour is essentially like that of a cantilever flexural beam.

2.5.3 Modal responses

Maximum responses in each natural mode of vibration can be expressed in terms of the modal properties and the earthquake response spectrum. For the *n*th mode, the base shear

$$V_{on} = \frac{A_n}{g} W_n \tag{2.4}$$

in which A_n is the ordinate, corresponding to the *n*th mode of vibration having the natural period T_n and damping ratio ξ_n, of the pseudo-acceleration response spectrum; g is the acceleration of gravity; and W_n the effective weight or the portion of the weight of the building that participates in the *n*th mode (see Ref. 2.11, p. 571),

$$W_n = \frac{\left[\sum_{i=1}^{N} w_i \varphi_{in}\right]^2}{\sum_{i=1}^{N} w_i \varphi_{in}^2} \tag{2.5}$$

where w_i is the mass lumped at the *i*th floor level, φ_{in} is the modal displacement of the *i*th floor, and N is the total number of floor levels. Equation 2.5 will give values of W_n that are independent of how the modes are normalized. The lateral force at the *i*th floor level in the *n*th mode of vibration is

$$f_{in} = V_{on} \frac{w_i \varphi_{in}}{\sum_{j=1}^{N} w_j \varphi_{jn}} \tag{2.6}$$

In applying the forces at the various floor levels to the building, their direction is controlled by the algebraic sign of φ_{in}. Hence, the modal forces for the fundamental mode will all act in the same direction; for the second or higher modes they will change direction as one moves up the structure. The form of Equations 2.4 and 2.6 is different from that usually employed[2.11], and is chosen here to highlight the relationship between the modal analysis procedure and the equivalent lateral force procedure (Section 2.6).

The subsequent calculation of internal forces — story shears and story moments — and deflections associated with the lateral forces for each mode does not involve any dynamic analysis. The lateral loads are applied at each floor level, statics is used to calculate the story shears and story moments, and a static deflection analysis to determine the floor deflections. The latter is not necessary, however, because floor deflections v_{in} due to the lateral forces f_{in} in the nth mode (Equation 2.6), are proportional to the mode shape, and the two are related rather simply:

$$v_{in} = \frac{1}{\omega_n^2} \frac{g}{w_i} f_{in} \tag{2.7}$$

where $\omega_n = 2\pi/T_n$ is the circular frequency of the nth natural mode of vibration.

The story shears and moments in individual modes will be combined to determine their total values, as presented in the next part of this section, and these total values are distributed to the various frames and walls that make up the lateral force resisting system (Section 2.7). Whereas it is convenient in all cases and satisfactory for many buildings to defer such distribution until after the modal values of story forces have been combined, the results may be in error for walls and braced frames. For better evaluation of shears and moments at various levels in walls and braced frames, individual modal values for these quantities should be determined, by appropriate distribution of the modal values of story shears and moments (Section 2.7), and combined in accordance with the procedures presented later in this section.

Similarly deformation quantities should not be determined from the total (after combining the modal values) floor displacements, but individual modal values should be determined and combined in accordance with the procedures presented later in this section. For example, Δ_i, the drift in Story i should be determined by combining the modal values

$$\Delta_{in} = v_{in} - v_{i-1,n} \tag{2.8}$$

2.5.4 Total responses

As mentioned earlier, total responses of an elastic structure are the superposition of responses in the natural modes of vibration of the structure, and the maximum responses in individual modes of vibration can be determined from the earthquake response spectrum. Because, in general, the modal maxima r_n do not occur simultaneously during the ground shaking, they cannot be directly superimposed to obtain r, the maximum of the total response. The direct superposition of modal maxima, however, provides an upper bound to the maximum of total response:

$$r \leq \sum_{n=1}^{N} |r_n| \tag{2.9}$$

This estimate is often too conservative and is therefore not recommended. A satisfactory estimate of the total response can usually be obtained from the root-sum-square of the modal responses:

$$r \simeq \sqrt{\Sigma r_n^2} \tag{2.10}$$

in which, as discussed earlier, only the lower few modes need to be included in the summation.

The maximum value of any response — story shear, story moment, shear and moment at various levels in braced frames and walls, floor displacement, story drift, etc. — can be estimated by combining the modal values for that response in accordance with Equation 2.10. The quality of this estimate is generally good for systems with well separated frequencies, a property typically valid for the building idealization adopted here, wherein only the lateral motion in a plane is considered. Improved combination formulas are available for systems lacking this property[2.10]; also see Chapter 3.

2.5.5 Application in inelastic systems

The modal analysis procedure described above is strictly valid only for systems in their linearly elastic range of behaviour. With the following modifications, it may, however, be employed as an approximate procedure for analysis of nonlinear responses. In Equation 2.4, replace A_n, the ordinate of the pseudo acceleration response spectrum for a linearly elastic system with vibration period T_n and damping ratio ζ_n, by A'_n, the corresponding value for a nonlinear system, with the same period of small amplitude vibration and damping ratio, which is determined from the design response spectrum (Section 2.3) for the selected ductility factor (Section 2.4). The base shear for the nth mode is now given as:

$$V_{on} = \frac{A'_n}{g} W_n \tag{2.11}$$

Multiply displacements calculated from Equation 2.7 by μ, the allowable ductility factor to obtain the total deflections in the nth mode:

$$v_{in} = \mu \frac{1}{\omega_n^2} \frac{g}{w_i} f_{in} \tag{2.12}$$

2.5.6 Earthquake design responses

Two independent analyses by the modal analysis procedure described above lead to the effects of lateral forces associated with ground motion in two orthogonal directions. The design forces and deformations due to earthquake effects are determined by combining the results of these independent analyses, and including the effects of torsional motions of the structure, of vertical motions of the structure due to horizontal ground motion, of the vertical component of ground motion, and the *P*-delta effects (Section 2.7).

2.6 EQUIVALENT LATERAL FORCE PROCEDURE

Although the building is idealized in the same manner as in the modal analysis procedure, the equivalent lateral force procedure requires less effort because, except for the fundamental period, the periods and shapes of the higher natural modes of vibration are not needed. The magnitude of lateral forces is based on an estimate of the fundamental period of vibration and their distribution on simple formulas appropriate for buildings with regular distribution of mass and stiffness over height. Situations where results from the equivalent lateral force procedure may not be satisfactory are discussed in Section 2.9. A number of versions of this procedure, differing in detail but based on the same underlying concepts, can be found in various building codes[2.15].

Planar models appropriate for each of two orthogonal lateral directions are analysed separately as described next; the results of the two analyses and the various effects (Section 2.7), including those due to torsional motions of the structure, are combined in accordance with Section 2.7.

2.6.1 Fundamental period of vibration

Methods of mechanics cannot be employed to calculate the vibration period before a design, at least a preliminary one, of the building is available. Simple formulas which involve only a general description of the building type — e.g., steel moment frame, concrete moment frame, shear wall system, braced frame, etc. — and overall dimensions such as height and plan size are therefore necessary to estimate the vibration period so that the base shear can be computed to provide the initial design. Because pseudo-acceleration values in design spectra for inelastic systems with moderate to large values of allowable ductility factor generally decrease with increasing values of vibration period, it is desirable to underestimate the fundamental period so that the computed base shear is conservative, except for very high frequencies or short periods.

A recently recommended formula for moment-resisting frame buildings is[2.13]:

$$T = C_T H^{3/4} \tag{2.13}$$

where $C_T = 0.035$ and 0.025 for steel and concrete frames, respectively, and H is the height in feet of the building. A commonly used formula for reinforced concrete shear-wall buildings and braced steel frames is[2.16]:

$$T = \frac{0.05H}{\sqrt{L}} \tag{2.14}$$

where L is the plan dimension in feet in the direction of analysis.

The fundamental vibration period of exceptionally stiff or light buildings may be significantly shorter than the estimate provided by formulas such as Equations 2.13 and 2.14. Especially for such buildings and as a general check, the period for an initial design of the building should be computed by established methods of mechanics[2.10,2.11]. An approximate formula based on Rayleigh's method is especially convenient:

$$T = 2\pi \left[\frac{\sum_{i=1}^{N} w_i v_i^2}{g \sum_{i=1}^{N} f_i v_i} \right]^{1/2} \tag{2.15}$$

in which the $v_i (i = 1,2,\ldots N)$ quantities are the static lateral displacements at the various floor levels, computed on linear elastic basis, due to a set of lateral forces f_i. Any reasonable distribution for f_i may be selected but it is convenient to use the lateral forces computed, as described later in this section, from the empirical estimate of the fundamental period.

Nonstructural elements participate in the behaviour of the structure even though the designer may not rely on them for contributing strength or stiffness to the structure. To ignore them would lead to longer periods and usually smaller design forces, and may cause underestimates of torsion; hence they should be considered in calculating the period and in the other aspects of the analysis.

2.6.2 Lateral forces

The distribution of lateral forces over the height of a building is generally quite complex because a number of natural modes of vibration contribute significantly to these forces. The contributions of the various vibration modes to the lateral forces and to the base shear depend on a number of factors, including shape of the design response spectrum, and natural vibration periods and mode shapes — which in turn depend on the mass and stiffness properties of the building. However, these forces are, in a large part, due to the first (fundamental) mode of vibration. Thus, in the equivalent lateral force procedure, they are determined from formulas similar to those for the first mode (Equations 2.4–2.6 with $n = 1$) but modified to approximately account for the effects of the higher modes.

The following formulas for V_0, the base shear, and f_i, the lateral force at each floor i, have been recommended[2.13].

$$V_0 = \frac{A_1'}{g} W \tag{2.16}$$

in which A_1' is the pseudo-acceleration corresponding to the estimated fundamental period and the appropriate damping, determined from the design response spectrum for the allowable ductility factor (Sections 2.3 and 2.4);

$$W = \sum_{i=1}^{N} w_i$$

is the total weight of the building;

$$f_i = V_0 \frac{w_i h_i^k}{\sum_{j=1}^{N} w_j h_j^k} \tag{2.17}$$

in which h_i is the height of the ith floor above the base and k is a coefficient related to the estimated fundamental vibration period as follows:

$$k = \begin{cases} 1 & T \leqq 0.5 \text{ s} \\ (T+1.5)/2 & 0.5 < T < 2.5 \text{ s} \\ 2 & T \geqq 2.5 \text{ s} \end{cases} \tag{2.18}$$

The basis for these formulas is presented in the following paragraph.

If W_1, the effective weight for the first mode, is replaced with W, the total weight, Equation 2.11 with $n=1$ will become identical with Equation 2.16 provided the same value of the fundamental period is used to determine A_1' in both cases. W_1 will always be smaller than W; typically values for W_1 are between 60 to 80% of W, depending on the distribution of weight over the height and the shape of the first mode. Equation 2.16 would therefore provide a value for the base shear which will be significantly larger than the first mode value; thus it indirectly and approximately accounts for the contributions of the higher modes of vibration.

If V_{01} is replaced with the total base shear V_0, Equation 2.6 will become identical with Equation 2.17 with $k=1$ if the first mode shape is a straight line; with $k=2$ if the first mode shape is a parabola with vertex at the base. Equation 2.17 with $k=1$ is appropriate for buildings having fundamental period of 0.5 sec or less, because the influence of vibration modes higher than the fundamental mode is small in earthquake responses of short-period buildings, and the fundamental vibration mode of regular buildings departs little from a straight line. Although earthquake responses of long-period buildings are primarily due to the fundamental mode of vibration, the influence of higher modes can be significant, and the fundamental mode lies between a straight line and a parabola with vertex at the base. The force distribution of Equation 2.17 with $k=2$ is therefore appropriate for buildings having fundamental period of 2.5 s or longer. Linear variation of k between values of 1 at a period of 0.5 s, and 2 at 2.5 s provides a simple transition between the two extreme values.

2.6.3 Story forces

Story shears are related to the lateral forces by equations of statics. The shear in any story is simply the sum of the lateral forces at floor levels above that story.

The story moments can be similarly determined from the lateral forces and heights of various stories by methods of statics. However, there are reasons for reducing the statically computed overturning moments to obtain design values:

(1) The distribution of story shears over height computed from the lateral forces of Equation 2.17 is intended to provide an envelope; because of the contributions of several modes, shears in all stories do not attain their maximums simultaneously during an earthquake. Thus the story moments statically consistent with the envelope of story shears will be overestimates.

(2) It is intended that the design shear envelope based on the simple distribution of lateral forces of Equation 2.17 be conservative. If the shear in some story is close to the exact value, the shears in almost all other stories are likely to be overestimated. Hence, the story moments statically consistent with the design shears will be overestimates.

(3) Under the action of overturning moments, one edge of the foundation may lift from the ground for short durations of time. Such behavior leads to substantial reduction in the axial forces in columns and other vertical members, caused by overturning.

Results of dynamic analysis taking into account the first two of the foregoing reasons[2.10] suggest that up to 20% reduction is, in general, reasonable for the story moments computed statically from the envelope of story shears, but no reduction should be permitted in the upper stories of a building. Rotational inertia due to axial deformations of columns and/or base rotation, in turn caused by soil–structure interaction or rotational components of ground motion[2.1], increases moments considerably, near the top. In any case, there is hardly any benefit in reducing the story moments near the top of buildings, because design of vertical elements near the top is rarely governed by these moments. Consequently, the following values have been recommended for the reduction factor by which the statically computed story moments should be multiplied: 1.0 for the top 10 stories; between 1.0

and 0.8 for the next 10 stories from the top, linearly varying with height; 0.8 for the remaining stories.

Formerly many building codes and design recommendations, including the 1968 SEAOC recommendations[2.17], allowed large reduction in overturning moments relative to their values statically consistent with the story shears. These reductions appeared to be excessive in light of the damage to buildings during the 1967 Caracas earthquake, where a number of column failures were primarily due to effects of overturning moment. In later (1973) versions of the SEAOC recommendations[2.18], no reduction was allowed. However, no reduction at all is unjustified in light of the reasons mentioned above and results of dynamic analysis[2.10].

Methods for distributing story shears and moments to the various frames and walls that make up the lateral force resisting system are presented in Section 2.7.

2.6.4 Deflections and drifts

A static deflection analysis that assumes linear behaviour of the building will lead to a set of floor deflections. In order to account for the inelastic effects, these deflections should be multiplied by the ductility factor (Section 2.4) that was used in establishing the design spectrum, resulting in the total deflections. The drift in a story is computed as the difference of the deflections of the floors at the top and bottom of the story under consideration.

2.6.5 Earthquake design responses

Effects of lateral forces associated with ground motion in two orthogonal directions can be determined from two independent analyses by the equivalent lateral force procedure presented above. The design forces and deformations due to earthquake effects are determined by combining the results of these independent analyses, and including the other effects mentioned in Section 2.7.

2.7 SPECIAL FACTORS

2.7.1 Torsion

Torsional responses in structures arise from two sources: eccentricity in the mass and stiffness distributions, causing a torsional response coupled with translational response; and torsion arising from accidental causes, including uncertainties in the masses and stiffnesses, the differences in coupling of the structural foundation with the supporting earth or rock beneath; and wave propagation effects in the earthquake motions that give a torsional input to the ground, as well as torsional motions in the earth itself during the earthquake.

In general, the torsion arising from eccentric distributions of mass and stiffness can be taken into account by ascribing an incremental torsional moment in each story equal to the shear in that story multiplied by the eccentricity between centres of mass and resistance for that story, perpendicular to the direction of applied ground motion. A precise evaluation of the torsional response is quite complicated because it is necessary to make essentially a two- or three-dimensional response calculation, taking into account the coupled lateral torsional vibration modes of the entire structure. However, it is probably adequate to approximate the response by summing from the top story the incremental torsional moments computed as described above, to obtain the total torsional moment in each story.

The 'static' torsional responses in each story are then determined by computing the twist in each story obtained by dividing the total torsional story moment by the story *rotational* stiffness. These twists are then added from the base upward to obtain the total twisting or torsional response at each floor level.

Since these are static responses, they should be amplified for dynamic response using the response spectrum amplification factor. It is probably adequate to use the factor corresponding to the response spectrum amplification factors in Table 2.1 for the fundamental torsional frequency of the structure. However, in many design codes no amplification whatsoever is used.

Accidental torsion may arise in various ways. One can take a value of accidental eccentricity of the order of about 5 to 10% of the plan dimension of the building story perpendicular to the direction of applied ground motion for the accidental torsional response. Most current building codes use a value of 5%. If one does this, one can consider the accidental torsion as an increase and also as decrease in the eccentricity corresponding to the distance between the centres of mass and resistances in the various stories, with consideration of increases in all levels or decreases in all levels to get two bounding values. The accidental torsion or the total torsion is computed in the same way as the 'real' torsion described above.

2.7.2 Distribution of shears and moments

The story shears arising from translational and from torsional response are distributed over the height of the building in proportion to the stiffnesses of the various elements in the building[2.13], with the translational shears being affected by the translational stiffnesses of the building, and the torsional shears being affected by the rotational stiffnesses of the building. The computed stiffness of the structure should take into account the stiffness of the floors and floor structure acting as a diaphragm or distributing element. The floor diaphragm can be considered infinitely stiff, and only the story stiffnesses are of importance. However, if the floor diaphragm is flexible and deforms greatly, the distribution of the forces becomes more nearly uniform than determined by the method discussed above. A simplified approach is possible by considering the relative displacements of the building due to translation, and that due to rotation of each story separately, as affected by the diaphragm or floor stiffness, with the stiffnesses being determined by the forces corresponding to a unit displacement in either translation or torsion, respectively. Then one distributes the shears due to translation or rotation in proportion to these stiffnesses.

The story moments are distributed to the various frames and walls that make up the lateral force system in a manner consistent with the distribution of story shears. In particular, the shears and moments in any frame or wall should be statically consistent.

2.7.3 Base or 'overturning' moments

The flexural moment about a horizontal axis at the base is of importance in connection with foundation design. The corresponding flexural moments at each floor level are important in connection with the calculation of vertical stresses in the columns and walls of the structure. These moments can be computed from modal analyses or from an effective lateral force analysis, as discussed in Sections 2.5 and 2.6. Modifications in the base flexural overturning moment can be made by use of the reduction factors given in Section 2.6 when the equivalent lateral force method is used. The modal analysis method takes account of these effects directly in the structure, although an additional reduction of approximately 20% of the moment may be allowed in the modal analysis for the foundation forces only. An additional reduction of approximately 10% may be allowed similarly, for the equivalent lateral force analysis.

2.7.4 Vertical component of ground motion

In highly seismic zones where the vertical acceleration of the ground may be large compared with the acceleration of gravity, the normal procedure of neglecting vertical accelerations in design may not be appropriate. Generally, the basis for neglecting these vertical accelerations is the fact that the building design usually provides for a high factor of safety in the vertical direction. The dynamic vertical response can be estimated by use of a vertical response spectrum to obtain the equivalent vertical forces for which the building must be designed. In making this computation, the response spectrum used for horizontal motion may be used as a basis, but with a scale factor of approximately two-thirds, which is generally the maximum ratio between vertical and horizontal accelerations in most ground movements due to earthquakes.

Elements which are particularly vulnerable to vertical components of ground motion are columns and walls in compression, and especially prestressed beams or other horizontal elements, and cantilevered elements, where the amplification factors for vertical response may be fairly large, or where there is a relatively small factor of safety for reversed or upward accelerations. In these instances, the amplified vertical acceleration as calculated from the response spectrum and from the period or the frequency of the particular element considered should be taken into account in the development of the design forces.

2.7.5 Combined effects of horizontal motions

Since the building responds in both horizontal directions at the same time and the stresses are caused by both motion inputs, as well as by the simultaneous vertical motion, it is necessary to consider the combined effects of the various directions of input earthquake motion. If these motions are computed separately and individually, one may make the combination in one of several ways. For example, let us define the effect at a particular point in a particular element, such as stress, moment, shear, etc., arising from the horizontal earthquake in one direction as F_1 and from the earthquake in the transverse horizontal direction as F_2. Let us also define the same effect arising from the vertical component of ground motion as F_3, in those instances where this cannot be neglected. In general, the combined effects may be computed as the square root of the sums of the squares of the individual effects, where the resultant effect F is given by the relation

$$F=(F_1^2+F_2^2+F_3^2)^{\frac{1}{2}} \tag{2.19}$$

When it is considered appropriate to neglect the vertical acceleration effect, F_3 may be taken as zero in Equation 2.19.

In some instances it is difficult to use this relationship, especially when the individual effects produce different components of stress, such as a maximum shear from the horizontal motion in the first direction, and a maximum normal stress from the horizontal motion in the second direction. In this case, the quantities to be combined by Equation 2.19 are not the same. For these cases, and for all cases as a general approximation, one may use instead the relationship that the effects of the different directions of motion are given approximately by assuming 1.0 times the effect of input motion in one direction, combined with $\frac{1}{3}$ of the effects in the other two directions. In order to use this relationship correctly, one must consider the maximum effect as being the one in which the factor of 1.0 is used, or alternatively one should take 1.0 times the inputs for each of the directions in turn, combined with $\frac{1}{3}$ of the inputs for the other two directions. The summation is made for the absolute values of the responses. It can be shown that this relationship is very close to that given by Equation 2.19. (See also Sections 1.10 and 3.4.3.)

2.7.6 Effects of gravity loads

The effect of gravity loads, when a structure deflects because of the horizontal motions transmitted to it, is to add a secondary moment owing to the eccentricity of the gravity loads acting through the lateral deflections corresponding to the horizontal responses. As a first approximation, one may compute the horizontal displacements and add the effects of the moments produced by the gravity loads directly. In making this calculation, the total horizontal deflection, including the inelastic portion of it, should be used, rather than the elastic component of the horizontal deflection. Consequently, the computed elastic deflection from the design shears and moments computed by the modal analysis or the equivalent lateral force procedure with a reduced design spectrum, must be increased by the factor μ, the ductility factor, in order to obtain the total deflections and the corresponding total gravity load moments or P-delta effects. A quick estimate of these moments can be made to determine whether the P-delta effect is important.

Unfortunately, when the P-delta effect is important, the above method underestimates the actual displacements because of the additional displacement caused by the additional moments accompanying the increment in deflection from the first step in the calculation described above. In effect, there is a series of corrections to be added, which requires the calculation of the successive increments in deflection caused by the P-delta effect and then the additional moments corresponding thereto in successive stages until convergence is reached.

One can make a very good approximation to this summation by considering the quantity θ as defining the relative increment in moment, stress or deflection, due to the first step in the P-delta calculation, and then computing the final moment by use of the factor $1/(1-\theta)$ as a multiplier times the effect computed without consideration of the P-delta effect. An alternative procedure is given in Ref. 2.19.

2.8 LIMITATIONS OF EQUIVALENT LATERAL FORCE AND MODAL ANALYSIS PROCEDURES

Following are the most important assumptions common to the equivalent lateral force procedure (Sections 2.6 and 2.7) and the modal analysis procedure (Sections 2.5 and 2.7): (1) Forces and deformations can be determined by combining the results of independent analyses of a planar idealization of the building for each horizontal component of ground motion, and including torsional moments determined on an indirect, empirical basis (Section 2.7). (2) Nonlinear structural response can be determined to an acceptable degree of accuracy by linear analysis of the building using the design spectrum for inelastic systems (Section 2.3). Both analysis procedures are likely to be inadequate if the dynamic response behaviour of the building is quite different from what is implied by these assumptions.

In particular, both methods may be inadequate if the lateral motions in two orthogonal directions and the torsional motions are strongly coupled. Buildings with large eccentricities of the centres of story resistance relative to the centres of floor mass, or buildings with close values of natural frequencies of the lower modes and essentially coincident centres of mass and resistance, exhibit coupled lateral-torsional motions. For such buildings independent analyses for the two lateral directions may not suffice, and at least three degrees of freedom per floor — two translational motions and one torsional — should be included in the idealized model. The modal method described in Section 2.5, with appropriate generalizations of the concepts involved, can be applied to analysis of the model. Because natural modes of vibration will show a combination of translational and torsional motions, in determining the modal maxima it is necessary to account for the fact that a given mode might be excited by both horizontal components of ground motion, and modes that are primarily torsional can be excited by translational components of ground motion. Because natural frequencies of a building with coupled lateral-torsional motions can be rather close to each other, the modal maxima should not be combined in accordance with the root-sum-square formula (Equation 2.10); instead a more general formula should be employed. (See, for example, Ref. 2.10, Chapter 10.)

The manner of combining the maximum responses due to the two horizontal components of ground motion depends on the correlation between these motions. As mentioned in Chapter 1: 'For earthquakes sufficiently intense to be of practical interest the intensities in all horizontal directions are comparable. It follows that there is little error in assuming that ground motions are uncorrelated in any two orthogonal horizontal directions'. With this assumption, the combined response considering both components can be estimated as the square root of the sum of squares of the responses to the two individual components.

The equivalent lateral force procedure (Section 2.6) and both versions of the modal method — the simpler version (Section 2.5) and the general version with three degrees of freedom per floor mentioned in the foregoing paragraphs — all are best applicable to analysis of buildings in which ductility demands imposed by earthquakes are expected to be essentially uniformly distributed over the various stories. For such buildings, the maximum ductility allowed for a particular structural system and material (Section 2.4) may be used in determining the inelastic response spectrum. If ductility demands are expected to be considerably different from one story to the next, a simple approach is to decrease the allowable ductility factors in establishing the inelastic response spectrum (Section 2.4), resulting in larger design forces. Whereas this simple approach is a step in the right direction, the foregoing analysis procedures may still err systematically on the unsafe side if the ductility demands are concentrated in a few stories of the building. For such buildings, the actual strength properties can be explicitly considered and the distribution of ductility demand determined, perhaps only by a nonlinear response analysis. However, as mentioned in Section 2.2, nonlinear analyses may not always be practical, nor lead to reliable results.

2.9 CHOICE BETWEEN EQUIVALENT LATERAL FORCE AND MODAL ANALYSIS PROCEDURES

Both procedures are based on the same basic assumptions and are applicable to buildings whose dynamic response behavior is in reasonable conformity with the implications of these assumptions (Section 2.8). The main difference between the two procedures lies in the magnitude of the base shear and distribution of the lateral forces. Whereas in the modal method the force calculations are based on computed periods and mode shapes of several modes of vibration, in the equivalent lateral force method they are based on an estimate of the fundamental period and simple formulas for distribution of forces which are appropriate for buildings with regular distribution of mass and stiffness over height. In what follows, a criterion to decide whether the equivalent lateral force procedure will be adequate in a particular situation is presented.

It would, in general, be adequate to use the equivalent lateral force procedure for buildings with the following properties: seismic force resisting system has the same configuration in all stories and in all floors; floor masses do not differ by more than, say, 30% in adjacent floors; and cross-sectional areas and moments of inertia of structural members do not differ by more than about 30% in adjacent stories. For other buildings, the following sequence of steps may be employed to decide whether the modal analysis procedure (Section 2.5) ought to be used:

(1) Compute lateral forces and story shears using the equivalent lateral force procedure (Section 2.6).
(2) Approximately dimension structural members.
(3) Compute lateral displacements of the structure as designed in step 2 due to lateral forces from step 1.
(4) Compute new sets of lateral forces and story shears by replacing h_i^k in Equation 2.17 with the displacements computed in step 3.
(5) If at any storey the recomputed storey shear (step 4) differs from the corresponding original value (step 1) by more than 30%, the structure should be analyzed by the modal analysis procedure. If the difference is less than this value the modal analysis procedure is unnecessary, and the structure should be designed using the story shears obtained in step 4; they represent an improvement over the results of step 1.

This method for determining whether modal analysis should be used is efficient, as well as effective. It requires far less computational effort than the use of the modal analysis procedure of Section 2.5, and in general, it will detect cases for which results from the equivalent lateral force procedure are in significant error; it will not, in general, indicate the need for modal analysis when its application would not significantly improve accuracy. It is of interest to note that steps 3 and 4 of this method are equivalent to a single cycle in an iterative procedure[2.10,2.11], based on the Stodola–Vianello iteration scheme and essentially the one proposed by Newmark[2.20], for calculating the fundamental mode of vibration.

The seismicity of the area and the potential hazard due to failure of the building should also be considered in deciding whether the equivalent lateral force procedure is adequate. For example, even irregular buildings, that may require modal analysis according to the criterion described, may be analyzed by the equivalent lateral force procedure if they are not located in highly seismic areas and do not house critical facilities necessary for post-disaster recovery or a large number of people.

ACKNOWLEDGEMENT

Much of the material contained in this chapter was derived from the authors' participation in the Applied Technology Council work[2.13] on the preparation of recommendations for earthquake-resistant design of structures in the United States. Acknowledgement is made to the various committee chairmen and committee members who participated in the ATC program and who had an influence on our views and concepts.

REFERENCES

2.1 Rosenblueth, E., 'Tall Buildings under Five-Component Earthquakes', *J. Struct. Div. Am. Soc. civ. Engrs*, Vol. 102, No. ST2, 455–459 (1976)

2.2 Chopra, A. K., Dibaj, M., Clough, R. W., Penzien, J. and Seed, H. B., 'Earthquake Analysis of Earth Dams', *Proc. Fourth World Conference on Earthquake Engineering*, Santiago, Chile (1969)

2.3 Bathe, K.-J. and Wilson, E. L., 'NONSAP–A Nonlinear Analysis Program', *Nuclear Engineering and Design*, Vol. 29, pp. 266–293 (1974)

2.4 Mondkar, D. P. and Powell, G. H., 'ANSR–I: General Purpose Computer Program for Analysis of Nonlinear Structure Response', *Report No. EERC* 75–37, Earthquake Engineering Research Center, Univ. California, Berkeley (1975)

2.5 Mahin, S. A. and Bertero, V. V., 'An Evaluation of Some Methods for Predicting the Seismic Behavior of Reinforced Concrete Buildings', *Report No. EERC* 75–5, Univ. California, Berkeley (1975)

2.6 Otani, S., 'SAKE — A Computer Program for Inelastic Response of Reinforced Concrete Frames to Earthquakes', *Structural Research Series No.* 413, Civil Engineering Studies, University of Illinois, Urbana-Champaign (1974)

2.7 Kanaan, A. and Powell, G. J., 'DRAIN–2D: A General Purpose Computer Program for Inelastic Dynamic Response of Plane Structures', *Report No. EERC* 73–6, Earthquake Engineering Research Center, Univ. California, Berkeley (1973)

2.8 Wilson, E. L. and Dovey, H. H., 'Three-Dimensional Analysis of Building Systems — TABS', *Repoert No. EERC* 72–8, Earthquake Engineering Research Center, Univ. California, Berkeley (1972)

2.9 Guendelman-Israel, R. and Powell, G. H., 'DRAIN–TABS: A Computer Program for Inelastic Earthquake Response of Three-Dimensional Buildings', *Report No. UCB/EERC*–77/08, Earthquake Engineering Research Center, Univ. California, Berkeley (1977)

2.10 Newmark, N. M. and Rosenblueth, E., *Fundamentals of Earthquake Engineering*, Prentice-Hall, Inc., Englewood Cliffs, N.J. (1971)

2.11 Clough, R. W. and Penzien, J., *Dynamics of Structures*, McGraw–Hill, New York (1975)

2.12 Thomson, W. T., *Vibration Theory and Applications*, Prentice-Hall, Inc., Englewood Cliffs, N.J. (1965)

2.13 'Tentative Provisions for the Development of Seismic Regulations for Buildings', prepared by Applied Technology Council, NBS Special Publication 510, U.S. Government Printing Office, Washington, D.C. (1978)

2.14 Bathe, K.-J. and Wilson, E. L., *Numerical Methods in Finite Element Analysis*, Prentice-Hall, Inc., Englewood Cliffs, N.J. (1976)

2.15 *Earthquake Resistant Regulations — A World List*, compiled by International Association for Earthquake Engineering, Gakujutsu Bunken Fukyu-Kai, Tokyo (1973)
2.16 *Recommended Lateral Force Requirements and Commentary*, Seismology Committee, Structural Engineers Association of California (1974)
2.17 *Recommended Lateral Force Requirements and Commentary*, Seismology Committee, Structural Engineers Association of California (1968)
2.18 *Recommended Lateral Force Requirements and Commentary*, Seismology Committee, Structural Engineers Association of California (1973)
2.19 Rosenblueth, E., 'Slenderness Effects in Buildings', *J. struct. Div. Am. Soc. civ. Engrs*, Vol. 91, No. ST1, 229–252 (1965)
2.20 Newmark, N. M., 'Numerical Procedure for Computing Deflections, Moments, and Buckling Loads', *Trans. Am. Soc. civ. Engrs*, Vol. 108, pp. 1161–1234 (1943)

3 Design: General

Luis Esteva

3.1 NATURE AND OBJECTIVES OF EARTHQUAKE RESISTANT DESIGN

Engineering design is rooted in society's need to optimize. It implies considering alternate lines of action, assessing their consequences and making the best choice. In earthquake engineering, every alternate line of action includes the adoption of both a structural system and a seismic design criterion, while assessing consequences implies estimating structural response and hence the expected cost of damage. The choice is based on comparison of initial, maintenance and repair costs for the various alternatives. However obvious these concepts may appear to the authors of design codes, they are often not explicit in those codes and hence they are not always present in the minds of those who apply design prescriptions to practical problems. Equally concealed within the regulations of seismic design codes are the approximations implicit in conventional criteria for the prediction of structural response: the accuracy of their predictions is often strongly dependent on the type of structural system considered. Base shear coefficients and design response spectra are taken as measures of response parameters, as the latter are usually expressed in terms of accelerations and equivalent lateral forces acting on linear systems. But these variables are no more than indirect measures of system performance during earthquakes: they serve to control the values of more significant variables, such as lateral deflections of actual nonlinear systems, global and local ductilities, and safety margins with respect to instability failure (second-order effects). Because the relations of control variables to actual response are affected by the type and features of the structural system, better designs will be obtained if these relations are understood and accounted for, in contrast with blindly applying codified recommendations. In seismic design more than in any other field of engineering, it is easy to fall on a strict — but blind — application of the most advanced regulations and yet to produce a structure bound to perform poorly. This chapter does not intend to summarize modern design specifications; it aims, instead, at discussing the main concepts on which they are based, analyzing their virtues and their weaknesses, and stating the conditions for which acceptable results are to be expected.

Codified values of design intensities and of allowable values of response control variables stem from formal or informal cost-benefit studies. As

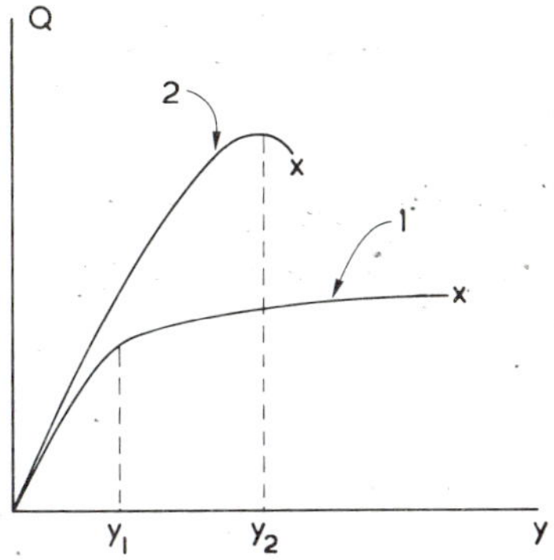

Fig. 3.1 Ductile and brittle systems

implicit in these studies, the general goal of optimization can be expressed in terms of direct, particular objectives: seismic design aims at providing adequate safety levels with respect to collapse in the face of exceptionally intense earthquakes, as well as with respect to damage to adjacent constructions; it also seeks to protect structures against excessive material damage under the action of moderate intensity earthquakes, to ensure simplicity of the required repair, reconstruction or strengthening works in case damage takes place, and to provide protection against the accumulation of structural damage during series of earthquakes. Finally, safety and comfort of occupants and of public in general is to be preserved by ensuring that structural response during moderate intensity earthquakes will not exceed given tolerance levels and that panic will not occur during earthquakes of moderate and high intensity, particularly in buildings where frequent gathering of people is expected.

Achievement of the foregoing objectives requires much more than dimensioning structural members for given internal forces. It implies explicit consideration of those objectives and of the problems related with nonlinear structural response and with the behavior of materials, members, and connections when subjected to several cycles of high-load reversals. It implies as well identifying serviceability conditions and formulating acceptance criteria with respect to them.

3.2 STRUCTURAL RESPONSE AND CONTROL VARIABLES

3.2.1 Ductility and strength

A structural system is said to be ductile if it is capable of undergoing substantial deformations at nearly constant load, without suffering excessive damage or loss of strength in face of subsequent load applications. Curves 1 and 2 in Figure 3.1 show typical load Q vs. deflection y relations for first load application in ductile and brittle systems, respectively. Curve 1 corresponds to the response under lateral load of an adequately

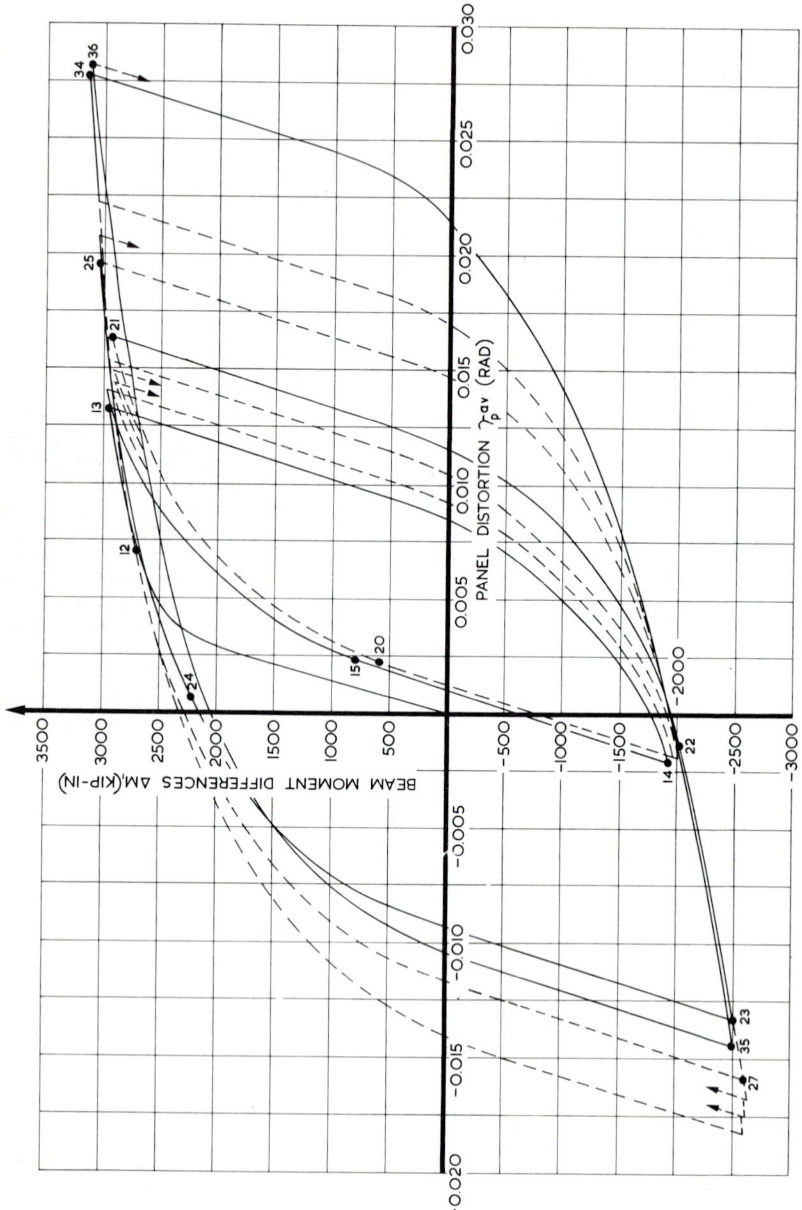

(a)

Fig. 3.2 (*a*) *Non degrading stiffness* (*after Krawinkler* et al.[3.2])

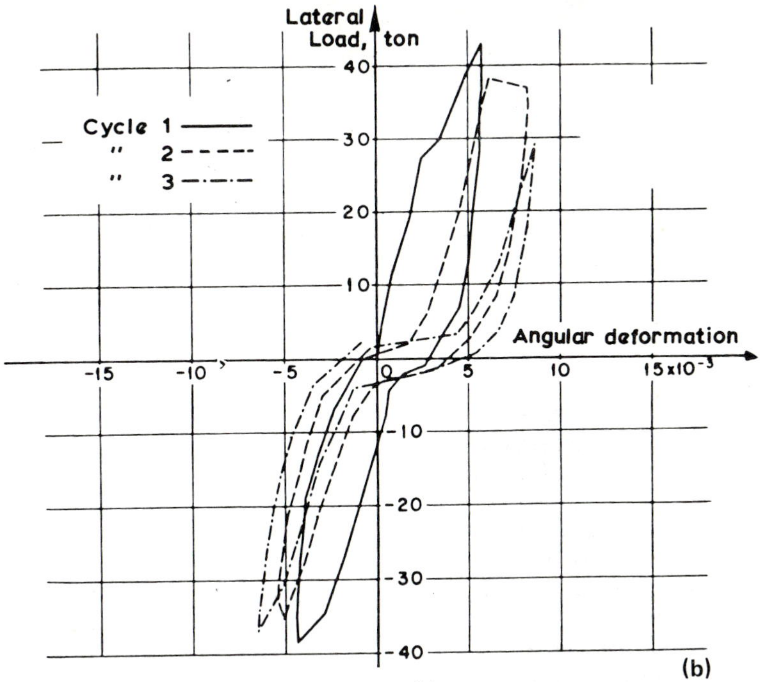

Fig. 3.2 (b) Degrading stiffness (after Esteva[3.1])

detailed reinforced concrete frame where slenderness effects are not significant; curve 2 is typical of weakly reinforced hollow block masonry. But when the effect of several loading cycles has to be considered, ductile behavior cannot be inferred from looking only at curves such as these for first load application: damage produced during the first cycles may impair the system's energy absorption capacity for subsequent cycles, and stiffness can degrade, as in Figure 3.2(b), typical of plain masonry shear walls confined by reinforced concrete frames[3.1]. In this case, stiffness degradation is associated with diagonal tension cracking of the infilling wall panel and the ensuing residual strains. Practically stable hysteretic cycles found for structural steel joints[3.2] as shown in Figure 3.2(a) are synonymous with negligible damage.

As shown in Section 3.2.2, the ability of structural systems to respond to dynamic excitations according to load deflection curves similar to Figure 3.2(a) provides support to conventional seismic design criteria, which require structures to sustain only a fraction of the lateral forces they should have to resist would they be demanded to remain within their linear range of behavior during strong earthquakes. Thus, safety against collapse can be provided by making a structure strong, by making it ductile, or by designing it for an economic combination of both properties. For some types of materials and structural members, ductility is difficult to achieve, and

Fig. 3.3 Instability failure

economy dictates designing for relatively high lateral forces; for others, providing ductility is much cheaper than providing lateral capacity, and design practice reflects this. But material ductility does not necessarily imply system ductility, as P-Δ effects (that is, interaction between lateral deflections and internal forces produced by gravity loads acting on the deformed structure) can lead to instability failure when the effective lateral stiffness is too low (see Figure 3.3).

Nonlinear ductile behavior of complex systems usually stems as a consequence of *local* or *concentrated* ductile deformations that take place at those particular sections of a given structure where yielding strains are reached (Figure 3.4). Numerically, local ductility can be expressed either as the ratio of total to yield-limit curvatures at a given section or as the ratio of total to yield-limit rotation at a member end[3.3]. *Global* or *overall* ductility is a property of a load-deformation curve expressed in terms of the resultant of external loads acting on a large portion of a given system. For instance, building frames are often dealt with as shear systems for the purpose of estimating their dynamic nonlinear response to seismic excitation. Global or overall ductilities can then be expressed in terms of the curves tying shear forces with lateral distortions. Numerical values of local ductilities determined by the above alternate criteria do not coincide among themselves, nor does overall ductility at a given story idealized as a segment of a shear beam coincide with the values of concentrated ductilities developed at the corresponding locations of the story, as overall ductility is a function of the ratio of the contributions to story distortion of concentrated ductile deformations and distributed elastic strains. Because beams are usually capable of developing larger ductilities than columns subjected to significant compressive loads, many building frames are designed under the 'strong column-weak girder' criterion, according to which different load factors are

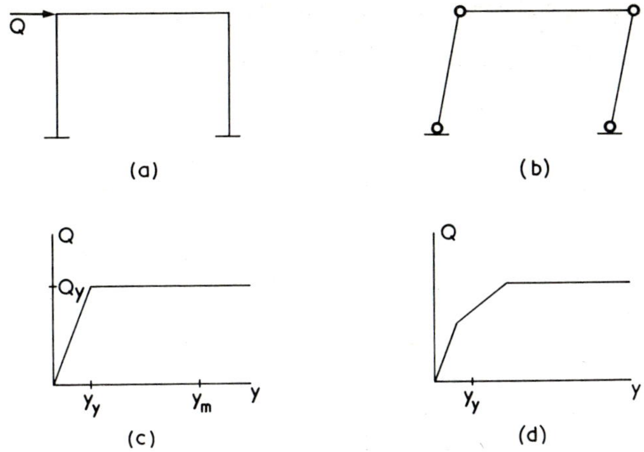

Fig. 3.4 Local and general ductility under lateral loads; (a) loads, (b) failure mechanism, (c) simultaneous yielding, (d) sequential yielding

adopted for different internal forces so as to make yielding much more likely at beam- than at column-ends. Under these conditions, significant coupling is introduced between nonlinear deformations of adjacent stories, and the shear-beam model may cease to apply. Whether the model in question strictly applies or not, nominal story ductilities are only indicators of their local values, and features contributing to ductility concentrations have to be held in mind while designing.

The relation between local and overall ductility is illustrated for a simple frame in Figure 3.4. Figure 3.4(b) shows the ideal case where moment-curvature graphs at critical sections are elasto-plastic and yielding is reached simultaneously at the four column-ends. If the frame is forced to undergo additional deformations at constant load, local curvatures at the yielding locations will increase and the lateral deflection of the frame will grow from y_y to y_m (Figure 3.4(c)). Local ductility can be measured by the ratio of the final and yielding values of the curvatures mentioned. Overall ductility is given by y_m/y_y and is a function of local ductility and of the lengths of the member segments along which curvatures will be greater than their values at yielding. Those lengths are functions of the type of material, the local details and the relative variation of bending moment ordinates and structural section strength.

Consider now a frame subjected to a constant system of vertical loads Q_2 (Figure 3.5) that produce an initial state of internal forces. If a system Q_1 of lateral loads is gradually applied, the ordinates of bending moment diagrams (b) and (c) will be additive at some locations and subtractive at others. Yielding will occur sequentially, say in the order DCBA, giving place to the load-deflection curve shown in Figure 3.5(d). Local ductilities will differ at the mentioned locations; they will depend, among other things, on the

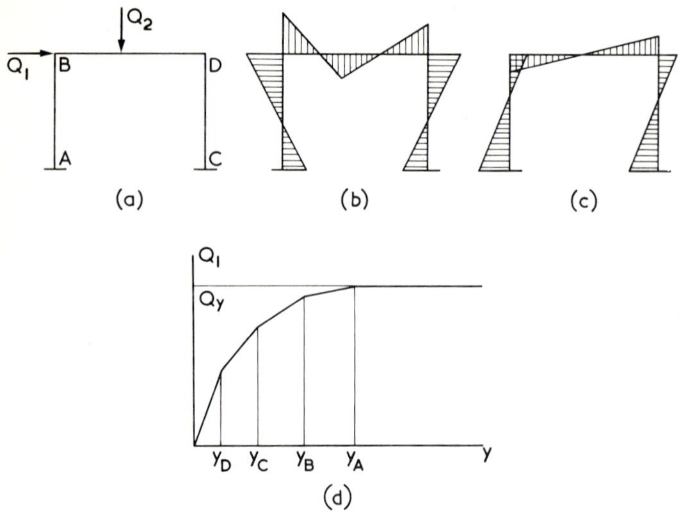

Fig. 3.5 Inelastic behaviour under vertical and lateral loads; (a) loads, (b) vertical-load moments, (c) lateral load moments, (d) resulting load deflection curve

order in which they reached their yield moments. Where axial loads are important, they can have a significant influence on these moments.

The following sections describe the quantitative relationships tying ductility demands with strength and stiffness in simple structural systems, as well as some problems found when trying to extrapolate those relationships to complex systems, representative of those encountered by engineers in their design practice.

Dynamic response of simple nonlinear systems A usual idealization of ductile structures is the elastoplastic system with load-deflection curve as shown in Figure 3.6(b), with stiffness k in the linear range of behavior, coefficient of viscous damping c, and top mass m. When the system responds to a strong earthquake, the maximum relative displacement D will exceed the yield deformation y_y, while the maximum lateral force will remain at the yield value Q_y if P-Δ effects are neglected. Failure is said to occur if the ductility demand D/y_y is greater than the available ductility μ. Figure 3.7 is a plot of yield deformations required to make ductility demands equal to available ductility for different values of this parameter, for the range of natural periods (computed in terms of the initial tangent stiffness of the elastoplastic system) most significant in practice, and for damping ratio $\zeta = 0.5c\,(km)^{-1/2}$ equal to 0.02. Pseudo-accelerations kD/m can be read on the proper scale in the same plot. Inspection of these curves shows that, provided the natural period is not too short, required yield deformations — and hence required base shear coefficients — vary inversely with ductility. The same conclusion is reached if one reads along the scale of spectral pseudoaccelerations. But this favorable influence of ductility in reducing the required base-shear coefficient is less

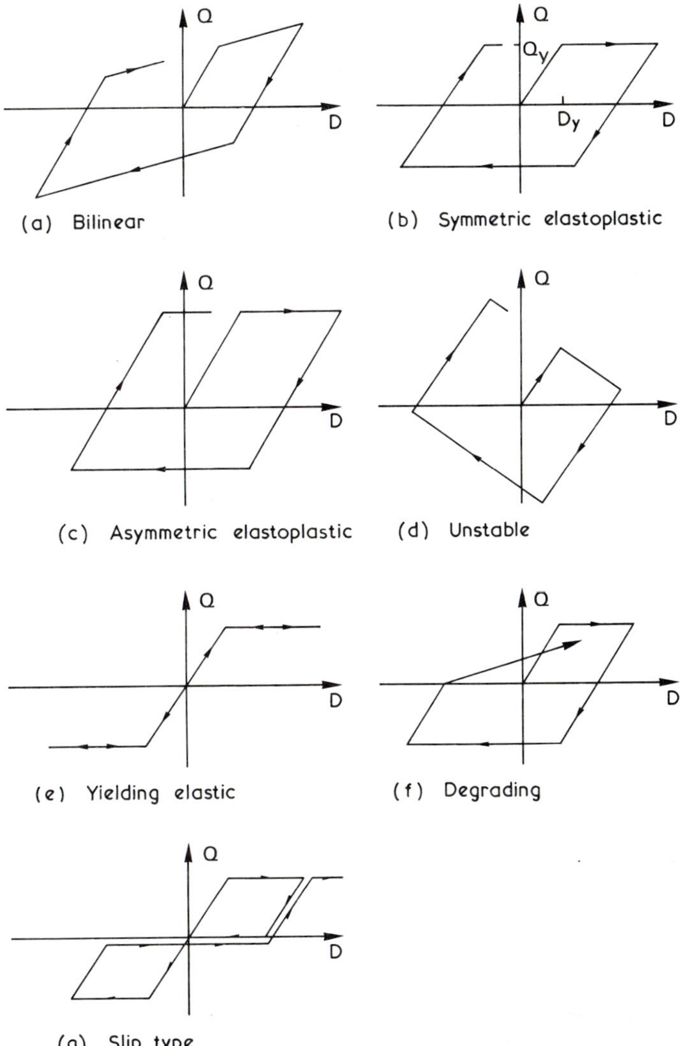

Fig. 3.6 Models of nonlinear behaviour

pronounced in the range of short natural periods, say shorter than $2\pi v/a$, where v and a are peak values of ground velocity and acceleration, respectively: as the system becomes stiffer, T tends to zero and spectral pseudo acceleration tends to a, regardless of μ, assuming that μ has to remain bounded. Actual values of lateral relative displacements are equal to μy_y, which means that, for moderate and long natural periods, those displacements are nearly insensitive to μ, while for very short natural periods they

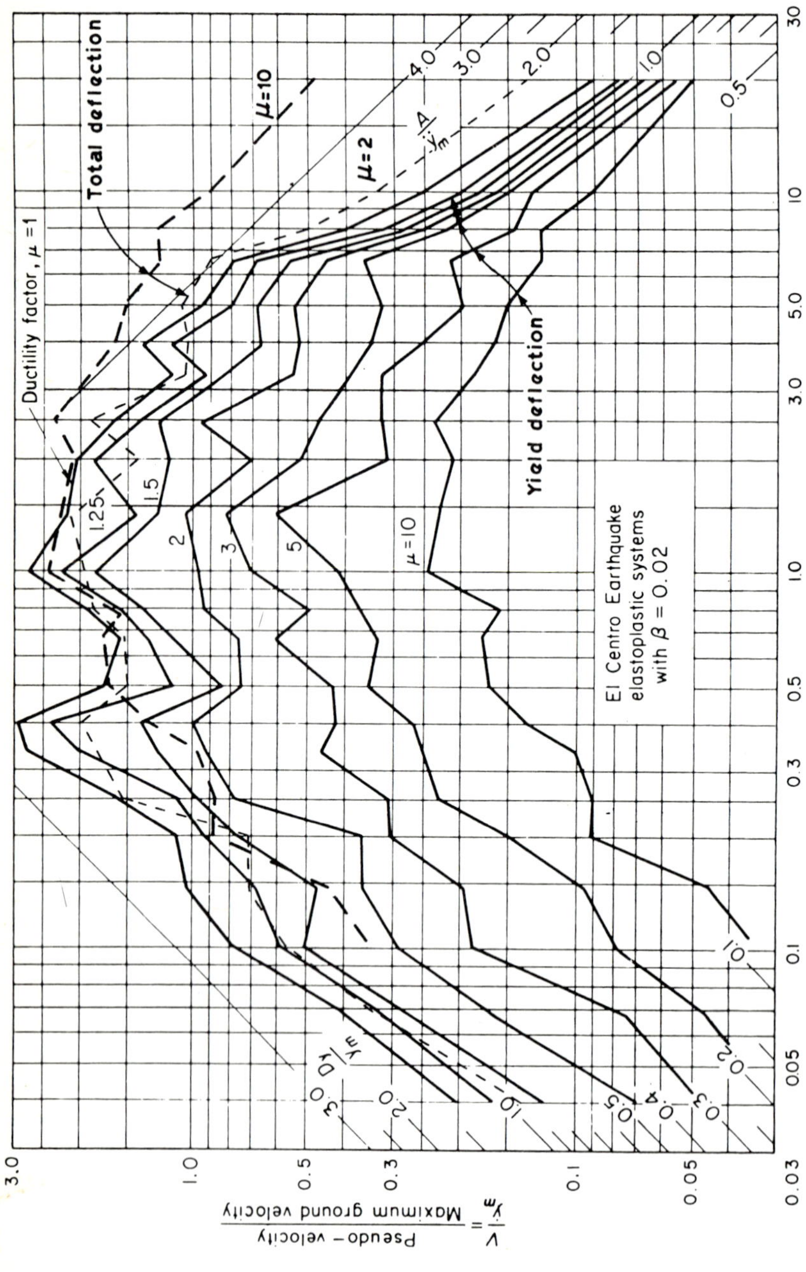

Fig. 3.7 Deformation spectra for elastoplastic systems with 2% critical damping subjected to the 1940 El Centro earthquake (after Newmark[3.33])

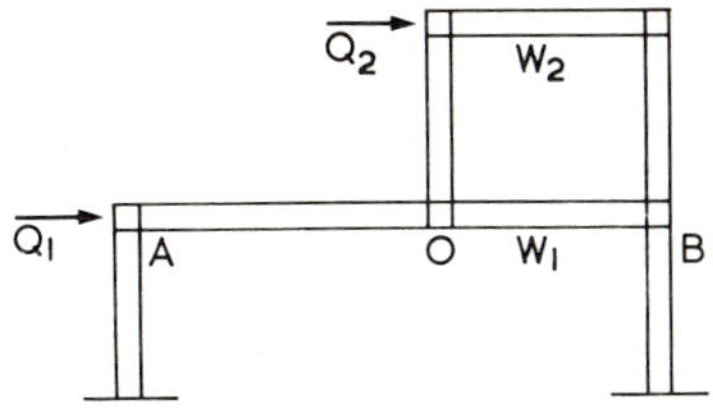

Fig. 3.8 Structure with asymmetric load–deflection curve

tend to be proportional to μ. The results just described can be expressed as follows: if a simple elastoplastic system with initial natural period T is to develop a ductility factor μ during an earthquake, the required base shear coefficient can be obtained by applying a reduction factor to the corresponding spectral value for an elastic system having equal natural period and damping; for moderate and long values of T, the reduction factor is approximately equal to μ^{-1}, while for short natural periods it will be comprised between μ^{-1} and 1. Relative displacements will equal μ times those of an elastic system subjected to the reduced base shear; that is, they will be approximately equal to those of the elastic system subjected to the actual, unreduced earthquake, if T is not too short, or to μ times the latter values if T is nearly zero. This is shown by a comparison of the dashed and full lines in Figure 3.7.

Similar conclusions have been derived from other earthquake records obtained on firm ground. Although these conclusions can be expected to be qualitatively valid for soft soil conditions, corresponding approximate quantitative rules are still to be derived.

The foregoing conclusions have to be modified when considering systems whose response cannot be idealized as elastoplastic. Other usual idealizations are depicted in Figures 3.6(e)–(f). Lateral strengths required for not exceeding given ductility demands in these systems are as a rule greater in 10 to 50% than those valid for the conventional elastoplastic system[3.5–3.9]. In the asymmetric elastoplastic case, yield strength is different for each direction of load application. It occurs, for instance, as a consequence of gravity loads giving place to increased or decreased lateral capacity of the second story of the system shown in Figure 3.8, depending on whether the vertical reaction to force Q_2, transmitted to beam AB at O, is directed upwards or downwards. Slip-type curves (Figure 3.9) usually stem as a result of lateral loads being carried by elements such as cross-braces or tie-cables, which can only carry tensile stresses. Yielding elastic curves are close approximations to the behavior of some prestressed concrete beams, subjected to antisymmetric end moments: these curves are often characterized by very narrow hysteretic loops. Degrading curves are frequently found in systems where a significant portion of the lateral capacity is due to members built with brittle materials and where no special precautions have been taken to prevent excessive damage in each cycle of load application. Such is the case, for

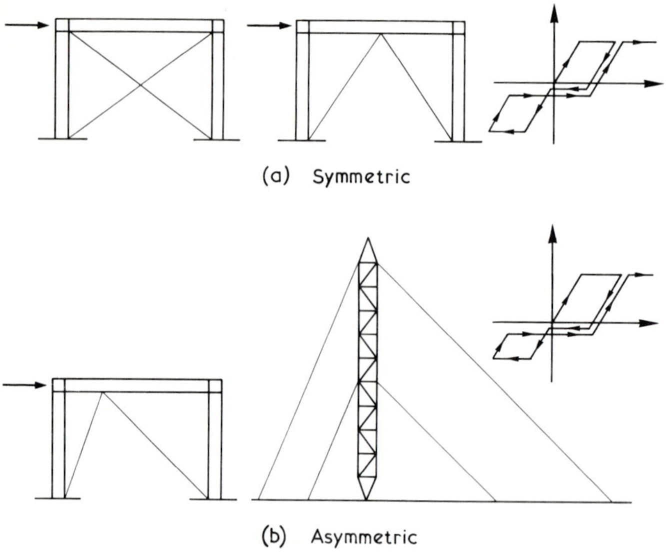

Fig. 3.9 Slip-type systems

instance, in masonry shear diaphragms or poorly detailed reinforced concrete frames.

Unstable curves (Figure 3.6(d) are produced by the influence of significant vertical loads acting on the displacements of the deformed structure. The influence of instability effects on ductility demands and on safety against collapse can be much more drastic than that associated with the features of the curves previously discussed, and is usually controlled in design practice by the specification of amplification factors for lateral deflections and internal forces that account for increments associated with second-order effects.

Ductility demands in complex systems Local ductility demands vary from point to point. Their distribution depends on that of local strength throughout the system, with significant interaction taking place between energy dissipation at different sections. The general patterns of ductility demands in complex systems have been studied almost exclusively in building frames, idealized either as shear beams or as assemblages of beams and columns where yielding is restricted to occur at plastic hinges located at the bar ends. Some results are plotted in Figures 3.10 and 3.11 for shear beams and frame systems, respectively. Each set of results corresponds to a different set of simulated earthquakes with frequency content similar to that observed under normal conditions on firm ground in the western coast of the United States. Structures were designed for the average ordinates, with respect to each set of motions, of the elastoplastic response spectrum corresponding to a ductility factor of 4. The systems in Figure 3.10 were

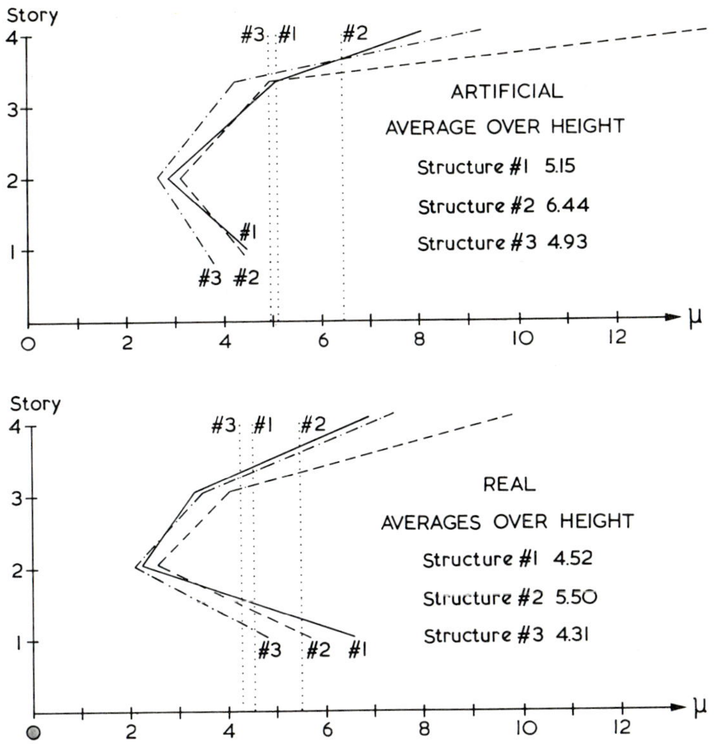

Fig. 3.10 *Ductility demands in shear systems subjected to simulated and real earthquakes (after Frank* et al.[3.3])

designed for the contribution of the fundamental mode of vibration alone, while that in Figure 3.11 was designed for the superposition of its four natural modes, in accordance with the criterion of square root of sum of squares advocated in Ref. 3.10. The load factor was in all cases taken as unity. Ductilities were expressed in terms of story sway for the shear beams and of local curvature at hinges for the framed system; thus, their absolute values cannot be compared. Their variability throughout the building is evident, however, as is the occurrence of large ductilities at the upper portion of systems for which the response associated with higher natural modes was neglected.

More pronounced variability in ductility demands has been observed in some shear systems with fundamental periods shorter than the dominant period of the ground motion, and in those whose safety factors with respect to design story shears vary significantly through the building height[3.11]. Such variability may be a consequence of architectural requirements, which often lead to some stories possessing elements stronger than they need to be in order to comply with the seismic coefficient adopted. When this happens,

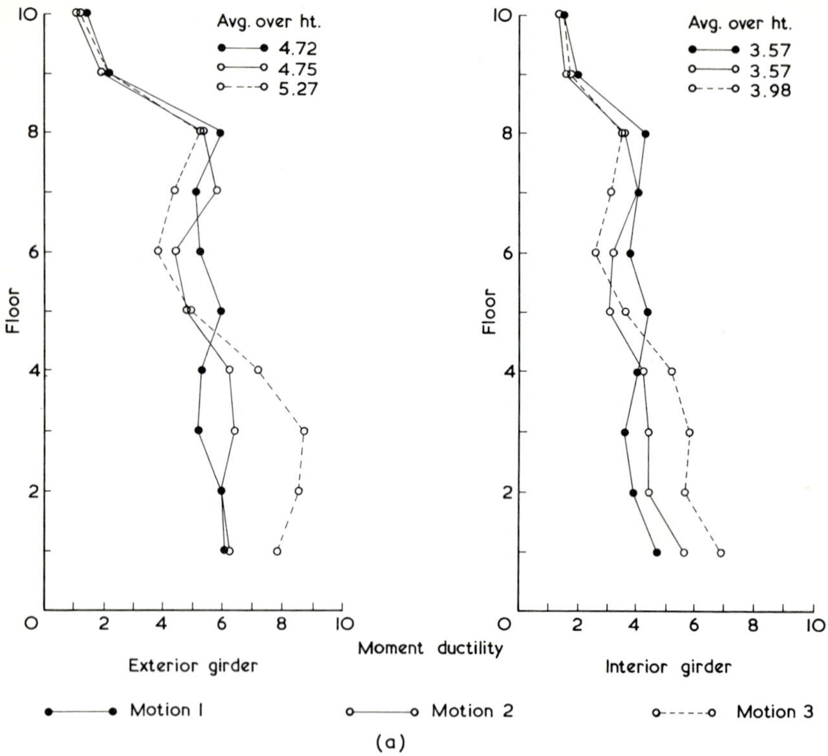

Fig. 3.11 (a) Maximum girder ductility factors for gravity and earthquake design

the relative contribution of each story to the hysteretic dissipation of kinetic energy changes, and those stories possessing the smallest safety factors are subjected to higher ductility demands than if the safety factor were uniform throughout the structure. When these increased ductility demands cannot be met with adequate yielding capacity, the lateral force coefficient has to be raised. Because of the large displacements implied, slenderness effects may become especially significant.

3.2.2 Stiffness and deformations

Structural stiffness controls natural period and hence seismic forces. The latter are lower for longer periods, that is, for small stiffnesses, but then displacements and deformations may become excessive. In addition to ensuring adequate safety factors against collapse, seismic criteria should aim at controlling deformations, because they are directly responsible for damage to nonstructural elements, impact with adjacent structures, panic and discomfort.

Stiffness is also the main variable controlling safety against instability.

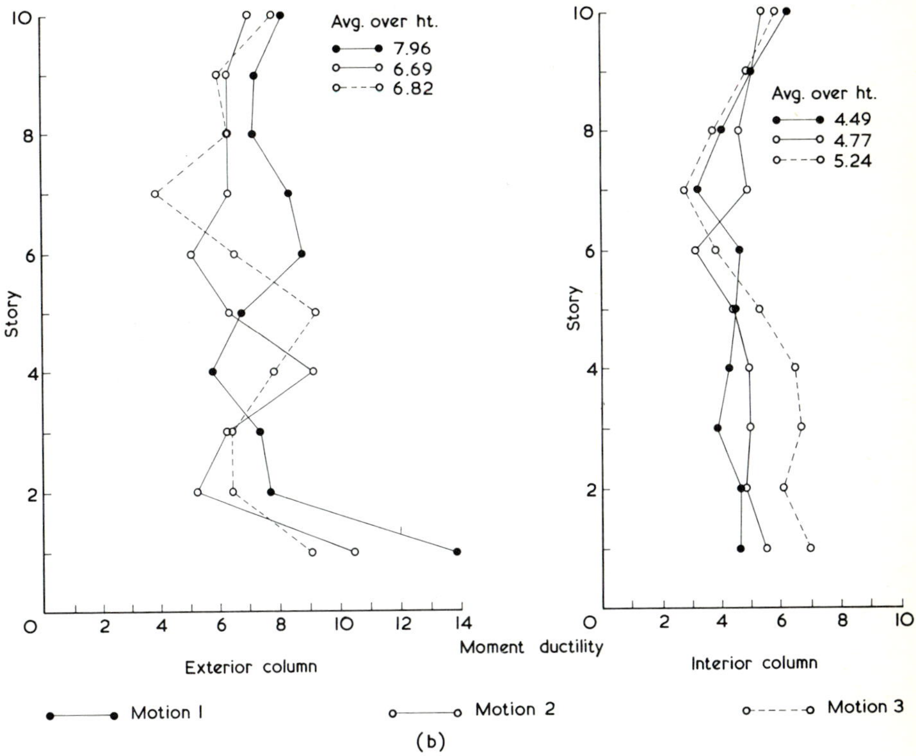

Fig. 3.11 (b) Maximum column ductility factors for gravity and earthquake design

Lateral displacements and internal forces produced by horizontal ground motion are amplified by interaction between gravity loads and the displacements mentioned. The amplification function varies in a nonlinear fashion with respect to lateral stiffness and reaches very high values when the latter variable approaches a certain critical value. In ductile structures, safety against instability failure is a function of effective stiffness, that is, of the slope of the line joining the origin of the force-deflection graph with the point representing the maximum deflection and the corresponding lateral force (in elastoplastic systems, this is the same as the value of the tangent initial stiffness divided by the ductility factor). The increasing rate of variation of the amplification function mentioned with respect to lateral stiffness when the latter is made to approach its critical value hinders the possibility of designing for very small lateral forces through the construction of very ductile structures (Figure 3.3).

3.2.3 Damage and energy absorption

Ductile hysteretic response provides a manner of transforming and

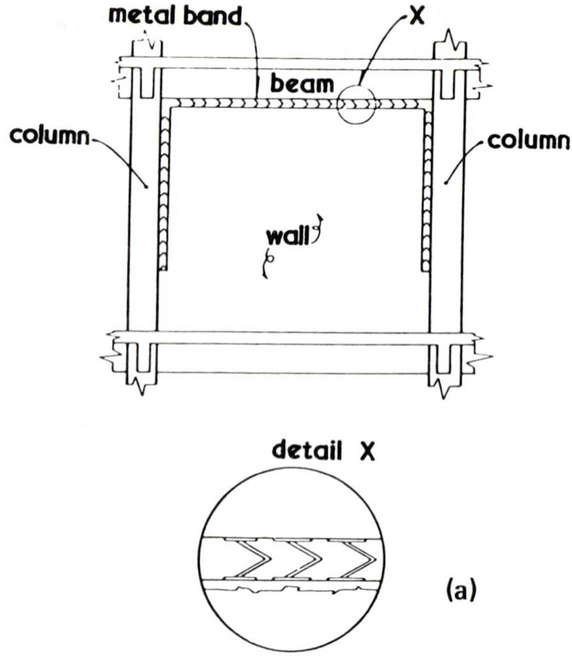

Fig. 3.12 Energy-absorbing and shock-isolating devices. (a) Metal band to protect partitions (after Newmark and Rosenblueth[3.10]*). (b) Roller support (after Ruiz,* et al.[3.15])

dissipating the kinetic energy imparted to a structure through its base. Such response usually implies some degree of damage, and possibly the deterioration of the system to withstand future severe earthquakes. Damage may accumulate during successive events, and the system's capacity may be seriously impaired. Decisions concerning the extent and level of damage that it is advisable to admit are mostly of an economic nature. In general, the degree of structural damage and its harmful effects on future performance can be controlled at some cost through selection of adequate materials and construction details, as described in Chapter 8. Damage to nonstructural members can be prevented through their isolation from the deformations of the structure. However, economy may dictate taking advantage of energy dissipation associated with damage. Architectural elements or ad hoc devices can be used for this purpose (Figure 3.12). In either case, considerations on facility of repair or replacement should form part of design.

The use of metal bands around partitions as shown in Figure 3.12(a) may serve the purposes of limiting the lateral forces that the structure will transmit to the partitions and at the same time taking advantage of the capacity of the partitions to resist such forces and making use of the energy absorbing capacity of the bands[3.10]. In other cases, designing for significant damage on partitions may prove to be attractive.

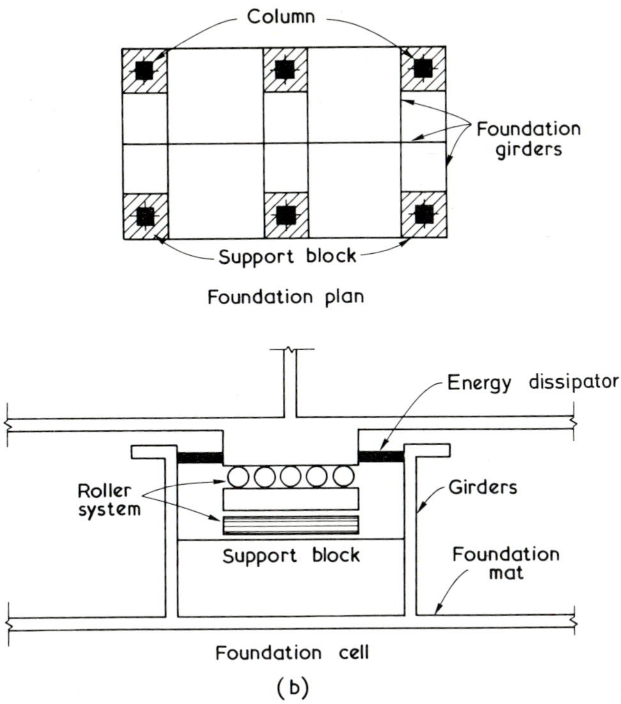

Fig. 3.12 (*contd.*)

Anchor bolts that yield during severe ground motion can provide protection to slender chimney stacks against local buckling or overall bending failure[3.12], at the expense of nonrecoverable elongations. Adequate performance of anchor bolts during sequences of earthquakes demands adjusting nuts after each event and replacing those bolts for which the sum of previous residual elongations is excessive. (See Section 4.8.)

Large concentrated deformations are frequent at spandrel beams connecting coupled shear-walls (Figure 3.16(b)) or at the ends of beams meeting shear-wall edges, and hence constitute adequate locations for energy-absorbing devices.

Partial isolation of building foundations from the ground motion has been advocated as a means to control structural response and nonstructural damage[3.13–3.15]. Isolating systems may consist of pads of very flexible material, assemblages of rollers or the like. Relative displacements between foundation and ground can be controlled by means of passive energy absorbing devices located at the ground-foundation interface (Figure 3.12(b)).

3.3 DESIGN PRINCIPLES

3.3.1 Design requirements and basic principles

The art of designing for earthquakes does not consist in producing structures capable of withstanding given sets of lateral forces, although that capability is part of a sound design. It involves producing systems characterized by an optimum combination of properties such as strength, stiffness, energy-absorption, and ductile-deformation capacities that will enable them to respond to frequent, moderate earthquakes without suffering significant damage, and to exceptional, severe earthquakes without endangering their own stability, their contents, or human life and limb. Achievment of this purpose means much more than application of codified rules; it demands understanding of the basic factors that determine the seismic response of structures, as well as ingenuity to produce systems with the required properties.

Codified requirements set optimum design levels in accordance with implicit cost–benefit analyses that balance initial construction costs with expected costs of damage and failure. They also recommend criteria and algorithms deemed adequate for the evaluation of the design actions tied to the optimum design levels. These recommendations serve the purpose of implementing sufficiently simple design criteria at the expense of narrowing the range of conditions where they give place to accurate predictions of response. It is the role of the engineer to recognize the possible deviations and to apply basic principles before trying to extrapolate general requirements to the particular problem at hand.

Static criteria of seismic design are stated in terms of the coefficients by which the masses of each structure have to be multiplied in order to produce the set of lateral forces to be designed for; but in most cases those coefficients stem from the dynamic response of linear shear beams possessing approximately uniform distributions of mass and stiffness. The meaning of the mentioned lateral forces must be clearly understood: they aim at providing a diagram of story shears that correspond to consistent safety levels; but they fail to predict other significant effects. Thus, reduction factors for overturning moment are required to account for the fact that maximum story shears do not occur simultaneously, and special algorithms have to be used to determine local effects, such as response of appendages and diaphragm stresses in floor systems, corresponding to safety levels consistent with those intended for story shear.

Dynamic criteria of design usually require performing a modal analysis, and hence variability in masses and stiffnesses is accounted for in the computation of the lateral force coefficients. Modal analysis however fails to predict the influence of nonlinear behavior except for some simple cases in which hysteretic dissipation of energy is distributed uniformly throughout the system and it is uncapable of predicting ductility-demand concentrations and nonlinear interactions for the simultaneous action of several ground motion components. Whatever design criterion is adopted, departures of actual conditions from those leading to uniform energy dissipation have to be recognized and their possible influence on behavior evaluated.

Given a set of design requirements and response control variables, a criterion of structural analysis capable of predicting with sufficient accuracy those variables must be applied to determine internal forces and deformed configuration. Simultaneous action of the significant components of ground motion has to be considered, including a scaling factor applied to each component in order to account for its probable value when the maximum absolute value of their combination takes place (see Chapter 2). The criterion of structural analysis adopted must be such as to recognize the possible concentrations of nonlinear behavior and to attain a sufficiently low probability that they occur at undesirable locations, as a consequence of inaccuracies of that criterion. This means that prediction of displacements and internal forces must account for stiffness and continuity, including all significant deformations; in particular, $P-\Delta$ effects must be considered at least by means of an approximate analysis intended to define the desirability of more refined studies. Some building code regulations state simple rules for deciding when $P-\Delta$ effects can be disregarded[3.16]. The contribution of the so-called nonstructural elements to stiffness should not be neglected, unless those elements are properly isolated from the structure or it is shown that they can not be harmful to its behavior.

Attention should be given to inertia forces associated with all significant components of local acceleration, namely angular acceleration (rotational inertia) of umbrella-like canopies or segments of stacks and vertical accelerations of long-span girders in bridges or industrial bents. Both types of acceleration are produced by horizontal, vertical, or rotational ground motion.

Adequate stress paths must be provided in order to guarantee that design forces can be transmitted down to the foundation. Deformability of the substructure and of the ground underlying it must be considered when defining the stiffness matrix of the whole system or the support conditions of the superstructure on the foundation. Distribution of contact pressures between ground and substructure should be computed on the bases that no tensile stresses can be transmitted at the interface, unless special provisions are taken, such as the construction of anchors or tension bearing piles.

Safety of structural and nonstructural elements to withstand the effects of local accelerations should be studied; in particular, overturning of walls and parapets produced by forces normal to their planes must be prevented by adequate reinforcement and anchorage.

3.3.2 Framing systems

Decisions concerning the selection of a framing system are influenced by many factors. Basic criteria are best illustrated by discussing some typical problems, as done in what follows.

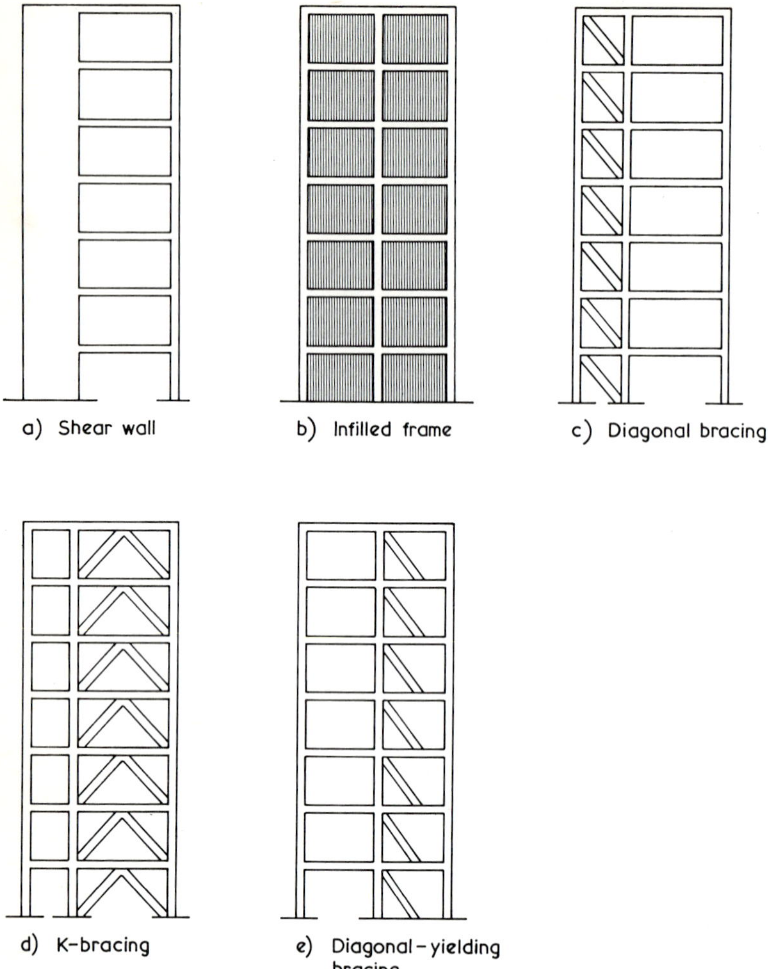

Fig. 3.13 Stiffening elements

Stiffening elements Continuous frames can usually resist seismic forces by developing rather uniform stress paths. Their main asset is that they can easily be designed and built so as to withstand large ductility demands. However, their efficiency, based on the bending capacity of beams and columns, is lower than that of systems that base their strength on that of elements subjected to simple shear or axial forces. Besides permitting the development of larger lateral capacities without excessive costs, stiffening systems can be decisive in the control of damage associated with lateral distortions. But economic and architectural considerations may preclude the use of these elements in some instances, and they may show significant

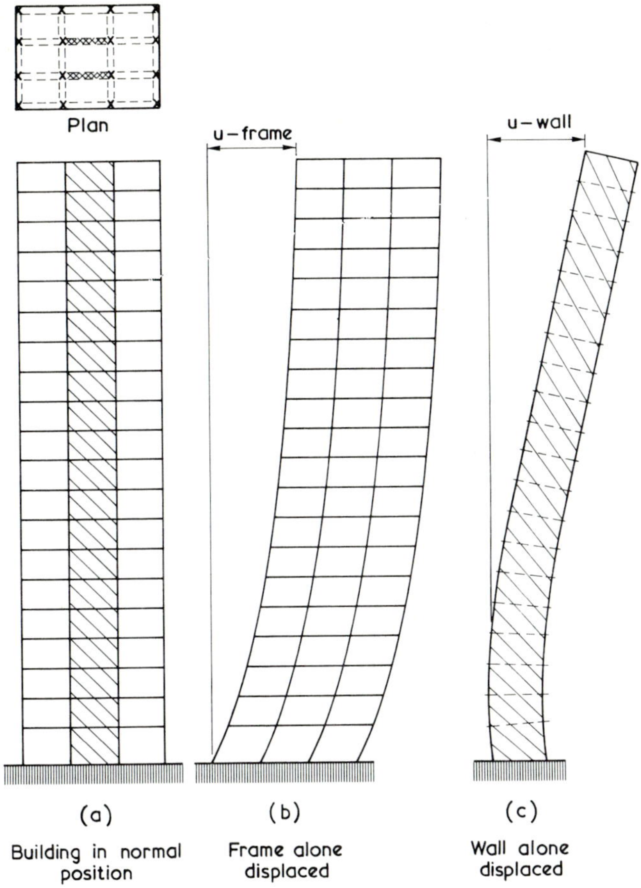

Fig. 3.14 Lateral load response of shear walls and frames (after Newmark[3.33]*)*

technical disadvantages in others. In tall buildings, enhanced stiffness is usually provided by diaphragms and cross braces; the former built in reinforced concrete or masonry, the latter in reinforced concrete or steel (Figure 3.13). Use of cross bracing is usually to be preferred over that of diaphragms in low buildings and industrial bents, except in those instances where the diaphragms are required for architectural reasons. In intermediate and tall buildings the reverse is usually true, mainly on account of the large cross-sectional dimensions that would be required for the bracing members and of the serious problems posed by their anchorage, particularly in reinforced concrete structures.

The efficiency of cross-braced bays and shear walls is reduced as their aspect ratio (height-to-width ratio) increases. The reason for this can be understood from Figure 3.14, which compares the deflected shapes of a

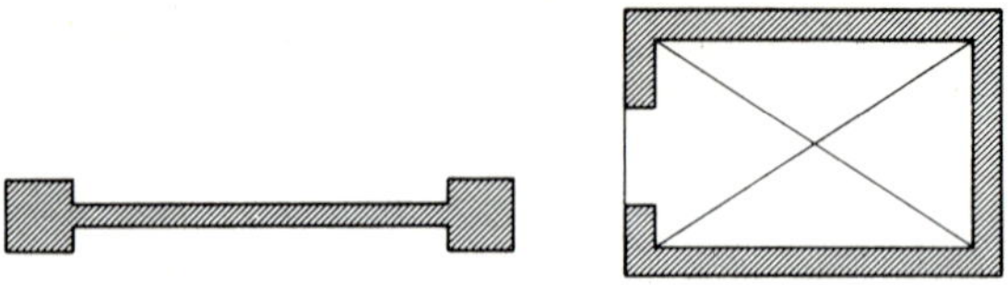

Fig. 3.15 Efficient shear wall sections

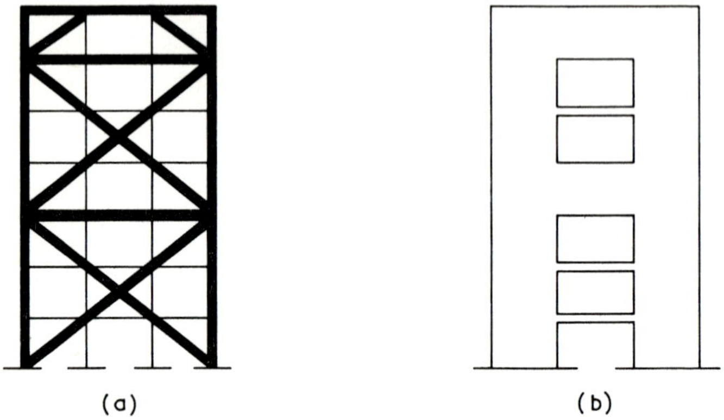

Fig. 3.16 Efficient use of stiffening elements; (a) cross braces, (b) shear walls and spandrel beams

braced bay or a wall acting as a flexural beam fixed at its base and a continuous frame acting essentially as a shear beam when both are subjected to a system of lateral forces. For equal top deflections, near the bottom the slopes of the flexural beam will be much smaller than those of the shear beam, but near the top the reverse will be true. The greater the aspect ratio of the flexural beam, the more important will be this effect. It follows that, when a system of lateral forces is resisted by the combination of a continuous frame and a slender wall, the latter will take a significant portion of the total story shear in the lower stories, but will fail to do so in the upper ones, as there the wall will tend to lean on the frame, instead of helping it to withstand the total story shear. The occurrence of large rotations of the wall horizontal sections gives place to excessive local deformations and ductility demands at the ends of beams connected to the wall edges. These problems can be aggravated by the occurrence of significant displacements associated with the flexibility of the foundation at the base of the wall. Adoption of cross sections as shown in Figure 3.15 can significantly enhance the efficiency of slender shear walls by increasing their flexural stiffness; but the most effective manner of reducing overall bending deflections is to get as wide a portion of a given bay to contribute to overall bending stiffness. In braced systems, this can be accomplished by adopting configurations as shown in Figure 3.16(a). Where architectural requirements force the adoption of a number of separate walls in the same plane, one solution consists in coupling several of them and making them act together by means of sufficiently stiff and strong spandrel beams (Figure 3.16(b)). Then stems the problem of attaining ductile behavior in these beams (see Chapter 5).

Use of stiffening elements may bring about other problems: the flexibility of the foundation and that of the floor diaphragms may be significant in comparison with that of the stiffening elements and have to be accounted for when obtaining the distribution of internal forces. In-plane deformability of horizontal diaphragms may become very important in buildings long in plan where lateral forces are resisted by shear walls located near the ends of the building plan. Not only stiffness, but also strength of the floor diaphragms in their own plane becomes then a relevant variable.

Symmetry The distressing influence of asymmetry in structural behavior has been recognized, and perhaps over-emphasized. Efforts to avoid its effects have concentrated on the problem of adjusting stiffnesses so as to avoid torsional eccentricities; but even if computed eccentricities are negligible, important torques can develop, for instance, when high stiffnesses of certain structural members on one end of the building plan are balanced by very dissimilar elements on the other, as the relative values of the computed stiffnesses may be little reliable. In addition, eccentricities .of variable magnitude may occur as a consequence of nonlinear behavior, even in those cases where conventional linear analysis predicts no torsional stresses. For this reason, it is desirable that structures be symmetric not only with respect to stiffnesses, but also to types of structural members.

Uniformity Adoption of very different spans in a given frame gives place to high shears and bending moments in the girders covering the shortest spans. These internal forces may reach excessive values in tall buildings, and even give place to objectionable variations on the axial loads of the adjacent columns; those variations can in turn affect foundation design. In low rise buildings these effects may be insignificant; the degree of uniformity which may be desirable in tall buildings may thus be objectionable in the lower ones, if it prevents taking advantage of the irregularly located points put forward by the architect. For instance, in the reinforced concrete frame structure whose floor plan is shown in Figure 3.17(a), the arrangement of service walls permits locating columns at points A, B, C, D. Such columns would reduce beam spans, and hence they would be desirable in a two- to five-story building; probably, they would be objectionable in a building having more than eight or ten stories. Nevertheless, it may in some cases be advantageous to locate columns at points that imply marked discrepancies between the spans of a tall building. It is then advisable to decrease the stiffnesses of the girders connecting those columns, mainly by reducing their depth, as shown in Figure 3.17(b).

As a consequence of vertical displacements produced by lengthening and shortening of columns, problems derived from excessive stiffness of short span beams tend to augment. Stiffness reduction called for by a good design for lateral loads might then be inconvenient because of limitations related

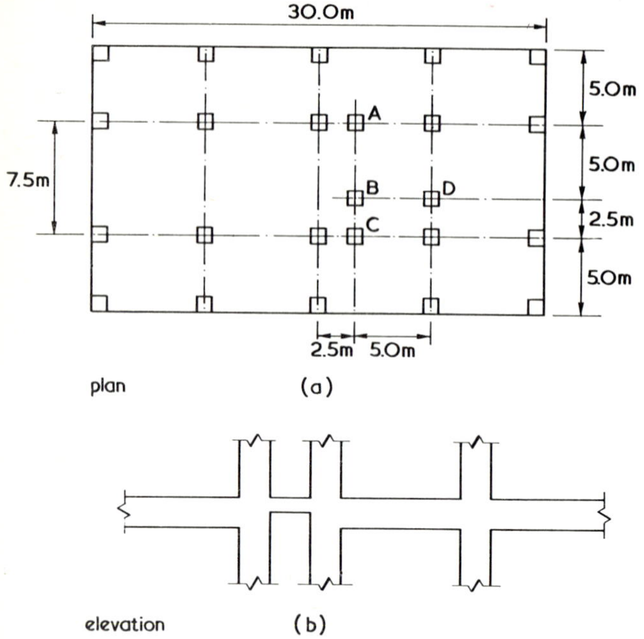

Fig. 3.17 Structural solution for building with uneven spans (after Newmark and Rosenblueth[3.10])

with vertical load deflections. It is then advisable to project plastic hinges at the ends of the elements under consideration.

Buildings having irregular plans that include two or more main sections interconnected by narrow corridors (Figure 3.18) pose special problems of analysis and design: excessive stresses in the corridor diaphragms and significant twisting forces in the building sections can result as a consequence of interaction among those sections. Evaluation of these effects is in general a difficult task including explicit consideration of diaphragm deformability. The problem can be successfully handled by means of properly located and carefully detailed vertical construction joints.

Staggered lines of defense A large number of buildings base their lateral strength on the contribution of brittle elements that crack while they absorb energy during the strongest phases of a shock. Design of such buildings is often done assuming no reduction in the elastic spectral ordinates on account of ductility, as cracking may be tantamount to collapse. Their performance is greatly improved, and the design forces for a given reliability sharply reduced, however, if the system is provided with a second line of defense, capable of developing a fraction of the lateral strength of the brittle elements and of showing ductile behavior after cracking of the first, stronger and usually stiffer system. This property has been recognized by some building codes[3.16,3.20], which specify nearly equal ductilities for moment resisting

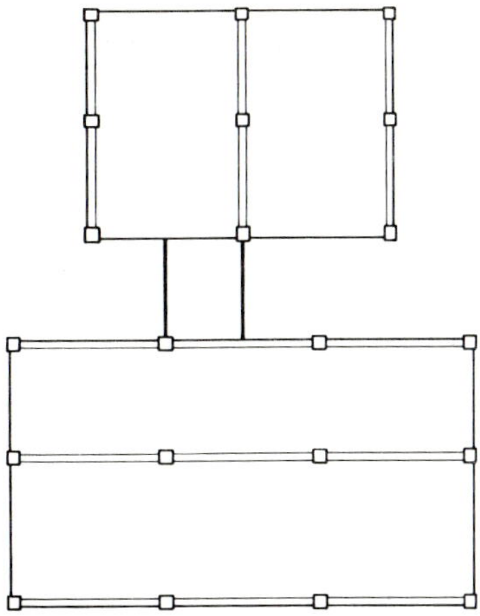

Fig. 3.18 Building sections interconnected by narrow corridors

structural frames as for dual systems that resist lateral forces by a combination of vertical bracing trusses, reinforced concrete or reinforced masonry shear walls and ductile moment resisting frames, provided the frames take at least 25% of the prescribed seismic forces.

3.3.3 Design for ductility and energy absorption

Neither local nor global ductilities can be ensured by use of a ductile material: both properties depend as well on the types of potential failure modes and on the relative values of the safety factors with respect to each of them. Thus, onset of instability precluded taking full advantage of the ductility inherent in the material used in the structure of Figure 3.3. While the stress-strain law for the material can be represented by Figure 3.6(b), the relationship between lateral force and displacement is better described by Figure 3.6(d), and this non-ductile curve will dominate system behavior unless the lateral stiffness is increased or the vertical load decreased; only the first of these actions is ordinarily feasible. Likewise, premature local buckling of a beam-flange may prevent the development of a ductile plastic hinge at the same cross section.

In order to attain ductile behavior, one must identify potential failure modes, determine those characterized by ductile behavior, and adopt a set of safety factors leading to a sufficiently low probability for the event that limit states with respect to brittle modes of behavior will be reached before those associated with ductile modes. For instance, reduction factors for lateral forces specified by Mexico City Building Code[3.16] for ordinary moment resisting reinforced concrete frames correspond to an assumed ductility of 4, but the code permits that parameter to be taken as 6 if some special requirements are satisfied. Among those requirements, load factors of 1.4 are specified for brittle-failure limit states, such as those associated with shear force, torsion and buckling, for the superposition of permanent loads and earthquake, instead of 1.1, applicable to ductile limit states.

Details and connections Because global ductility of usual structures depends as a rule on local ductilities of small regions, careful design and detailing of those regions is mandatory. In building frames, yielding is usually restricted to occur at plastic hinges located at the sections where the ratios of capacity to action are lowest. As a rule, it is feasible and convenient to have those sections at the member ends. Chapters 4 and 5 deal with the specific design criteria intended to ensure that sufficiently ductile plastic hinges will form at predetermined locations.

Brittle modes of behavior are often the consequence of exceedance of structural capacity at some particular regions where drastic changes in mechanical properties of the structural members take place. As a rule, brittleness of behavior can in those cases be ascribed to local nonlinear buckling or to stress concentrations usually unaccounted for in ordinary design. Typical among the vulnerable regions are connections between structural members. In steel structures, local brittle behavior usually results from local buckling or welding failure, while in reinforced concrete, problems of bond, diagonal tension, and stress transfer between reinforcement of different members dominate. On account of the complexity of the stress patterns usually involved, the problem is in general not only one of brittleness but also one of ignorance or carelessness in the evaluation of the structural capacity of the joint. Practical recommendations for evaluation of this capacity are provided in Chapters 4 and 5. The condition that the probability of brittle failure is sufficiently smaller than that of ductile failure is attained by adopting larger safety factors with respect to capacity of the joint than to that of the members it connects; but often the difference in safety factors is insufficient to override the wide uncertainties associated with joint behavior.

Ductility of members and subsystems In members and subsystems, ratios of safety factors with respect to brittle and ductile modes depend on the capacities of critical sections with respect to various combinations of internal forces and on the ratios between those internal forces when the member or subsystem deforms beyond the failure limit states of the critical sections. Thus, a reinforced concrete beam acted on its ends by moments M_1 and M_2 produced by permanent loads and by seismic couples M'_1 and M'_2 which grow from zero to their final values, will attain its bending capacity if either $M_1+M'_1$ or $M_2+M'_2$ reach the corresponding strength. Failure will be ductile if the beam is under-reinforced, i.e. if tensile bending failure governs. Brittle failure will take place if the member is over-reinforced or if development of the bending capacity is precluded by premature failure in diagonal tension. As couples M'_1 and M'_2 grow, end shears vary as $[V=V_0 \pm(M'_1+M'_2)/L]$, where V_0 is the effect of permanent loads and L is the member span, and the member fails prematurely in diagonal tension if the

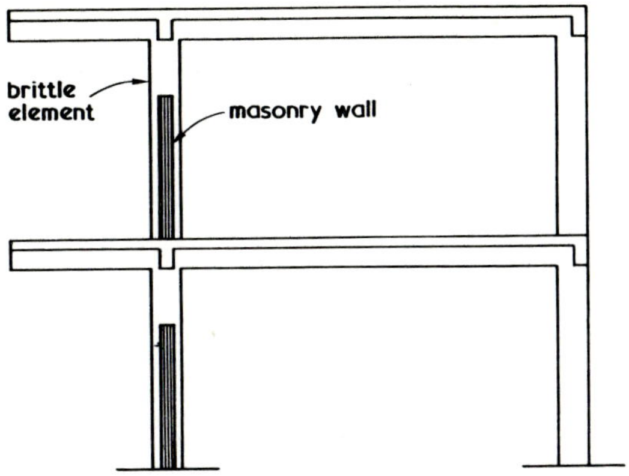

Fig. 3.19 *Low ductility structure*

shear at either end reaches the beam capacity before moments $M_1 + M'_1$ and $M_2 + M'_2$ reach the corresponding bending strengths.

Large values of L imply small values of the shear force for given values of M'_1 and M'_2, and bending failure is likely to dominate; ductile behavior will take place at ordinary under-reinforced members. For small values of L the opposite will be the case: brittle-type diagonal tension failure will be reached before bending failure, unless special precautions are taken to ensure that the safety factor with respect to the former mode is greater than that applicable to the latter.

The condition is often encountered in buildings with irregular plan, as shown in Figure 3.17(a); adoption of a smaller beam depth can lead to a ratio of shear to bending strengths capable of ensuring ductile behavior. The same problem is characteristic of the structural system shown in Figure 3.19, typical of school buildings in some countries: the clear height of some columns is reduced by their interaction with masonry panels lower than the story height. This leads on one hand to shear concentrations and torsional response, and on the other to large ratios of shear force to bending moments, and hence to brittle failure, under usual conditions. All these problems can be avoided if the columns are liberated from restrictions throughout the full story height, either by placing a flexible joint between wall panels and columns, or by locating frame and wall on different, parallel planes. Alternatively, ductile behavior can be accomplished in this case by designing the free-standing portion of a given column for a shear capacity equal to or larger than the sum of the bending capacities at the ends of the mentioned portion divided by its height. Interaction with axial forces must not be forgotten. In the extreme case of very short spandrel beams used for providing coupled action of adjacent shear walls (Figure 3.20), special reinforcement has to be furnished in order to attain ductile behavior under diagonal tension.

Axial loads reduce available ductility at columns ends; the larger the axial

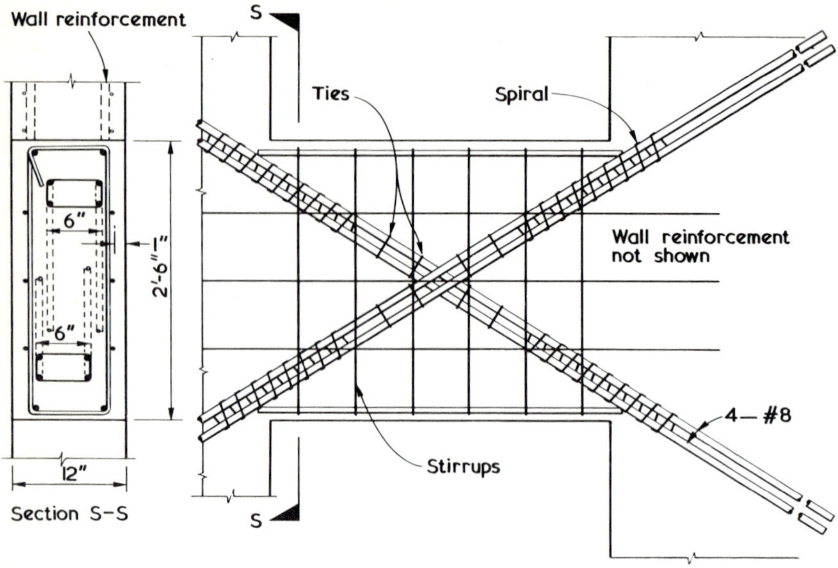

Fig. 3.20 Suggested reinforcement for coupling beam (after Paulay[3.34])

stress, the larger the reduction, as shown in Figure 3.21 for a reinforced concrete column of given characteristics. Hence the criterion that suggests that plastic hinges occur at the end of beams, rather than of columns; this can be accomplished with reasonable reliability by adopting slightly higher load factors (say 10 or 20%) for column than for beam design.

The consequences of designing exclusively for strength, with neglect of ductility considerations, can be as serious as displayed in Figure 3.23, which shows the brittle failure of a large number of columns of a building having the cross section shown in Figure 3.22, during the Caracas earthquake of 1967[3.17]. Axial loads due to gravity forces and to seismic response impaired the capacity of the otherwise strong columns to develop sufficient ductility; the situation may have been aggravated because the upper stories, being much stronger in shear than the lower ones, must have given place to the occurrence of specially higher ductility demands at the columns under consideration.

3.4 SAFETY CRITERIA

3.4.1 Structural safety

Uncertainty and safety in seismic design Neither loads acting on buildings nor strengths of structural members can be predicted with sufficient accuracy that uncertainty can be neglected in design. Nominal values of loads and strengths are most unfavorable values only in the sense that the probability

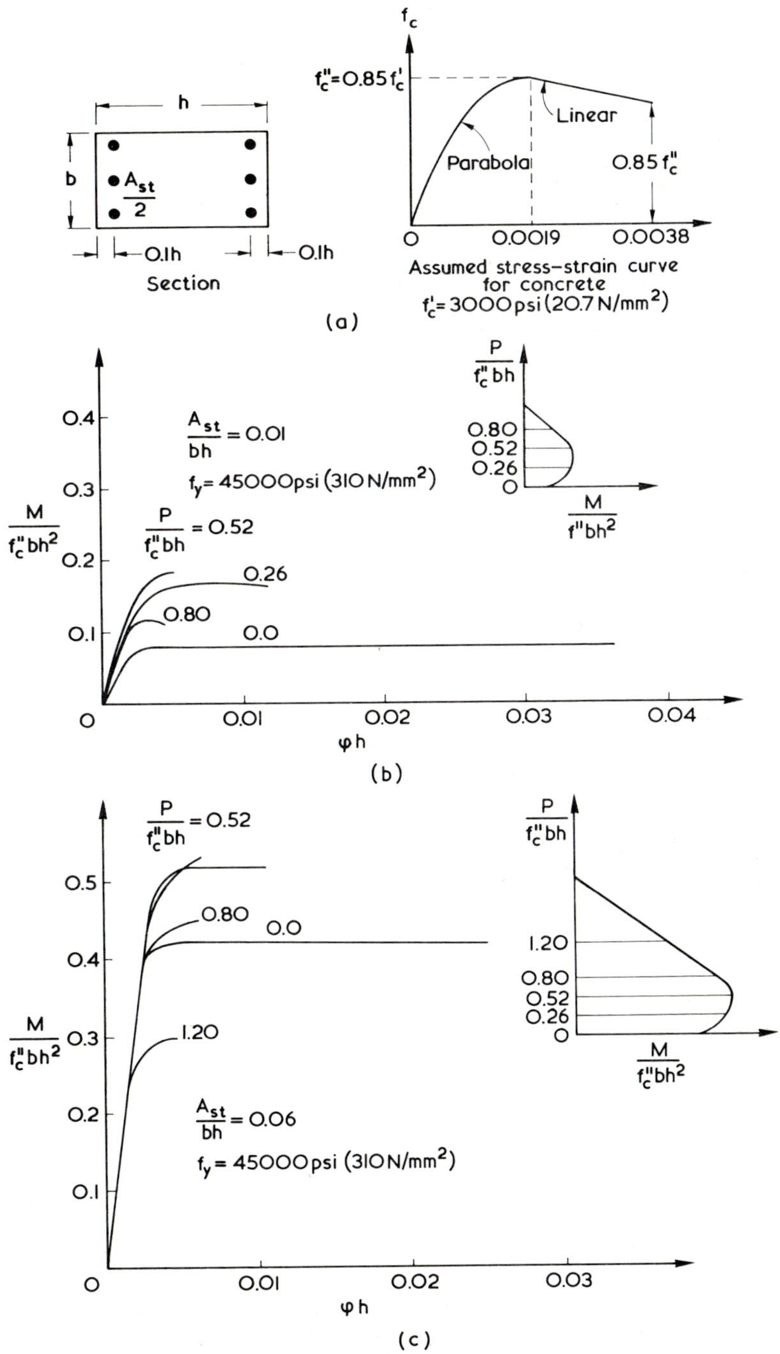

Fig. 3.21 Influence of axial load on column ductility (after Park and Paulay[3.35])

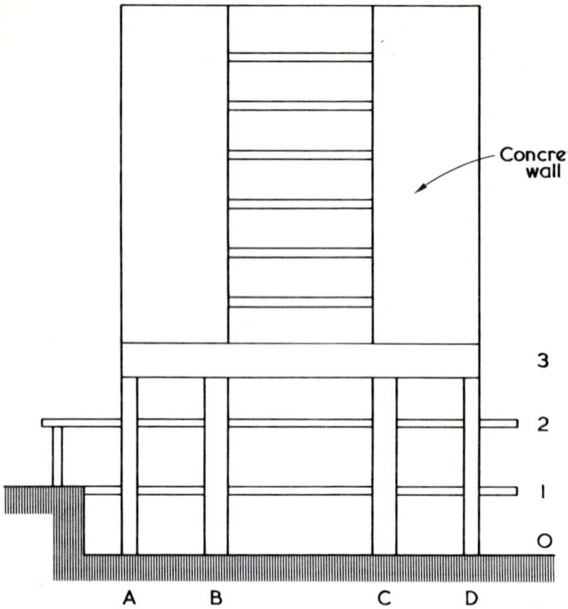

Fig. 3.22 Schematic cross-section of building damaged during Caracas 1967 earthquake[3.17]

that those loads and strengths adopt values more dangerous for the performance of a given system is sufficiently small. If the actual value of the internal force acting on a critical section or subassemblage of a structure exceeds the actual value of the corresponding strength, failure occurs. Structural safety is measured by the probability of survival, that is, that failure does not take place. When only a single load application is contemplated the probability of survival is determined by the probability distributions of load and strength at the instant when the load is applied, provided the safety margin, i.e. the difference between strength and load does not decrease with time. Seismic excitation however consists of a random number of events of random intensities taking place at random instants in time, and seismic safety cannot be described by a single probability of survival under a given load application, but rather by a time-dependent reliability function $L(t)$, equal to the probability that the structure survives all combinations of dead, live and seismic actions that affect it during an interval of length t starting at the same time as construction.

Limitation of material losses and other forms of damage is as important an aim of earthquake resistant design as is safety against collapse. For the sake of simplicity, these two objectives are usually pursued by design codes through the specification of a design earthquake for which collapse safety and deformation restrictions have to be verified. Some special structures are analyzed for two different design earthquakes; safety requirements with respect to collapse limit states are established for an extreme intensity event,

Fig. 3.23 Failure of reinforced concrete columns

while limitation of non-structural damage is aimed for through the control of stresses and deformations for shocks of moderate intensity, likely to be exceeded several times during the structure's life.

Complying with collapse safety design conditions does not mean that failure probability is annulled: it is rarely possible to set sufficiently low upper bounds to seismic intensity at a site or to structural response that designing for them will be economical or even feasible. Besides, neither structural strength nor performance for a given intensity can be predicted with certainty. Establishment of design conditions follows cost-benefit studies, where the initial costs required to provide given safety levels and degrees of protection with respect to material losses are compared with the present value of the expected consequences of structural behavior. This is obtained by adding up the costs of failure and damage that may occur during given time intervals, multiplied by their corresponding probabilities and by actualization factors that convert monetary values at arbitrary instants in the future into equivalent values at the moment of making the initial investment.

Evaluation of failure and damage probabilities implies an analysis of the uncertainties associated with structural parameters, such as mass, strength, stiffness and damping[3.18], and with those defining seismic excitation, such as motion intensity and relation of the latter to the ordinates of the response spectra for given periods and damping values, or to other variables closely correlated with structural response. Conversely, attainment of given safety levels and degrees of protection for material losses is accomplished through the specification of nominal values of design parameters used to compute structural capacity and response and of safety factors that must relate the latter variables.

Optimum safety The formal application of cost-benefit studies to decision making in earthquake engineering is often hindered by problems that arise in the evaluation of expected performance of structures. Prominent among them is the difficulty to express different types of failure consequences in the same unit or, more specifically, to assign monetary values to concepts such as panic, injury, death and even loss of prestige of designers, contractors or regulating agencies responsible for safety policies. Those difficulties can be overcome through adoption of decision-making models that account for uncertainty in the mentioned concepts and of policies for assessing that uncertainty. An important asset of decision oriented cost-benefit studies, however informal they may be, is their providing of insight into the relevant variables and the manner in which optimum design intensities and safety factors should vary with respect to those variables. Thus, it is concluded that optimum design intensity is an increasing function of the ratio of the derivative of initial cost with respect to capacity to the expected cost of failure, and is a decreasing function of seismic activity at a site.

The latter conclusion means that the higher the activity the higher the optimum level of risk to be accepted in design[3.19]. This is often neglected, as it contradicts the widely extended concept that in seismic design consistent safety means design for intensities having a given return period, regardless of initial costs.

The benefits of adopting safety levels that depend on the consequences of failure have been recognized in some modern design regulations. For instance, structures are classified in Mexico City Building Code[3.16] in three categories according to their usage, namely provisional, ordinary and specially important. The second category includes apartment and office buildings, and the third includes structures the failure of which would have especially important consequences, the good performance of which is critical just after an earthquake (hospitals, fire stations), or the contents of which are very valuable (museums). Structures in the first category do not require formal earthquake resistant design, while those in the third category are designed for 1.3 times the spectral ordinates specified for the second group.

In the recently proposed Recommended Comprehensive Seismic Design Provisions for Buildings[3.20], structures are classified into three main groups according to their seismic hazard exposure, that is, the relative hazard to the public based on the intended use of the building. In decreasing order of importance, these groups include, respectively, buildings housing critical facilities which are necessary to post-disaster recovery, those which have a high density of occupancy or which restrict the movements of occupants, and other structures. Seismic design spectra are based in all seismic regions on intensities that may be exceeded with 10% probability in 50 years. Differences in the optimum safety levels for different building usages are not recognized in the adoption of different seismic coefficients, but only in the restrictions concerning height and types of structural systems and in the refinement of the criteria for structural analysis and design, which are made to depend on the seismic zone and the seismic hazard exposure.

3.4.2 Design values

Nominal values of design variables and safety factors — and hence of implicit safety levels — have been traditionally established by trial and error and engineering judgement. Although explicit optimization as described above seems the ideal framework for design, its direct application by designers is at present impractical, with the exception, perhaps, of extremely expensive structures, such as nuclear reactors, or structures built in large numbers from the same design, such as offshore drilling platforms. Design values specified in a building code should be based on optimization studies covering the types of structures contemplated by that code, and optimization should be referred to the expected population of those structures. The fact that explicit optimization is not directly applied to each individual structure implies that we are dealing with suboptimization, that is, optimization within given restrictions: design formats must be kept simple, and the number of relevant variables small. As a consequence, what is optimum for a population of structures may not be optimum for every individual member.

Nevertheless, the theory of structural reliability has provided the framework for recent attempts to attain consistency between those rules and to extrapolate them to more general conditions; simplified formulations derived from the basic concepts have led to design criteria that approach consistency while not departing from the simplicity required for practical applications[3.21–3.24]. Nominal values of the design variables are chosen such that the probability that each variable will adopt a more unfavorable value does not exceed a certain limit; often, the probability limit specification is substituted with a criterion stating a number of standard deviations above or below the mean value of each variable. Consistent safety levels based on cost–benefit studies are approached through proper handling of load factors and strength reduction factors[3.25].

Permanent loads Dead and live loads affect seismic design conditions in various manners: they give place to internal forces produced by gravity — thus reducing capacity available to resist seismic forces — and they influence seismic response, both with regard to the structure's vibration periods and to the relation between mass, acceleration and force. The influence on natural periods is usually disregarded when specifying design loads, but can be accounted for by stating probable ranges of variation of those periods with respect to their computed values. Because dead loads are essentially constant in time, their design values for the combination of permanent and accidental loads coincide with those valid for the action of the former alone. Design values for live loads to be used in combination with earthquake must be obtained from the probability distributions of their value at an arbitrary instant in time, rather than of their maximum during a relatively long interval; the fact that the cost of failure in case it occurs is a function of the acting live load has been accounted for in some recent cost–benefit studies[3.26]. These considerations substantiate the requirements of some design codes that state different design live loads for their combination with permanent and accidental loads or with permanent loads alone[3.16].

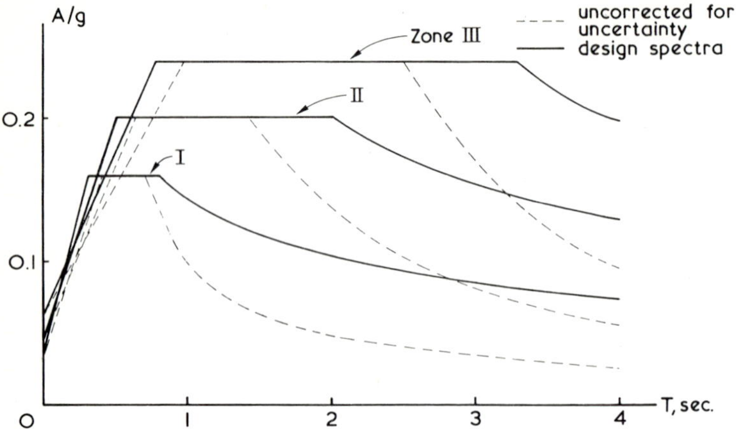

Fig. 3.24 Design spectra corrected for uncertainty in natural period

Natural periods Uncertainty in natural periods stems from that associated with mass and stiffness as well as with soil-structure interaction; its significance arises from the sensitivity of spectral ordinates to this parameter. That uncertainty can be taken into account by adopting unfavorable values derived either by applying corrective factors to those computed in terms of nominal values of the relevant parameters or by covering those uncertainties by means of suitable modifications to the ordinates of the nominal design spectra. As a rule, corrective factors greater than unity are applied to periods lying on the ascending branch of the acceleration spectrum, and values smaller than unity are applied otherwise. For instance, Figure 3.24 shows design spectra for three microzones in Mexico City both for deterministically known and uncertain natural periods[3.16]. For multidegree of freedom systems this criterion errs on the safe side, as it neglects probabilistic correlation among natural periods.

Design spectra Detailed characteristics of earthquakes are only approximately specified when a design intensity is adopted. Specification of design spectra for linear systems involves making decisions with respect to the design intensity and to the probability of exceedance of the proposed spectral ordinates given that intensity. Because the frequency content of ground motion varies with magnitude, focal mechanism, and site-to-source distance, earthquake intensity by itself does not determine the probability distribution of spectral ordinates for all ranges of natural periods. Unless seismic risk at a site can be ascribed exclusively to shocks that may generate at the same source, design spectra can not be made to correspond to the 'worst probable earthquake' to be expected at the site; rather, they should be obtained from the probability distributions of maximum response for different natural periods, regardless of the seismic source where every particular shock may have originated.

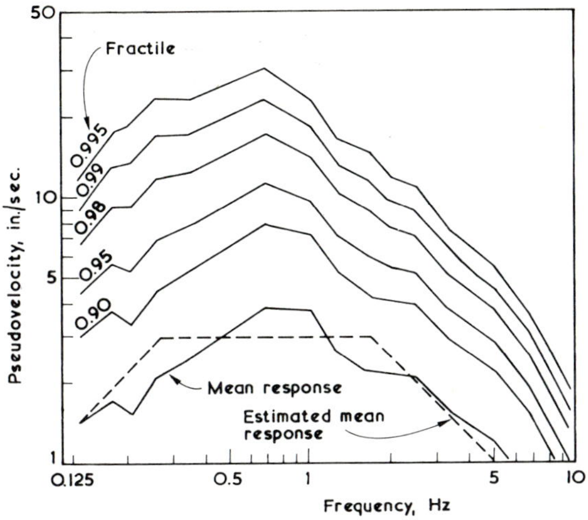

Fig. 3.25 Response spectra for different exceedance probabilities (after McGuire[3.29])

As a rule, the probability distributions of maximum spectral ordinates referred to in the foregoing paragraphs cannot be directly inferred from strong-motion records obtained at the site of interest, as only exceptionally is a large enough sample of those records available for the site. Instead, those distributions are usually generated from stochastic process models of local seismicity in the near-by seismic sources and the transformation of magnitudes and source locations into intensities at the site by means of attenuation laws that relate the pertinent variables with site-to-source distance[3.27,3.28]. Spectral ordinates corresponding to given probabilities of exceedance for a given magnitude and distance are shown in Figure 3.25, obtained from Ref. 29. If peak ground acceleration and velocity are given, *mean* values of design spectra or values corresponding to given exceedance probabilities for different damping ratios can be readily estimated, as shown in Figure 2.1.

Use of elastic spectra on firm ground as the basis for constructing inelastic design spectra is illustrated in Figure 2.3. The solid line represents an elastic design spectrum constructed according to the criterion of Ref. 3.30; ordinates are pseudovelocities and abscissas are natural frequencies, and both scales are logarithmic. The dashed line represents the nonlinear spectrum for the same damping as the elastic spectrum and a ductility factor μ; spectral accelerations can be directly read from the dashed-line plot by referring it to an adequate system of straight lines sloping down from the left, and total displacements of the inelastic system are obtained by multiplying those corresponding to the dashed line by the ductility factor μ (dash-point line). The relation between the various segments of the reduced acceleration spectrum $D'V'A'A'_0$ and their counterparts for the elastic case is as follows[3.30]. The extreme right-hand portion of the spectrum, where the response is governed by the maximum ground acceleration, remains at the same acceleration level as for the elastic case, and therefore at a corresponding increased total displacement level; the ordinates of segments D and V in the small and intermediate frequency ranges, respectively, are divided by μ, and the ordinates of segment A are reduced according to an equal-energy criterion, which for elasto-plastic systems is tantamount to dividing by $(2\mu-1)^{\frac{1}{2}}$.

The accelerograms of some earthquakes recorded on the surface of thick sediments of soft soil are characterized by their long duration and by their nearly harmonic nature. These properties are reflected in their linear response spectra, which show very narrow and pronounced peaks at one or more dominant periods (Figure 3.26). The validity of the foregoing rules for transforming linear spectra into their nonlinear counterparts has not been assessed yet, but some significant features have been qualitatively applied in the formulation of design spectra for the soft soil region in Mexico City. As Figure 3.26 shows, the design spectra uncorrected for uncertainty in natural period shows a wide plateau of constant ordinates, which is intended to cover the tendency of structures possessing natural periods shorter than those dominant in the ground motion to show increased responses as their effective periods grow as a consequence of nonlinear behavior.

Figure 3.26 also shows a correction for uncertainty in natural periods on both sides of the region of maximum ordinates; it also shows that on the long period side, specified ordinates are made to decay at a significantly slower rate than in the recorded spectrum. The latter requirement stands for the convenience of covering the decrease in reliability due to the possibility of occurrence of a large number of failure modes, and of providing additional protection with respect to unfavorable behavior caused by phenomena typical of long period structures and not normally considered in analysis, such as some forms of soil-structure interaction, concentrations of ductility demand, and slenderness effects in excess of computed values.

Damping and ductility The recommendations of some modern building codes[3.16,3.20] are formulated as though design spectra were actually based on linear response spectra for 5 to 10% viscous damping, with correction factors intended to account for ductilities in the approximate range of 1 to 6. But structural damping at small strains is much smaller than openly recognized in design specifications. Thus, while linear response spectra that provide the basis for the recommendations of Ref. 3.16 correspond to a damping ratio of 0.05 of critical, tests on actual structures subjected to small amplitude vibration show that this value should not exceed 2 to 3% for reinforced concrete structures or 0.5 to 1% for welded steel structures with low density of nonstructural elements. Apparent inconsistencies are rather a matter of tradition and of nomenclature than of actual safety, as most damping, even at low strains, must be ascribed to nonlinear response and deterioration rather than to viscous, non-damaging behavior. Nominal ductile capacity for given structures has not been derived from probabilistic

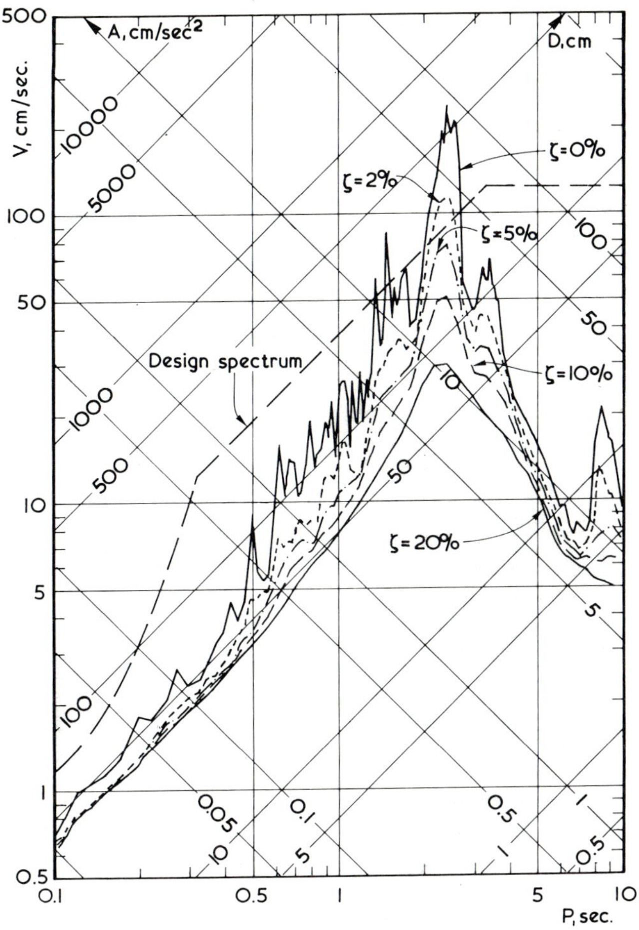

Fig. 3.26 Typical response spectra on soft clay in Mexico City

analysis of the measured ductilities developed by various structural systems subjected to dynamic excitation, but rather by semi-empirical adjustment of design coefficients based on engineering judgement, economic considerations, and study of the observed response of structures of known properties to severe ground shaking.

3.4.3 Reliability analysis in seismic design

As mentioned above, the reliability function of a system in a given environment is the probability that the system survives all the actions exerted upon it by the environment during a given time interval. Its computation is based on the probability distribution of the minimum safety margin during the given interval, and this probability is in turn dependent on the probability distributions of system strength and environment action at every instant within the interval. In seismic reliability problems the environment is described by stochastic models of dead, live and seismic loads, while system strength is described by probabilities of occurrence of given failure modes for given combinations of the mentioned loads. Uncertainty in seismic loads arises from randomness in earthquake origin, magnitude, rupture mechanism and wave propagation path, as well as from uncertainty in dynamic response for a given earthquake intensity. A brief description of the basic concepts of seismic reliability analysis is presented in the following, with the intention that it will provide a conceptual framework for the rational determination of safety levels and hence of pertinent design values and safety factors. More complete studies can be found in Refs. 3.21, 3.27 and 3.31.

Seismicity Let Y be earthquake intensity, expressed in terms of a set of parameters (such as peak ground acceleration or velocity, spectral response for given period and damping) that can be directly correlated with structural response or performance. Seismicity will be described by the stochastic process of occurrence of significant events, that is earthquakes having at the site of interest an intensity sufficiently high as to affect engineering structures, and by the conditional probability distribution of intensity given the occurrence of an event. Let T be the interval between occurrence of successive significant events, T_1 the time from the instant observations are started to the first event, and $f(t)$, $f_1(t)$ the respective probability density functions. The probability density function of the time to the rth significant event is obtained recursively as follows:

$$f_r(t)=\int_0^t f_{r-1}(t-\tau)f(\tau)\mathrm{d}\tau, \qquad r>1 \tag{3.1}$$

thus, the probability density function of the time to first exceedance of intensity y equals

$$g_1(t)=\sum_{r=1}^{\infty} f_r(t)PQ^{r-1} \tag{3.2}$$

where $Q(y)$ (assumed independent of t and r) is the conditional cumulative probability distribution of intensity given that a significant event has occurred, and $P=1-Q$. The probability density function of the time to failure for a structure having a deterministically known strength Y_R, can be obtained by means of Equation 3.2, making $Q=Q(Y_R)$.

Different expressions have been proposed for f_T and f_{T_1}[3.28]. The simplest of them, although not the most realistic, assumes significant events to take place in accordance with a Poisson process, that is,

$$f_{T_1}(t)=f_T(t)=\nu \exp(-\nu t)$$

where v is the mean rate of occurrence of the mentioned events. Under this assumption, Equation 3.2 leads to

$$g_1(t) = \nu P e^{-\nu P t} \quad (3.3)$$

The discussion that follows will be confined to this assumption. A more general treatment can be found in Ref. 3.31.

Structural response Let D be the cost of damage caused by an earthquake on a structure. It can vary between 0 and $D_0 + A$, where D_0 is the total cost of the structure and A that of its contents, as well as all other consequences (such as loss of human lives and indirect effects) expressed in monetary terms, diminished by the salvage value. A probability density function of D conditional to every possible value of intensity can be established[3.32]. If that function is denoted by $f_{D|Y}(d|y)$, the probability density function of D every time a significant event takes place is

$$f_D(\mathrm{d}) = \int \frac{\mathrm{d}Q(y)}{\mathrm{d}y} f_{D|Y}(\mathrm{d}|y)\,\mathrm{d}y \quad (3.4)$$

It may be advantageous to express the domain of possible damage levels of a given structure by a set of potential failure modes. If $p_i(y)$ is the probability of failure in mode i given an intensity equal to y, and D_i is the corresponding cost of damage, then the marginal probability of failure in mode i given the occurrence of a significant event is

$$\bar{p}_i = \int \frac{\mathrm{d}Q(y)}{\mathrm{d}y} p_i(y)\mathrm{d}y \quad (3.5)$$

and the expected cost of damage for each event is

$$\bar{D} = \sum_i D_i \bar{p}_i \quad (3.6)$$

the $\bar{p}_i$'s are functions of acting permanent loads, design parameters, and safety factors with respect to all relevant failure modes. By changing relative values of those safety factors it is possible to make failure modes with the highest consequences (in general, brittle modes) much less likely than those leading to lower damage levels. Thus, adoption of higher load factors for column than for beam bending moments may be advisable when significant axial loads hinder the development of enough ductile capacity at column ends; or it may be advantageous to make a structure safer with respect to overturning moment than to lateral yielding. Quantitative assessment of adequate increments of load factors can be established from economic considerations within the cost–benefit framework advocated in the sequel.

Optimum design Let $C(x_1, \ldots, x_n)$ be the initial cost of a given structure, and $x_1, \ldots, x_n$ a set of design parameters (resistances, stiffnesses, ductilities). Optimal values of those parameters are those maximizing the function

$$V = B - C - Z \tag{3.7}$$

where B and Z, also functions of the set of design parameters, are present values of the expected benefits and failure consequences, respectively. In other words; if $b(t)$ is the expected value of benefits at time t derived from performance of the structure, and γ is a discount rate such that present values of future losses or benefits can be obtained through multiplication of the latter by $\exp(-\gamma t)$, then

$$B = \int_0^\infty b(t) e^{-\gamma t} L(t) dt \tag{3.8}$$

and

$$Z = \int_0^\infty v\bar{D} e^{-\gamma t} L(t) dt \tag{3.9}$$

where $L(t)$ is the reliability function defined above. The meaning of $L(t)$ in Equations 3.8 and 3.9 is that production of benefits and losses is subjected to the condition that the structure has survived all previous loads. For the case of deterministically known strength Y_R, Equation 3.3 leads to

$$L(t) = \exp(-vP_R t) \tag{3.10}$$

where $P_R = P(Y_R)$.

From Equations 3.8, 3.10 and the assumption that $b(t) = b$ is constant,

$$B = \frac{b}{\gamma + vP_R} \tag{3.11}$$

Likewise, from Equation 3.9,

$$Z = \frac{v\bar{D}}{\gamma + vP_R} \tag{3.12}$$

and the expression for utility becomes

$$U = \frac{b}{\gamma + vP_R} - C - \frac{v\bar{D}}{\gamma + vP_R} \tag{3.13}$$

If Y_0 is the minimum intensity of significant events, that is, an intensity below which no damage can occur, then ν can be approximately expressed as KY_0^{-r}, where K and r depend on the activity of seismic sources near the site[3.28]. Under these conditions, $P_R=(Y_0/Y_R)^r$.

Expressing b, P_R, C and $\bar{D}$ in terms of the set of design parameters and differentiating with respect to them, a system of equations is obtained from which optimum values of those parameters can be determined.

If structural strength for a given set of design parameters is uncertain, Equations 3.10 and 3.13 become respectively

$$L(t)=E[\exp(-\nu P_R t)] \tag{3.14}$$

$$U=E\left[\frac{b}{\gamma+\nu P_R}-C-\frac{\nu\bar{D}}{\gamma+\nu P_R}\right] \tag{3.15}$$

The expectations in the above equations are taken with respect to the probability density functions of structural strength Y_R.

Different expressions for U can be obtained, depending on the policy adopted *a priori* with regard to repair and reconstruction measures to be taken after every damaging event[3.31]. Optimum design parameters may be strongly influenced by that policy.

Specification of safety in codified design According to the optimization criteria described above, determination of design resistances, stiffnesses and ductilities is not based on the expected response to a single event, defined by a given spectrum and assumed to correspond to a given return period. Instead, design parameters are optimum in the sense that they lead to the best investment of resources taking into account long term expected behavior under the action of a random number of random loads. However, by comparison with safety requirements for permanent loads, it is usually advantageous to specify seismic safety in terms of a design earthquake, assumed to correspond to a given return period, a set of rules to define minimum probable resistances from their expected values and variation coefficients, and a set of load factors. Safety under the action of the design earthquake is not significant by itself, but because it is an indirect measure of the reliability function $L(t)$.

Member and system reliability In the applications of the theory of structural reliability to the formulation of consistent safety design criteria for a single load application, nominal capacities of members or critical sections are often defined by either of the following expressions[3.21,3.22]

$$R^*=\bar{R}\exp(-\alpha V_R) \tag{3.16}$$

$$R^*=\bar{R}/(1+\alpha V_R) \tag{3.17}$$

Here, R^* denotes nominal value of the random strength R, $\bar{R}$ its expected value, V_R its coefficient of variation and α a constant that depends on the probability that R is smaller than its nominal value. It is clear that the ratio $R^*/\bar{R}$ is smaller than unity and decreases when V_R increases.

The capacity with respect to some failure modes in ductile systems can be expressed as the sum of the contributions of the capacities of a number of critical sections. Take for instance the shear capacity of a given story of a frame building and consider that capacity to be made up of the contributions of the moment capacities at all column ends. The coefficient of variation of the story shear capacity is equal to

$$V = \left(\sum_i \sum_j \rho_{ij} V_i V_j \bar{R}_i \bar{R}_j \right)^{1/2} \Big/ \sum_i \bar{R}_i$$

where R_i is the strength at the ith critical section, $\bar{R}_i$ and V_i respectively its expected value and coefficient of variation and ρ_{ij} the correlation coefficient between R_i and R_j. If the latter variables are stochastically independent,

$$V = \left\{ \sum_i (V_i \bar{R}_i)^2 \right\}^{1/2} \Big/ \sum_i \bar{R}_i,$$

and if all V_i's are equal to v,

$$V = v\left(\sum_i \bar{R}_i^2 \right)^{1/2} V = v\left(\sum_i \bar{R}_i^2 \right)^{1/2} \Big/ \sum_i \bar{R}_i$$

hence, $V \leqq v$ and the nominal value of R that would be obtained by direct application of Equation 3.17 with the adequate value of V will exceed that obtained by simple addition of the nominal values R_i^* of the contributions of all critical sections. This result is an analytical way of expressing an often intuitively derived principle: that under similar safety conditions for individual critical sections the reliability of ductile systems with respect to failure modes that require the development of the capacity of n critical sections decreases with decreasing n. Because design criteria for the revision of safety conditions are usually stated in terms of the ratio of structural capacity to internal load at each individual critical section, the effect under study has to be accounted for by making required safety factors vary with the number of critical sections involved in a failure mode. This is the basis for the prescription in the 1976 Mexico City Building Code stating that the generalized force acting on every shear wall or column that takes up more than 20% of the story generalized force (shear, torque or overturning moment) be increased 20%; or by the prescription concerning nonredundant systems in ATC recommendations[3.20] stating that when a building system is designed or constructed so that the failure of a single member, connection or component would endanger the stability of the building, that member, connection or component should be provided with a strength at least 50% greater than otherwise required.

Superposition of modal contributions Maximum contributions of all natural modes to a given response — internal force at a critical section, displacement or deformation — do not take place simultaneously. The design value of a response parameter is assumed proportional to its standard deviation at the end of the earthquake. After some simplifications[3.10], this criterion leads to the following expression:

$$S=\left(\sum_i\sum_j\frac{S_iS_j}{1+\varepsilon_{ij}{}^2}\right)^{1/2} \tag{3.18}$$

in which

$$\varepsilon_{ij}=\frac{\omega_i'-\omega_j'}{\zeta_i'\omega_i+\zeta_j'\omega_j} \tag{3.19}$$

and S_i is the maximum absolute value of the contribution of the ith mode to the response of interest; it is to be taken with the sign adopted by the unit impulse response function of the response of interest to a ground velocity step-increment when the mentioned function attains its maximum numerical value.

In Equation 3.19, ω_i = undamped circular frequency of ith natural mode, $\omega_i'=\omega_i\sqrt{1-\zeta_i'^2}$ = damped circular frequency of ith natural mode, $\zeta_i'=\zeta_i+2/\omega_i s$, ζ_i = damping ratio of ith natural mode (assumed equal to 0.05 unless a different value is justified), and s = duration of segment of stationary white noise equivalent to the family of actual design earthquakes; s may range from 15 to 40 seconds for ground conditions ranging from firm ground to thick deposits of very soft material. The influence of stochastic correlation between the instants when the response associated with each mode reaches its maximum is reflected in Equation 3.18 through the participation of ε_{ij}: when ω_i' differs significantly from ω_j', ε_{ij} is large and S^2 approaches the sum of the squares of the individual mode contributions, $\sum_i S_i{}^2$. However, when ω_i' is close to ω_j', ε_{ij} tends to zero and the cross-product tc ms S_iS_j, for $i\neq{}^*j$, become significant. The fact that each of these terms can be either positive or negative accounts for the possibilities of strongly correlated modal responses taking place with phase angles close to either 0 or 180°.

In buildings, cross-product terms are usually negligible. Exceptions occur, for instance, in the modal analysis of buildings possessing small torsional eccentricities, when torsional degrees of freedom are taken into account, or in the analysis of any type of structure when the response of an appendage (portion characterized by a mass much smaller than the others into which the system has been discretized) is taken as a degree of freedom in the computation of modal shapes and frequencies[3.16].

Superposition of ground motion components It has been customary to design structures so that they resist the envelope of effects of the various components of earthquake motion as though these components acted one at a time. There is growing consciousness that design should recognize the simultaneous action of all the components, as a number of conditions have been identified where superposition of those components significantly affects safety. Take, for instance, a building possessing continuous frames in two orthogonal directions, another with an asymmetrical plan, and a long continuous bridge with several supports. If the columns in the first structure are built in reinforced concrete and possess a square cross section, the most unfavorable direction of application of seismic forces will be along their diagonal, rather than parallel to either system of orthogonal frames. In addition, if the nonlinear response of the structure is analyzed and substantial ductility is developed at the column ends, effective stiffnesses of the frames in one direction will depend at any instant on the simultaneous state of deformation of the other system of frames; in other words, significant interaction will exist between ductility demands in both directions. Frames normal to the direction of asymmetry in the second case are subjected to the effects of direct shear produced by the horizontal ground component parallel to them, and to the torsional effects associated with the other horizontal component. Out of phase motion of the various supports in the third structure affect qualitatively and quantitatively the distribution of internal forces.

An approximate criterion to account for the foregoing effects has been recently developed; it evolved from a simplification of a second moment formulation of structural safety[3.36], and consists of the following[3.37]:

(1) Compute the responses to gravity loads and to the components of ground motion regarded as potentially significant. Let those responses be arranged into vectors $R = R_0$ and R_i respectively, with $i = 1, 2, \ldots, n$.

(2) Obtain vectors

$$R = R_0 + \sum_{i=1} \alpha_i R_i,$$

assigning plus and minus signs to $\alpha_i R_i$, ordering the R_i's in all possible permutations, and giving the α_i's the values in Table 3.1.

(3) If the problem is one of analysis, find out whether all points fall within the failure surface. If the problem is one of design, assign the design parameters such values that the safe domain will contain all the points.

In the analysis and design of towers and chimney stacks it is advisable to take α_i equal to 0.5 instead of 0.3 for $i \geqq 2$. This recommendation stems from two considerations: in towers having square or rectangular plan, supported on four equal columns, application of the foregoing criterion with $\alpha_2 = 0.3$ to safety checking with respect to axial stresses produced by overturning moment leads to systematic errors on the unsafe side; and in structures nominally having radial symmetry, such as chimney stacks, an apparently insignificant asymmetry causes an appreciable degree of coupling between modes of vibration involving orthogonal horizontal displacements. (See Sections 1.10 and 2.7.4.)

TABLE 3.1 Values of α_1 and maximum errors in amplitude of seismic-response vector (after Ref. 3.27)

i or *n*	α_1	*max. error* (%)	α_1	*Max. error safe side* (%)	*Max. error unsafe side* (%)
1	1.000	0.0	1.0	0.0	0.0
2	0.336	5.5	0.3	4.4	8.1
3	0.250	8.4	0.3	8.6	7.6
4	0.206	10.4	0.3	12.7	5.0
5	0.179	11.8	0.3	16.6	1.6
6	0.160	13.0	0.3	20.4	−2.1
7	0.146	13.9	0.3	24.1	−5.8
8	0.135	14.7	0.3	27.7	−9.6
9	0.126	15.4	0.3	31.1	−13.3
10	0.118	16.0	0.3	34.5	−17.0

3.4.4 Repair and strengthening of existing structures

Historical monuments, damaged structures and those to be remodeled or the use of which is modified, often pose the problem of deciding about adequate safety levels and compliance with current building codes. In some regions, large portions of important buildings have been designed and built according to standards that were afterwards deemed insufficiently strict, and there are large numbers of unengineered dwelling units. Adoption of standards applicable to new structures is cumbersome and expensive in most cases mentioned above. The situation must be coped with having in mind that the objective of engineering design is to optimize for society. Decision models dealing with these cases have recently been developed[3.32].

REFERENCES

3.1 Esteva, L., 'Behavior Under Alternating Loads of Masonry Diaphragms Framed by Reinforced Concrete Members', *Proc. International Symposium on the Effects of Repeated Loading of Materials and Structures*, RILEM, Mexico City (1966)

3.2 Krawinkler, H., Bertero, V. V. and Popov, E. P., 'Inelastic Behavior of Steel Beam–to Column Subassemblages', *Earthquake Engineering Research Center, College of Engineering, Univ. California*, EERC 71–7, Berkeley, Cal. (1971)

3.3 Frank, R. A., Anagnostopoulos, S. A., Biggs, J. M. and Vanmarcke, E. H., 'Variability of Structural Response Due to Real and Artificial Ground Motions', MIT Department of Civil Engineering, *Research Report R*76–6 (1976)

3.4 Havilane, R. W., Biggs, J. M. and Anagnostopoulos, S. A., 'Inelastic Response Spectrum Design Procedures for Steel Frames', MIT Department of Civil Engineering *Research Report R*76–40 (1976)

3.5 Veletsos, A. S., 'Maximum Deformations of Certain Nonlinear Systems', *Proc. Fourth World Conference on Earthquake Engineering*, Santiago, Chile (1969)

3.6 Bielak, J., 'Dynamic Response of Single-Degree-of-Freedom Bilinear Systems', Master's Thesis, Rice University, Houston, Tex. (1966)

3.7 Spencer, R. A., 'The Nonlinear Response of Multistory Prestressed Concrete Structures to Earthquake Excitation', *Proc. Fourth World Conference on Earthquake Engineering*, Santiago, Chile (1969)

3.8 Penzien, J. and Liu, S. C., 'Nondeterministic Analysis of Nonlinear Structures Subjected to Earthquake Excitations', *Proc. Fourth World Conference on Earthquake Engineering*, Santiago, Chile (1969)

3.9 Ramírez, J., 'Efectos de Inestabilidad en la Respuesta Sísmica de Estructuras de Cortante Inelásticas', Master's Thesis, *Faculty of Engineering, National University of Mexico* (1977)

3.10 Newmark, N. M. and Rosenblueth, E., 'Fundamentals of Earthquake Engineering', Prentice-Hall, Englewood Cliffs, N.J. (1971)

3.11 Guerra, O. R. and Esteva, L., 'Equivalent Properties and Ductility Requirements in Seismic Dynamic Analysis of Nonlinear Systems', *Proc. Sixth World Conference on Earthquake Engineering*, New Delhi, India (1977)

3.12 Rosenblueth, E., 'Temblores Chilenos de Mayo 1970: Sus Efectos en Estructuras Civiles', *Ingenieria*, **31**, 1, Mexico, D.F. (1961)

3.13 González-Flores, M., 'Sistema para Elimar los Esfuerzos Peligrosos que los Temblores Causan en las Estructuras', *Quinto Congreso Mexicano de la Industria de la Construcción*, México, D.F. (1964)

3.14 Skinner, R. I., Kelly, J. M. and Heine, A. J., 'Energy Absorption Devices for Earthquake Resistant Structures', *Proc. Fifth World Conference on Earthquake Engineering*, Rome, Italy (1973)

3.15 Ruiz, S. E., Esteva, L. and Guerra, O. R., 'Desarrollo y Evaluación de Sistemas para Limitar las Acciones Sísmicas sobre Edificios', *Ingeniería*, **47**, 2, Mexico, D.F. (1977)

3.16 'Diseno estructural', Mexico City Building Code, Mexico, D.F. (1977)

3.17 Esteva, L., Díaz de Cossío, R. and Elorduy, J., 'El Temblor de Caracas, Julio de 1967', *Ingeniería*, **38**, 3, Mexico, D.F. (1968)

3.18 Haviland, R., 'A Study of the Uncertainties in the Fundamental Translational Periods and Damping Values for Real Buildings', MIT, Department of Civil Engineering Research *Report R*76–12 (1976)

3.19 Esteva, L., 'Seismic Risk and Seismic Design Decisions', *Seminar on Seismic Design of Nuclear Power Plants*, MIT Press, Cambridge, Mass (1969)

3.20 *Recommended Comprehensive Seismic Design Provisions for Buildings*', Applied Technology Council, Palo Alto, Calif. (1977)

3.21 Rosenblueth, E. and Esteva, L., 'Reliability Basis for Some Mexican Codes', American Concrete Institute, *Special Publication* 31 (1971)

3.22 Meli, R., 'Bases para los Criterios de Diseno Estructural del Proyecto del Reglamento de Construcciones para el Distrito Federal', *Institute of Engineering, National University of Mexico*, Report 375 (1976)

3.23 Esteva, L., 'Regionalización Sísmica de México para Fines de Ingeniería, Institute of Engineering, National University of Mexico, *Report* 246 (1070)

3.24 Rosenblueth, E., 'Code Specification of Safety and Serviceability', State of Art Report 2, Technical Committee 10: Structural Safety and Probabilistic Methods, *ASCE-IABSE Joint Committee on Planning and Design of Tall Buildings*, Fritz Engineering Laboratory, Lehigh University, Bethlehem, Pa. (1972)

3.25 Lind, N. C., 'Deterministic Format for the Probabilistic Design of Structures', in *An Introduction to Structural Optimization*, Solid Mechanics Division, Univ. Waterloo, Waterloo, Ont., Canada, SM Study 1 (1969)

3.26 Rosenblueth, E. and Mendoza, E., 'Optimum Seismic Design of Auditoriums', *Proc. Fifth World Conference on Earthquake Engineering*, 2, Rome, Italy (1973)

3.27 Esteva, L., 'Criterios para la Construcción de Espectros de Diseno Sísmico', Institute of Engineering, National University of Mexico, Report 178 (1968)

3.28 Esteva, L., 'Seismicity', Chapter 6 of *Seismic Risk and Engineering Decisions* (Edited by C. Lomnitz and E. Rosenblueth), Elsevier, Amsterdam (1976)

3.29 McGuire, R. K., 'Seismic Structural Response Risk Analysis, Incorporating Peak Response Regressions on Earthquake Magnitude and Distance', Ph.D. Thesis, *MIT Department of Civil Engineering* (1974)

3.30 Newmark, N. M. and Hall, W. J., 'Procedures and Criteria for Earthquake Resistant Design', *Building Practices for Disaster Mitigation, Building Science Series* 46, National Bureau of Standards, Washington, D.C. (1973)
3.31 Rosenblueth, E., 'Optimum Design for Infrequent Disturbances', *Proc. Am. Soc. civ. Engrs*, **102**, ST9 (Sept. 1976)
3.32 Rosenblueth, E., 'Decisiones Optimas para Estructuras Existentes en Zona Sísmica. Primera Etapa', Institute of Engineering, National University of Mexico (in press)
3.33 Newmark, N. M., 'Current Trends in the Seismic Analysis and Design of High-Rise Structure', in *Earthquake Engineering* (Edited by R. Wiegel), Prentice Hall, Englewood Cliffs, N.J. (1970)
3.34 Paulay, T., 'Coupling Beams of Reinforced Concrete Shear Walls', *Proc. Am. Soc. civ. Engrs*, **97**, ST3 (March, 1971)
3.35 Park, R. and Paulay, T., 'Reinforced Concrete Structures', John Wiley and Sons, New York (1975)
3.36 Hasofer, A. M. and Lind, N. C., 'Exact and Invariant Second-Moment Code Format', *Proc. Am. Soc. civ. Engrs*, **100**, EM1 (1974)
3.37 Rosenblueth, E. and Contreras, H., 'Approximate Design for Multicomponent Earthquakes', Submitted for publication in *J. Engng Mech. Div. Am. Soc. civ. Engrs*

4 Steel structures

Oscar de Buen

4.1 INTRODUCTION

Most modern buildings are designed to support much smaller seismic forces than those corresponding to an unlimited elastic behavior. Hence under strong quakes they suffer localized plastic deformations which dissipate an important amount of energy. In buildings having few partitions and other nonstructural elements, the structure alone has to dissipate almost all the energy.

A ductile frame can undergo important inelastic deformations, localized in the neighborhood of sections of maximum bending moment. These eventually lead to formation and rotation of plastic hinges and redistribution of bending moments, allowing the structure to resist higher loads than those predicted by an elastic analysis. If the structure is highly redundant and the number of plastic hinges associated with the collapse mechanism is high, the energy dissipated as they form and rotate is quite important. The energy absorption capacity demand in elastic regions is drastically reduced.

Structural steel is a very ductile material and can be used to build statically indeterminate structures fulfilling the foregoing conditions. It is, therefore, very adequate for earthquake resistant construction. But intrinsic ductility is not necessarily retained in the finished structure. Care must hence be had in design and construction to avoid losing this property.

4.2 BEHAVIOR OF STEEL STRUCTURES UNDER STRONG EARTHQUAKES

Behavior of steel building structures under strong earthquakes has generally been satisfactory from a strength point of view. Structures have had enough strength even in buildings 10 to 12 stories high designed only for vertical permanent loading and having flexible connections. On the other hand, their lateral stiffness has frequently been inadequate, and windows and nonstructural partitions have suffered considerable damage.

Two cities with an important number of tall steel buildings have suffered strong earthquakes: San Francisco in 1906 (Richter magnitude 8 1/4, 14.5 km from downtown to failure) and Mexico City in 1957 (magnitude 7.5, 275 km from epicenter).

The behavior of tall buildings will complete steel structure during the San Francisco earthquake was satisfactory. Some had rigid frames or X-bracing; the rest had partition walls, which were almost non-existent in the ground story; in this story lateral resistance depended on semi-rigid beam-column joints. Structural damage was negligible[4.1].

Buildings with exterior bearing walls and interior steel frames suffered some damage, but none of them collapsed.

In 1957 most of the buildings having steel structure in Mexico City were riveted; there were several in the range of 10 to 22 stories, and one 45 stories tall; the excellent behavior of this building, the Latin–American Tower, is well known the world over.

The only high-rise completely welded building, 20 stories tall, also behaved satisfactorily. Designed to resist a seismic coefficient of only 0.025, it did not suffer damage of any kind.

Those buildings were designed elastically, and no precautions were taken to secure a good behavior in the plastic range. On the other hand, building codes employed at that time lead probably, under equal circumstances, to stronger structures than today's codes.

Many buildings designed for permanent vertical loading only, or with very small horizontal loads, survived the earthquake without structural damage, but often there was much nonstructural damage. Some buildings were stiffened afterwards, and at least one was demolished.

Tall steel buildings have been affected by later earthquakes (Anchorage, 1964; Caracas, 1967; San Fernando, 1971; Managua, 1972; Guatemala, 1976; Mexico City, 1979). Their behavior has consistently been adequate.

Nevertheless, it must be kept in mind that most of those buildings had many nonstructural partitions that provided lateral strength and stiffness and high energy absorption capacity.

4.3 MOMENT RESISTANT FRAMES

4.3.1 Behavior

Behavior of a story in a multistory frame under constant vertical loading and monotonically increasing horizontal forces can be described by its $Q-\Delta$ (horizontal load-lateral displacement) curve. The frame remains in the elastic range until the first plastic hinge forms (to simplify the discussion, residual stresses are neglected, shape factors are assumed equal to one, and the story-shear configuration in the building is assumed to remain constant save for a proportionality factor). Additional load increments are resisted by a deteriorated structure, and lateral displacements increase as the first plastic hinge rotates under constant plastic moment, M_p. Lateral stiffness decreases again upon formation of a second plastic hinge. The process continues, as the number of plastic hinges increases, until the structure becomes a mechanism which sways under decreasing lateral load.

The plastic collapse load corresponding to an overall mechanism is the maximum load which can theoretically be resisted by a steel frame, and the one which allows the maximum energy absorption before failure. There are, however, several factors that can lead to failure under a smaller load.

4.3.2 Causes of premature failures

A premature failure can be triggered by any of several factors:

(1) Incapacity to reach the M_p moment.
(2) Insufficient rotation capacity.
(3) Failure of members or connections.
(4) Overall or partial structure instability.

A section where a plastic hinge is supposed to develop can fail under a bending moment smaller than its plastic moment or when the rotation has not reached the value associated with the collapse mechanism. This can be due to local or lateral buckling or to fracture because of insufficient deformation capacity. Ductility can be drastically reduced by inadequate fabrication procedures, low working temperatures, triaxial states of stress, or fatigue. High strength steels are generally less ductile than ordinary structural steels.

A beam or column can fail between plastic hinges by local or lateral buckling or a combination of both. Connections may be unable to resist the forces and moments transmitted to them by beams and columns.

Finally, if lateral stiffness is not high enough, the structure as a whole, or one or more stories, can fail by instability due to second order P–Δ effects.

Extensive theoretical and experimental research has been done on every possible type of failure of structural members, subassemblages, and rigid frames under monotonically increasing loading. Their behavior under cyclic load, though, has been studied only in the last few years, and much information is still needed.

4.4 BEHAVIOR AND DESIGN OF STRUCTURAL MEMBERS

If the complete load-displacement curves of a structure are known, the load factor against collapse and the lateral displacements under working loads can be determined. If either of these is unacceptable, beams and columns are modified until an adequate response is obtained, consistent with the assumptions made. In multistory frames a curve for each story is theoretically necessary but in most cases a few curves suffice.

Q–Δ curves obtained for monotonically increasing lateral deformations give a good indication of the frame seismic behavior. To obtain the Q–Δ curve of a structure the characteristics of the component members must be known; beams, columns and joints must therefore have been tentatively designed before plotting the curve.

Member and connection behavior and design are discussed in the next paragraphs.

4.4.1 Beams

Beam behavior can be understood studying the curves relating loads and deflections in and outside the plane of bending. Information to plot these curves has been obtained experimentally. Figure 4.1 shows curves relating bending moment to vertical and lateral deflections. Curves correspond to the beam's central segment, under uniform bending[4.2,4.3]. Supports and points of load application are laterally and torsionally braced. Lateral displacements of the compression flange start as soon as the bending moment reaches the M_p value. They increase gradually. Beam sections lose their original shape (Figure 4.1(b)). Resistance is reached when the critical half of the compression flange buckles.

This behavior corresponds to adequately braced beams with sufficiently low width-to-thickness ratios in web and flanges. Lateral and local buckling start after important plastic deformation has taken place under constant plastic moment.

There are other possible types of behavior, some of them shown in Figure 4.2. Strength and rotation capacity can be reduced because of lateral or local buckling.

Lateral buckling The moment-end rotation curve $M_o - \theta$ (Figure 4.3) describes the behavior of a beam in the plane of symmetry and loading. Lateral buckling is shown by curves $M_o - u$ or $M_o - \varphi$, moment-lateral displacement or moment-torsional rotation. If the beam has no geometrical imperfections and there are no eccentricities in the loads, curves $M_o - u$ and $M_o - \varphi$ will be as depicted by the full line, with a bifurcation at A; after this point the beam can theoretically remain in the original plane as loading increases, or deform laterally under constant moment.

Bifurcation of equilibrium never takes place in actual beams because of inevitable initial imperfections. Failure is never due to buckling but to instability under continuously increasing lateral deformation. Nevertheless, the theoretical critical load is an upper limit of the actual resistance and is employed for design purposes.

Very short beams yield completely before buckling. They can resist moments equal to M_p, or higher because of strain hardening. Strength of intermediate-length beams is reduced because partial plastification takes place before buckling. The load carrying capacity of long beams is determined by elastic buckling. A given shape can exhibit any one of these behaviors, as shown in Figure 4.4, which gives the resisting moment as a function of the distance between laterally supported sections.

Near sections of plastic-hinge formation in fully rigid frames to be constructed in strongly seismic areas, beams should be in the first interval of Figure 4.4. Lateral buckling should not prevent them from reaching the plastic moment, nor should their rotation capacity be significantly impaired.

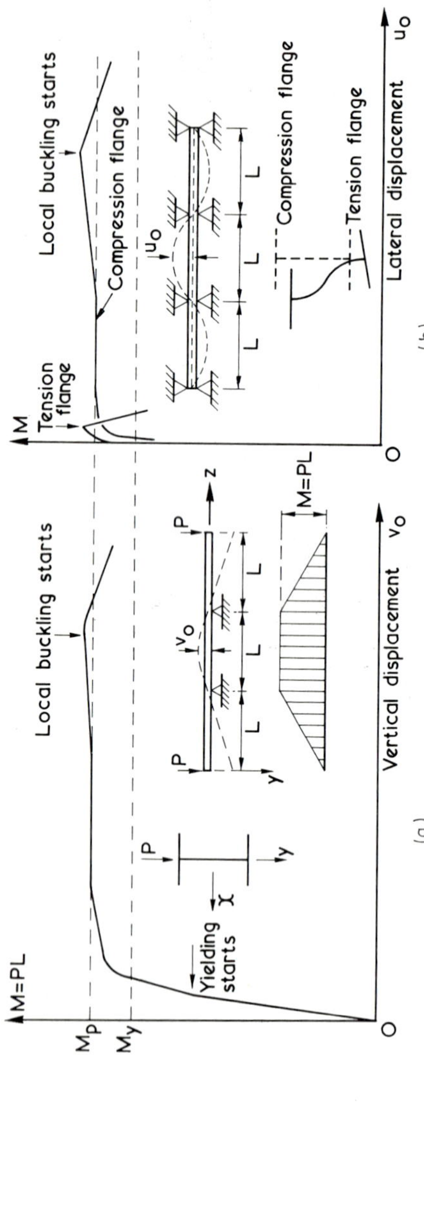

Fig. 4.1 Beam under uniform moment; moment–deflection curves[4.3]

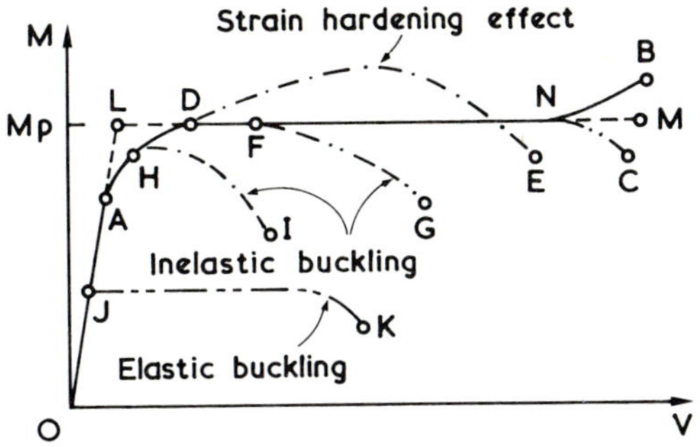

Fig. 4.2 Beam moment–deflection curves

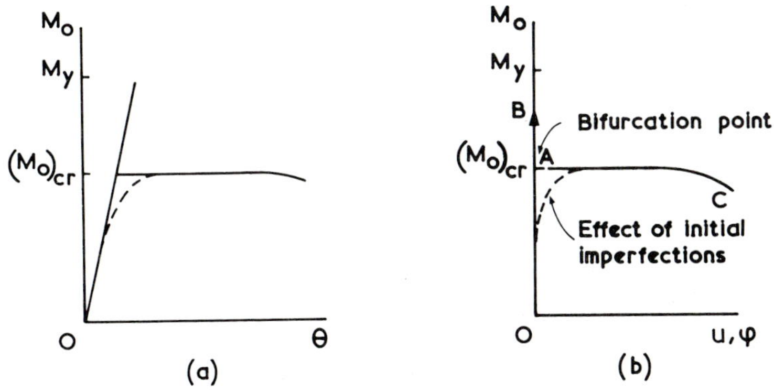

Fig. 4.3 Behaviour of a beam under uniform moment

Away from plastic hinges, beams may be in the second or third interval, but they have to be designed with safety factors high enough to avoid instability before attaining the plastic collapse load.

Instability of completely yielded beams Lateral and local buckling are unavoidable in I or wide-flange beams in strong axis bending. However, if the depth-to-thickness ratios of their flanges and webs are small enough and adequate lateral bracing is provided, buckling can be sufficiently delayed to attain satisfactory behavior under static as well as dynamic loading.

Curves in Figure 4.5 show adequate and inadequate behaviors from the viewpoint of rotation capacity. If unloading starts when the moment is smaller than M_p, behavior is even less adequate.

In many wide-flange beams tested in strong-axis bending, lateral deflection starts as soon as M_p is reached, but when distances between laterally supported sections are equal to or smaller than a certain critical length, initiation of lateral buckling does not correspond to failure. Unloading begins when the compression flange buckles locally under important lateral

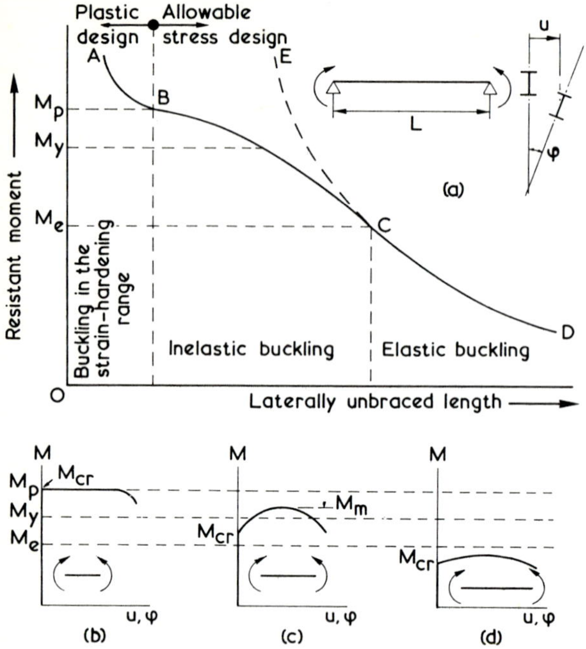

Fig. 4.4 Influence of unbraced length on beam behaviour

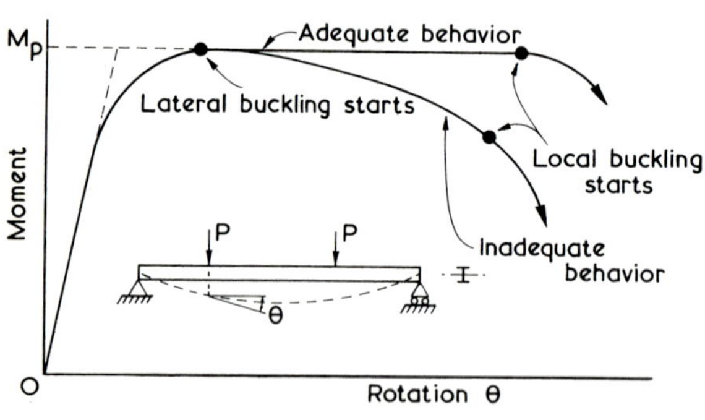

Fig. 4.5 Beam under uniform bending; moment–rotation curves

deformations. Failure is always due to a combination of both types of buckling[4.4,4.5].

Beam behavior is strongly dependent upon values of end moments in the unbraced length. If bending moment is uniform or changes slowly, the compression flange yields over a considerable length and its stiffness diminishes markedly. Lateral deflections and additional compression stresses grow rapidly. Eventually a local buckling wave develops, due to

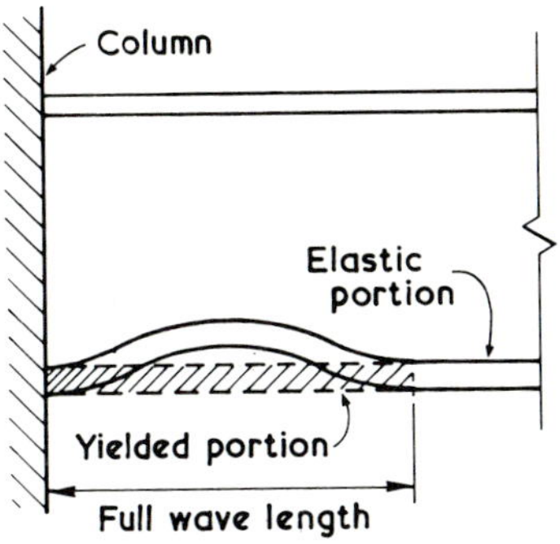

Fig. 4.6 *Compression flange local buckling at the end of a beam*

combined primary and lateral bending compression stresses[4.13]. On the other hand, if the slope of the bending moment diagram is steep, only a small length of the compression flange adjacent to the plastic hinge yields, loss of stiffness is small and lateral deflections do not grow rapidly; failure begins by local buckling under uniform compression, and collapse eventually takes place by lateral buckling[4.6,4.7].

The slope of the bending moment diagram near the end of frame beams, where plastic hinges develop, is very steep. The yielding portion of the compression flange is highly restrained at both ends. Under static loading, the compression flange buckles locally when the length of the yielded portion equals that of a wave of local buckling[4.6,4.7] (Figure 4.6).

Under cyclic loading, beam flanges adjacent to columns are alternately in tension and compression. The compression flange buckles locally, but buckling waves appear and disappear in every cycle, as compression changes into tension. If lateral buckling is not critical, failure sets in eventually by formation of a crack in the zone of maximum inelastic deformation[4.8].

Test results reported in Ref. 4.8 clearly show that under cycling, alternating loads, local buckling of beams is more critical than material low-cycle fatigue.

There are indications that flange local buckling does not significantly lower frame capacity. Accordingly, use of H shapes, having depth-to-thickness flange ratios almost 25% in excess of limits specified for plastic design under static load, has led to satisfactory behavior[4.9,4.10]. Still, load reversals enhance effects of local imperfections, hastening initiation of torsional and local buckling and reducing strength and stiffness[4.11,4.12,4.14]. Rotation capacity improves when depth-to-thickness ratios in webs and flanges are reduced.

Reference 4.11 suggests that the maximum story drift index*, corresponding to the most severe design earthquake, should be limited to 0.03 to avoid

* The story drift index is the relative lateral displacement between consecutive levels divided by the story height.

Table 4.1 Maximum allowable width-to-thickness ratios and laterally unsupported distances

	Maximum allowable width-to-thickness ratios			*Maximum allowable laterally unsupported distance*	
		Webs			
Yield stress	*Flanges*	$P/P_y \leqq 0.27$	$P/P_y > 0.27$	$1.0 \geqq M/M_p > -0.5$	$-0.5 \geqq M/M_p > -1.0$
36 ksi (24.84 kN/cm^2)	17.0	68.7–96.1 P/P_y	42.8	63.2 r_y	38.2 r_y
50 ksi (34.50 kN/cm^2)	14.0	58.3–81.6 P/P_y	36.6	52.5 r_y	27.5 r_y
65 ksi (44.85 kN/cm^2)	12.0	51.1–71.5 P/P_y	31.9	46.2 r_y	21.2 r_y

instability. The maximum necessary plastic hinge rotation capacity of the beam's ends can then be conservatively taken as 0.03 rad. Beams satisfying the depth-to-thickness ratios specified for plastic design under static loads, with continuous lateral support of the top flange, can apparently develop at least those rotations without loss of strength.

Design The rotation capacity necessary in structures under strong earthquakes and the rotation capacity which they can supply are still plagued with uncertainties. Until additional information becomes available it is wise to satisfy requirements on lateral bracing and depth-to-thickness ratios specified for plastic design of rigid frames[4.15], which are shown in Table 4.1. This recommendation should be followed even in structures designed by allowable-stress methods.

Cross sections corresponding to plastic hinges associated with the failure mechanism must be adequately braced against lateral and torsional displacements. The distance from such braced hinge locations to similarly braced adjacent points should not exceed the value given in the last two columns of Table 4.1. M/M_p, ratio of end moments in the unbraced length, is negative for single curvature; M is always smaller than M_p.

Effects of shear upon beam bending strength can almost always be disregarded. The plastic moment needs not be modified if the shear force is not greater than $(\sigma_y/\sqrt{3})\ wd_w$, where σ_y = steel yield stress, w and d_w = web thickness and height[4.16].

Design of beams in regions not adjacent to plastic hinges is normally done by allowable stress methods.

4.4.2 Beam-columns

Columns in rigid frames work generally under axial compression and biaxial bending.

Failure of a beam-column can be due to one or a combination of several of the following causes.

(1) Plastic hinge formation under axial compression and bending.
(2) Instability by axial load-moment interaction.
(3) Lateral-torsional buckling.
(4) Axial compression buckling by weak axis bending.
(5) Local buckling.

Building columns usually fail through a combination of the first two factors, by development of a mechanism under axial loads and bending moments magnified by second-order effects. Dimensions of these members are such that lateral-torsional buckling is seldom critical; and, as bending almost always plays an important role, axial compression buckling is not critical either.

In braced frames or frames with shear walls of adequate stiffness columns fail by formation of three plastic hinges (Figure 4.7), or development of hinges in the column and the neighboring beams, which produces an equivalent effect.

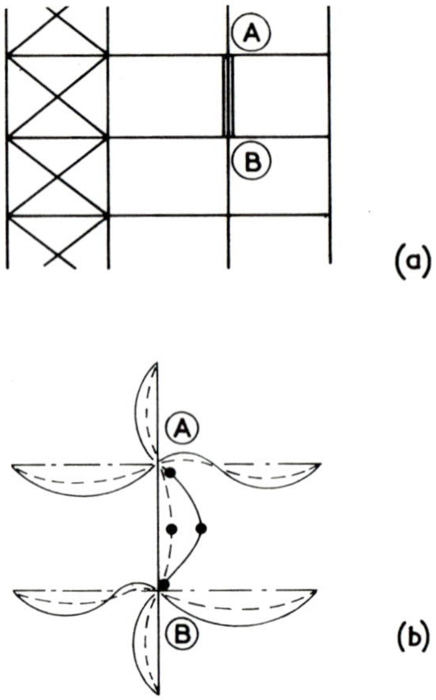

Fig. 4.7 Column failure in a braced frame

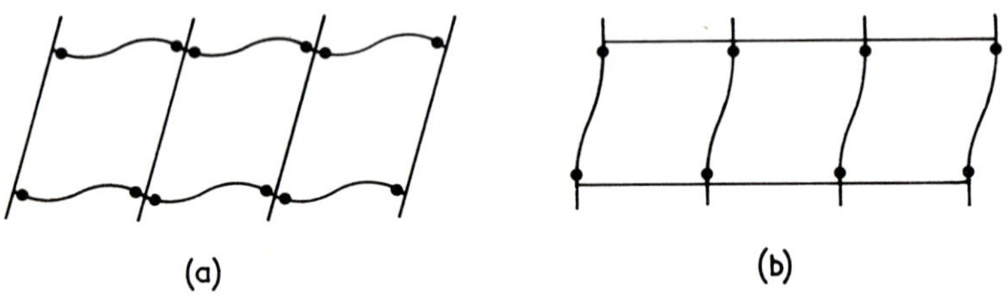

Fig. 4.8 Column behaviour in unbraced frames

Unbraced frames behave quite differently. Overall stability is primarily a function of beam stiffnesses; if beams are very flexible, columns behave basically as rigid bodies (except at their bases, where plastic hinges develop), and lateral displacements are due to bending of beams. Every story becomes a mechanism when plastic hinges form at the beam ends (Figure 4.8(a)).

When beams are very stiff, sway is resisted by column bending. The mechanism condition is reached when plastic hinges appear in both ends of every column. Lateral stability is a function of the properties of all columns in each story. Figure 4.8(b) shows the theoretical behavior of a story with infinitely rigid beams.

If the columns in a story of an unbraced frame have different strengths some of them can fail as shown in Figure 4.7(b) when the story's lateral resistance is not yet exhausted; this situation should generally be avoided.

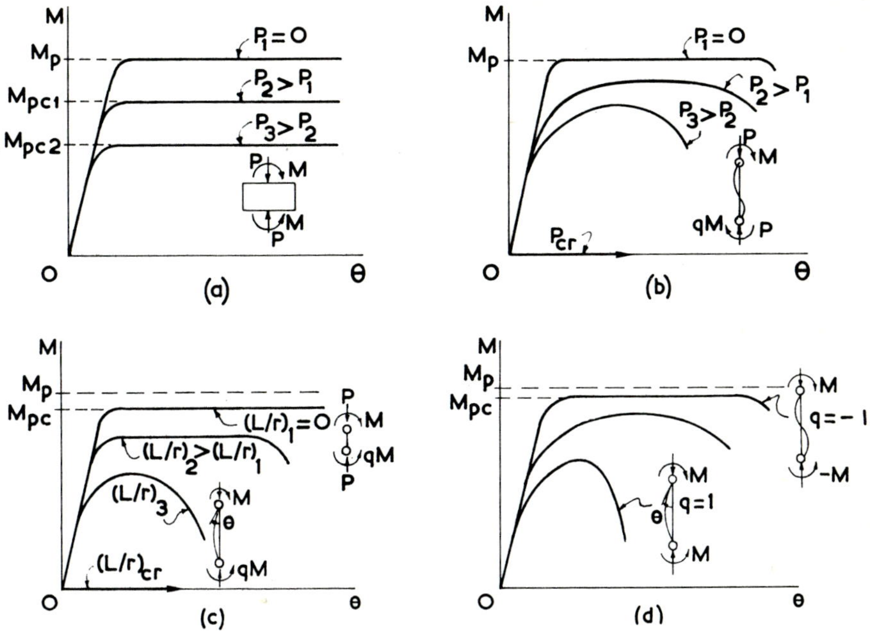

Fig. 4.9 Moment–rotation curves of beam-columns which fail by excessive bending in the end moments plan. (*b*) L/r and q *constant*, P *variable*, (*c*) q *and* P *constant*, L/r *variable*, (*d*) P *and* L/r *constant*, q *variable*

According to current practice, columns of frames in strongly seismic areas are usually designed to remain basically in the elastic range, except at their bases, as plastic hinges form in the beams. This is one reason why information on cyclic behavior of beam-columns is scanty.

Behavior of an isolated beam-column with couples applied at the ends can be represented by an end moment–end rotation curve. Member response is determined by several parameters: member slenderness and curvature, compression load. The possibility of lateral buckling and linear displacements of the column ends makes the problem more involved. Figure 4.9 shows several $M-\theta$ curves corresponding to beam-columns with linearly supported ends, assuming that there is no local or lateral buckling.

The strength of a beam-column that fails by excessive strong axis bending can be computed using elastic or elastoplastic methods. In the first case it is assumed that the limit of structural usefulness is reached when the yield stress appears at the critical section; it does not correspond to the maximum strength.

Disregarding residual stresses, elastic behavior ends when

$$\sigma_{max}=\frac{P}{A}+\frac{M_{max}}{S_x}=\sigma_y \tag{4.1}$$

A and S_x are the column's area and section modulus, P and M_{max} the compression load and maximum bending moment, including second-order effects, and σ_y is the steel yield point.

Dividing both members by σ_y, recalling that $A\sigma_y = P_y$ and $S_x\sigma_y = (M_x)_y$, and expressing M_{max} as an amplification factor times the larger end moment, M_o, Equation 4.1 can be written

$$\frac{P}{P_y} + \varphi \frac{M_o}{(M_x)_y} = 1.0 \tag{4.2}$$

Exact theoretical expressions for the amplification factor φ are too involved for design. Approximate equations are generally employed:

$$\frac{P}{P_y} + \frac{C_m}{1 - P/P_E} \frac{M_o}{(M_x)_y} \leqq 1.0 \tag{4.3}$$

$$\frac{P}{P_y} + \frac{M_o}{(M_x)_y} \leqq 1.0 \tag{4.4}$$

Both formulas must be simultaneously satisfied.

$C_m M_o$ is a fictitious uniform moment approximately equivalent to the actual one, and $1/(1 - P/P_E)$ is an amplification factor; P_E is the Euler critical load in the plane of bending; C_m can be computed using a formula given in Reference 4.15.

Methods have been developed to evaluate beam-column maximum strength and to draw complete end moment-end rotation curves[4.17–4.20]; these curves are fundamental in earthquake resistant design because they provide the member energy absorption and rotation capacities. Such curves have been plotted for members under constant compression and monotonically increasing end moments, but they still provide a qualitative indication of beam-column behavior under dynamic load.

Figure 4.10 shows moment rotation curves corresponding to three beam-columns with slenderness ratios $L/r_x = 30$ and compression loads $P = 0.30\,P_y$, under increasing end moments[4.21]. None can resist the complete M_p moment. Columns (a) and (b) have significant rotation capacity under M_{pc} moment (plastic moment reduced by axial compression), and second-order effects are more decisive in column (c). This beam-column is unable to reach M_{pc}, and its rotation capacity decreases dramatically. This phenomenom is more important when L/r_x and P increase: slender columns under high compression load resist maximum moments much smaller than M_{pc}; their rotation capacity is quite small or nil.

Columns of frames built in strongly seismic areas are usually in the low P/P_y range because they are strongly influenced by bending. Besides, they have small slenderness ratios and bend in double curvature under vertical and earthquake loads. Their rotation capacity is generally high, about half that predicted for beams[4.21].

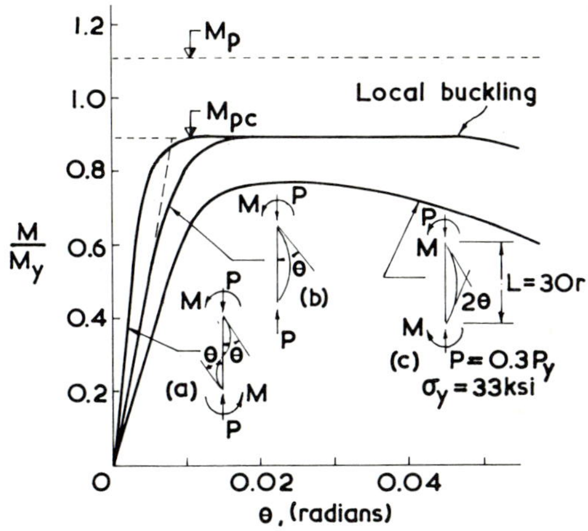

Fig. 4.10 *Typical beam–column moment rotation curves (after Galambos and Lay*[4.21])

Equations 4.3 and 4.4 can also be used to determine the load condition corresponding to the beginning of elastic lateral-torsional buckling; P_y and $(M_x)_y$, Equation 4.3, are replaced with P_{cr}, Euler's buckling load, and $(M_{cr})_x$, elastic buckling critical moment of the member under uniform strong axis bending; Equation 4.4 does not change.

The strength of beam-columns that fail by excessive bending without buckling, given in non-dimensional form, is almost independent of the member cross section dimensions. A family of curves corresponding to a given ratio of smaller to larger end moments can be used for all rolled or welded H-shapes. On the other hand, lateral-torsional buckling resistance depends markedly on cross-sectional geometry. Reference 4.22 shows that the strength of most H-shapes employed as beam-columns in buildings is not significantly reduced by lateral-torsional buckling under static load.

If P_y and $(M_x)_y$ in Equation 4.3 are respectively replaced with the inelastic critical load of the column under axial compression, P_{cr}, and the maximum moment which could be resisted by the member under uniform bending, taking into account lateral-torsional buckling when significant, M_m, we obtain Equation 4.5 to compute pairs of values of P and M_o which produce failure by inelastic instability; this equation is accurate enough for design purposes.

Equation 4.6, corresponding to the formation of a plastic hinge in one end, replaces Equation 4.4.

Equations 4.5 and 4.6 can be adapted to beam-columns belonging to frames subject to joint translation (sidesway).

$$\frac{P}{P_{cr}} + \frac{C_m}{1 - P/P_E} \frac{M_o}{M_m} = 1.0 \qquad (4.5)$$

$$\frac{P}{P_y}+\frac{M_o}{1.18M_p}=1.0 \tag{4.6}$$

M_m, maximum moment that can be resisted by the member in the absence of axial load, is given by the empirical formula[4.15]

$$M_m=\left[1.07-\frac{(L/r_y)\sqrt{\sigma_y}}{2625}\right]M_p\text{, but not greater than } M_p \tag{4.7}$$

σ_y is in kN/cm², and M_m has the same units as the plastic moment M_p.

Equation 4.7 was developed for beam-columns under uniform primary bending moment, but can be adjusted to different load conditions.

Equations 4.5 and 4.6 provide two values of the moment M_o; the smaller one is the moment that can be resisted by the beam-column:

$$M_o=\frac{1}{C_m}\left(1-\frac{P}{P_E}\right)\left(1-\frac{P}{P_{cr}}\right)M_m \tag{4.8}$$

$$M_o=1.18\left(1-\frac{P}{P_y}\right)M_p \tag{4.9}$$

Biaxial bending Reference 4.23 presents an approximate method for the elastic design of H-shaped beam-columns with couples applied at both ends, about the two principal axes. The method is recommended for design purposes in References 4.24 and 4.25.

Maximum inelastic strength of H-shaped isolated beam-columns under biaxial bending has been evaluated by determining their complete load-deformation curve, using numerical methods with loads applied in a sequence of sufficiently small increments. Interaction equations have been obtained to express failure conditions by formation of a plastic hinge in a cross section or by column instability[4.26–4.30]. The following condition must be satisfied at braced cross-sections (both ends of the column);

$$\left(\frac{M_x}{M_{pcx}}\right)^{\xi}+\left(\frac{M_y}{M_{pcy}}\right)^{\xi}\leq 1.0 \tag{4.10}$$

M_x and M_y are the bending moments at the section and M_{pcx} and M_{pcy} the plastic moments modified to include the effect of axial compression.

$$M_{pc_x}=1.18[1-P/P_y]M_{px}\text{, but not greater than } M_{px}$$

$$M_{pc_y}=1.19[1-(P/P_y)^2]M_{py}\text{, but not greater than } M_{py}$$

In H-sections in which $0.5 \leqq D/H \leqq 1.0$, ξ is given by

$$\xi = 1.6 - \frac{P/P_y}{2\ln(P/P_y)}$$

D is the flange width, H the web depth, and ln means natural logarithm. Column stability between braced points is checked with

$$\left(\frac{C_{m_x} M_x}{M_{uc_x}}\right)^{\eta} + \left(\frac{C_{m_y} M_y}{M_{uc_y}}\right)^{\eta} \leqq 1.0 \tag{4.11}$$

M_x and M_y are the larger of the moments applied at one or the other end of the beam-column, and M_{uc_x}, M_{uc_y}, the ultimate symmetrical single-curvature moment capacities of the beam-columns, about the x or y axis, respectively, when there is zero moment about the other axis, including the effect of lateral torsional buckling, if critical.

$$M_{uc_x} = M_{u_x}[1-(P/P_u)][1-(P/P_{E_x})]$$

$$M_{uc_y} = M_{p_y}[1-(P/P_u)][1-P/P_{E_y})]$$

P_u is the ultimate load of axially loaded column, P_{E_x} and P_{E_y} the Euler buckling loads, M_{u_x} the plastic moment capacity about the strong axis, reduced for lateral torsional buckling, if necessary, and η is given by

$$\eta = 0.4 + P/P_y + D/h \geqq 1.0, \text{ when } D/h \geqq 0.3$$

$$\eta = 1.0, \text{ when } D/h < 0.3$$

h is the depth of the W or I section.

To use Equations 4.10 and 4.11 to design beam-columns in moment-resistant sway frames, second order moments corresponding to ultimate load must be taken into account by an analysis including P–Δ effects in two orthogonal directions[4.30].

Reference 4.30 recommends that if Equations 4.10 and 4.11 are used to evaluate the strength of beam-columns in biaxial bending, then the reversible actions produced by earthquake or wind, and the variable actions due to wind or earthquake plus vertical dead and live loads (with a load factor of 1.5 multiplied by 0.7, or so, to account for probability of occurrence) should not produce stresses above the nominal yield point of the material.

Safety against instability failures of beam-columns under biaxial bending can be evaluated employing a generalized form of equation 4.5[4.16]:

$$\frac{P}{(P_{cr})_y} + \frac{C_{m_x} M_x}{M_{cr_x}(1-P/P_{E_x})} + \frac{C_{m_y} M_y}{M_{p_y}(1-P/P_{E_y})} = 1.0 \tag{4.12}$$

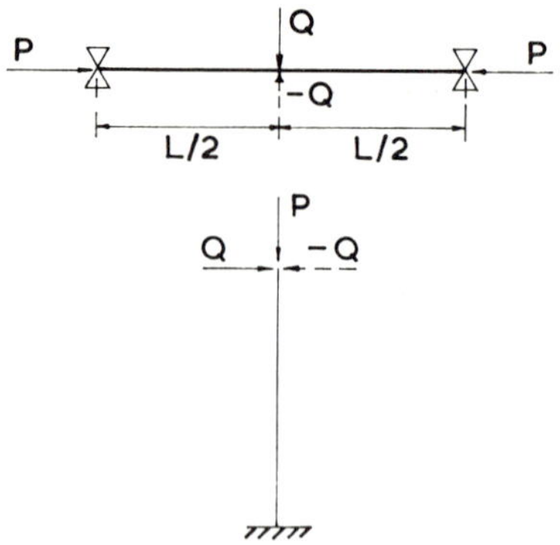

Fig. 4.11 Beam-columns under cyclic lateral load tested by several researchers

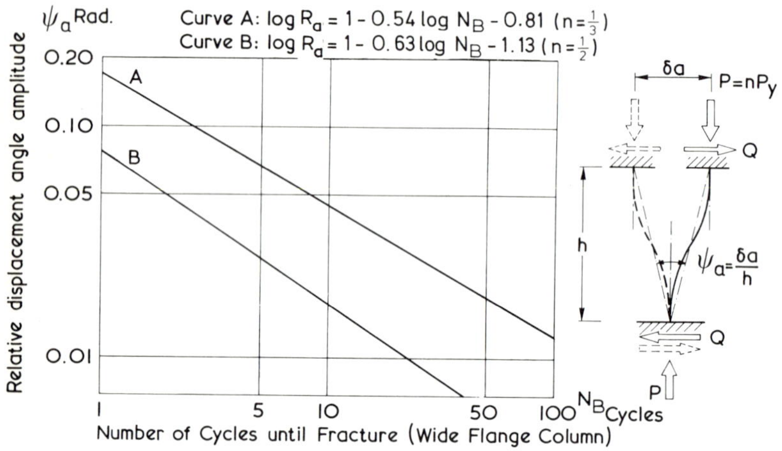

Fig. 4.12 Low cycle fracture limit of wide flange columns (after Yamada[4.32])

Beam-columns under cyclic loading Available experimental information on the behavior of beam-columns under cyclic loading is scarce, and is limited to uniaxial bending.

Isolated H-shape specimens supported and loaded as shown in Figure 4.11 have been tested under constant axial and transverse cyclic loads[4.31–4.35]. Tests have been run with different magnitudes of the axial load to establish its influence upon member behavior.

Column response deteriorates under increasing axial load. If lateral displacements are prevented, failure starts by flange local buckling, which increases as the number of load cycles increases and spreads to the web. A combination of flange and web local buckling eventually produces collapse.

Figure 4.12 shows that as axial load increases, both deflection amplitude and number of load cycles to failure decrease[4.31,4.32].

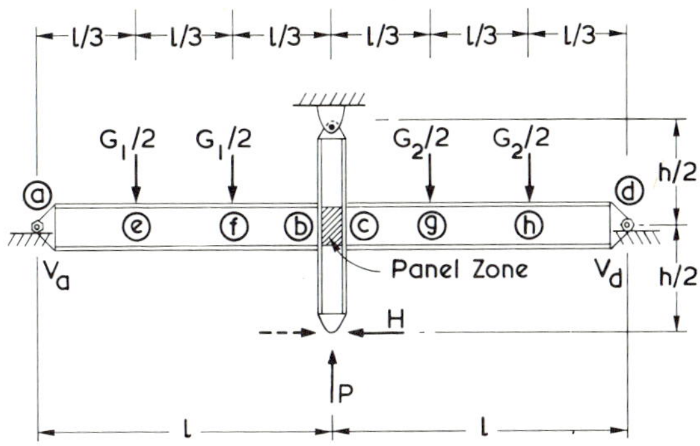

Fig. 4.13 *Test subassemblage* (*after Krawinkler* et al.[4.42])

Rotation capacity decreases also when the width-to-thickness ratios of flanges and web, or the axial force, increases[4.31].

Inelastic behavior of beam-columns under cyclic load and its influence upon rigid frame response are not well understood. Design of rigid frames is done by forcing plastic hinges to appear in the beams, as the columns remain basically in the elastic range up to frame collapse, except at their bases. Actual response of a structure under simultaneous gravity and seismic loads is so complex that we cannot be sure that all plastic hinges will be confined to the beams, even when we designed with such condition in mind.

Design is based on handbook dimensions and nominal yield point. Geometry and material properties in actual members differ, often substantially, from their theoretical values. Actual strengths can significantly differ from those computed. A combination of higher beam strength with lower column resistance can shift plastic hinges to the columns. Also, vertical accelerations are not usually taken into account in the design stages, disregarding the column axial loads they produce and the corresponding moment resistance diminution. It is important to recognize the possibility of column plastic hinging and to design accordingly[4.36].

Six subassemblages of the type shown in Figure 4.13 were tested under constant vertical load and cyclic horizontal forces[4.36]; beams were deliberately overdesigned to force plastic hinges to appear in the columns. Based on this limited number of tests, it appears that plastic hinges can be accepted in columns in strong or weak axis bending if P/P_y is not greater than about 0.5. Performance is satisfactory, mainly because of strain hardening. On the other hand, plastic hinges should be avoided when axial load exceeds 0.5 P_y: drastic strength reductions were observed in specimens with P/P_y equal to 0.6 and 0.8. Adequate behavior was also obtained testing two frames with column P/P_y ratios of 0.20 and 0.41 [4.10].

It can be concluded that beam-columns with slenderness ratios usual in buildings behave correctly in frames under seismic loading. They allow formation and rotation of plastic hinges if their flanges and webs satisfy the

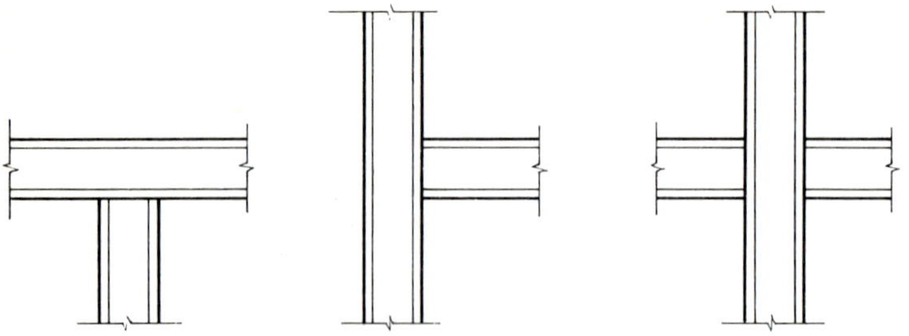

Fig. 4.14 *Typical beam-to-column connections*

width-to-thickness ratios specified for plastic design and the compression load does not exceed about $0.5P_y$.

4.4.3 Connections

To insure moment-resisting frame behavior, beam-to-column connections must have enough strength and rigidity to hold virtually unchanged the original angles between connected members. Three types of common beam-to-column connections are shown in Figure 4.14.

Although three or four beams generally connect to each column, most analytical and experimental studies made until now belong to plane frames, with beams connected only to the column flanges. A limited number of tests have also been done with beams connected to the column web.

When the column is continuous through the joint, as usually happens in building frames, joint design consists of:

(1) Design of connecting media to transmit bending moments, shears and axial forces from beams to column, using welding, rivets, or high-strength bolts (rivets are rarely used in modern steel structures). In welded joints load transmission can be achieved directly or using plates and angles.

(2) Column revision to check if its strength and rigidity are adequate to resist beam loads.

(3) Design, if necessary, of stiffeners, doubler plates*, etc.

Connection design according to American practice[4.15,4.16] is based on the results reported in Ref. 4.38, which delves in behavior and plastic design of beam-to-column moment-resisting frame joints under static load.

Behavior is adequate if the connection has enough rotation capacity, as compression acts simultaneously upon the column, to permit the formation of plastic hinges at the beams central region, or has sufficient strength to allow plastic hinges to develop at the beams ends and to rotate until formation of hinges at the central part of the span.

To find out if the joint behavior will be adequate, the following items must be investigated:

* Doubler plates are reinforcing plates welded to one or both sides of the column web.

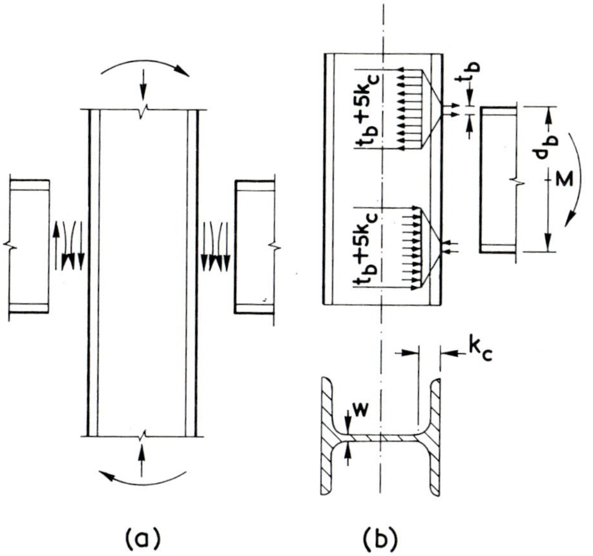

Fig. 4.15 *Forces in beam-to-column connection*

(1) Column strength at the regions adjacent to the beam flanges, when column is not reinforced.
(2) Column strength increment produced by horizontal stiffeners between flanges.
(3) Possibility of column failure by combination of compression and shear stresses.
(4) Influence of web-connected beams.
(5) Rotation required and connection rotation capacity.

According to Ref. 4.38, the effect of axial load upon connection behavior can be neglected, and four-beam connections can be safely designed disregarding the beams connected to the column web, as their stiffening action more than offsets the weakening effect due to the triaxial stresses that they produce.

The rotation required of a connection varies with structural geometry and loading. Typical rotations, higher than usually required, have been computed, and all tested joints were able to admit them under practically constant moment. Besides, if the connection strength is high enough, plastic hinges form at the end of the beams, which have enough rotation capacity if designed correctly.

Figure 4.15(a) shows moments and forces in an interior joint of a frame under vertical loading. In Figure 4.15(b) the beam has been replaced with the forces it applies to the column, disregarding web forces, which are of secondary importance.

As the distance from the column outer face increases, the area resisting the

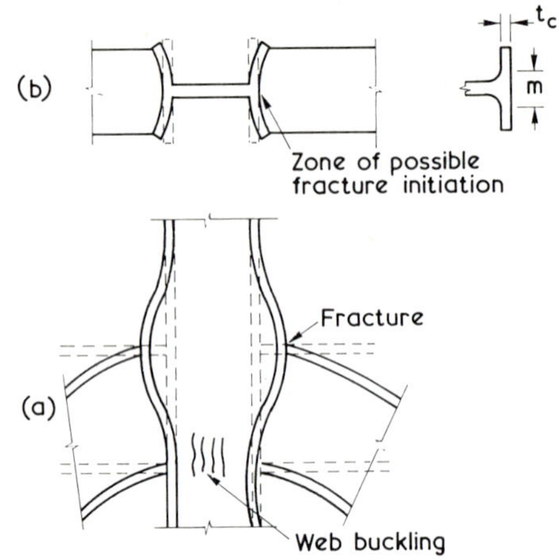

Fig. 4.16 Column flange distortion

forces increases and the corresponding stresses decrease. At the web toe of the fillets each force is resisted by a web length equal to $t_b + 5k_c$ (Figure 4.15(b)); t_b = beam flange thickness and k_c = distance from the column flange outer face to the web toe of fillet.

If the product $\sigma_y w(t_b + 5k_c)$, w being the column web thickness, is smaller than the beam flange force $A_f \sigma_y$ (A_f = area of the beam flange), the column web opposite the beam flanges yields in tension or compression. Besides, the web can buckle in the compression area or the weld in the beam tension flange can fracture (Figure 4.16).

To avoid yielding of the column web opposite each beam flange and web fracture in the tension region, Equation 4.13 must be satisfied:

$$\sigma_y \; w(t_b + 5k_c) \geqq A_f \sigma_y \tag{4.13}$$

Horizontal stiffeners are not necessary if

$$w \geqq A_f/(t_b + 5k_c) \tag{4.14}$$

If beam and column steels are different, σ_y does not cancel in Equation 4.14.

To avoid web buckling in the area adjacent to the beam compression flange, condition 4.15 must also be satisfied[4.15]:

$$w \geqq d_c \sqrt{\sigma_y}/150 \tag{4.15}$$

d_c = column web depth clear of fillets; σ_y = yield stress in kN/cm^2.

The most frequent type of failure opposite the beam tension flange is by bending of the column flanges (Figure 4.16) and excessive deformation of the central part of the welds and the adjacent column flange area. Eventually one of these regions, usually the weld, cracks after exhausting its deformation capacity.

Equation 4.16 provides the smallest column flange thickness for which stiffeners adjacent to the beam tension flange are not yet required:

$$t_c = 0.4\sqrt{A_f} \tag{4.16}$$

Coefficient 0.4 has been obtained for H rolled shapes[4.38]. It might require modification if Equation 4.16 is to be applied to welded sections.

In summary, stiffeners opposite the beam compression flange are not required if Equations 4.14 and 4.15 are both satisfied, nor adjacent to the tension flange if the column flange thickness is equal to or greater than that given by Equation 4.16. If these conditions are not satisfied, stiffeners must be used[4.15].

When beam-end moments in an interior joint have different signs and magnitudes, or when both have the same sign, as happens in low levels of buildings during strong earthquakes, the column web panel is under important shear forces. Sometimes it must be reinforced.

Figure 4.17(a) shows the moments and axial and shear forces that act on a typical interior connection. Figure 4.17(b) is the free body diagram of the upper pair of stiffeners. V_{col} is the upper column shear force, T_b and T_c the beam flanges forces, which can be approximately evaluated dividing bending moment by beam depth, and τ_{wd_c} is the panel zone shear. By equilibrium,

$$\tau_{wd_c} = \frac{M_b}{d_b} + \frac{M_c}{d_b} - V_{col}$$

The web column thickness required to resist the shear is obtained solving for w and making $\tau_w = \tau_y = \sigma_y/\sqrt{3}$:

$$w = \frac{\sqrt{3}}{\sigma_y d_c}\left(\frac{M_b + M_c}{d_b} - V_{col}\right) \tag{4.17}$$

If its thickness is smaller than this, the web must be reinforced by diagonal stiffeners, doubler plates, or vertical plates parallel to but separated from the web. In exterior column joints M_b and M_c reduce to only one moment.

The commentary to Reference 4.15 recommends Equation 4.18 to compute the smallest column web thickness that does not require reinforcement:

$$w = \frac{\sqrt{3}}{\sigma_y} \; \frac{\Delta M}{0.95 d_b \times 0.95 d_c} \tag{4.18}$$

This equation provides thicknesses 20 to 30% higher than Equation 4.17

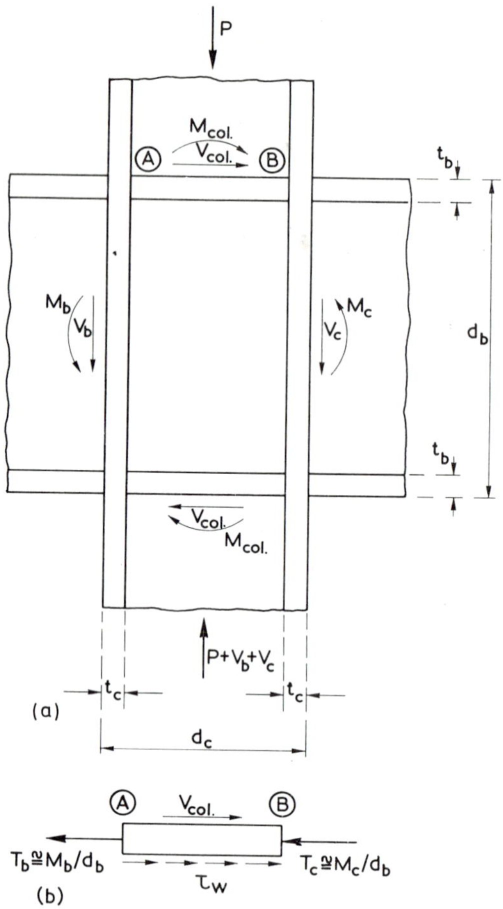

Fig. 4.17 Bending moments, axial and shear forces in an interior beam-to-column connection

because of the 0.95 factors that multiply beam and column depths, d_b and d_c, and because the beneficial effect of the upper column shear force is disregarded.

Recent studies have confirmed the adequacy of the AISC rules[4.15] for designing welded moment-resisting frame connections loaded statically. Reference 4.39, for instance, shows that joints so designed can be employed in plastically designed structures because they allow the beams to reach their plastic load and have adequate stiffness and rotation capacity.

Beam-to-column connections under cyclic loading Behavior of the several types of connecting media employed in beam-to-column joints has been studied experimentally. Cantilever beams connected to very rigid stub columns with welds, high-strength bolts or a combination of both have been

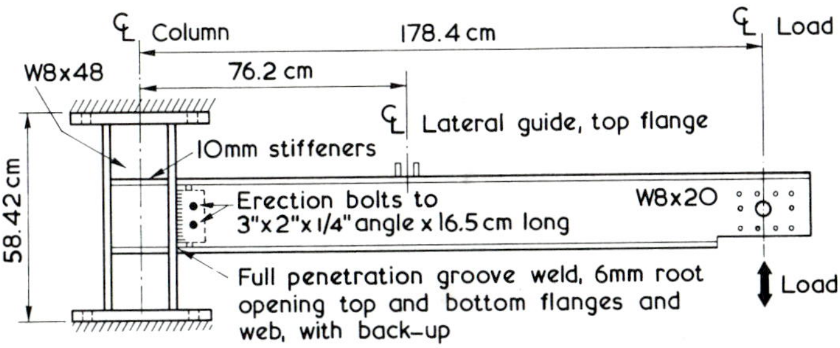

Fig. 4.18 *Welded beam-to-column connection specimen* (*after Popov and Pinkney*[4.9])

tested under cyclic transverse loads applied at their free end. In most cases beams have been connected to the column flanges (Figures 4.18 and 4.19)[4.9,4.40–4.42]. The panel zone has also been extensively studied[4.11,4.12,4.36,4.37]. The main conclusions obtained from the tests were:

(1) All the joints, including an exploratory one unwelded at the web*, were able to develop strengths higher than predicted by simple plastic theory, transmitting simultaneously appreciable shear forces. Strain hardening contributes significantly to this very satisfactory behavor.
(2) Connections are quite reliable. They absorb a definite amount of energy in every cycle for a prescribed displacement, as shown by their stable load-deflection hysteresis loops (Figures 4.20 and 4.21).
(3) Hysteresis loops for specimens connected with high strength bolts in flanges and web have a characteristic shape (Figure 4.22) due to slippage at failure surfaces. Joints with welded flanges and high-strength bolted webs also show bolt slippage.
(4) Connections adequately designed and built appear to be able to sustain severe cyclic loading. Their intrinsic energy absorption capacity is high. They can safely resist a number of loading cycles apparently higher than the number to be expected during the life of actual structures.
(5) Good workmanship and careful inspection during and after fabrication are of paramount importance. This was evidenced by premature failure of two specimens with defective welds.

Behavior of the column web at the joint (*panel web*)[4.11,4.12,4.37,4.42] Failure of beam-to-column connections can be due to crushing or buckling of the column web, column flange distortion, or panel zone shear yielding or buckling. Design recommendations to avoid the first three types of failure have already been given. They were developed for joints under static loading, but it has been experimentally verified that connections designed following

* This is in disagreement with the behavior of a similar connection reported in Reference 4.39. Under static load, this joint failed at 51% of the maximum load predicted by simple plastic theory. Although one test is not conclusive, the result seems to point out that joint behavior under cyclic loading is more favorable than under static load.

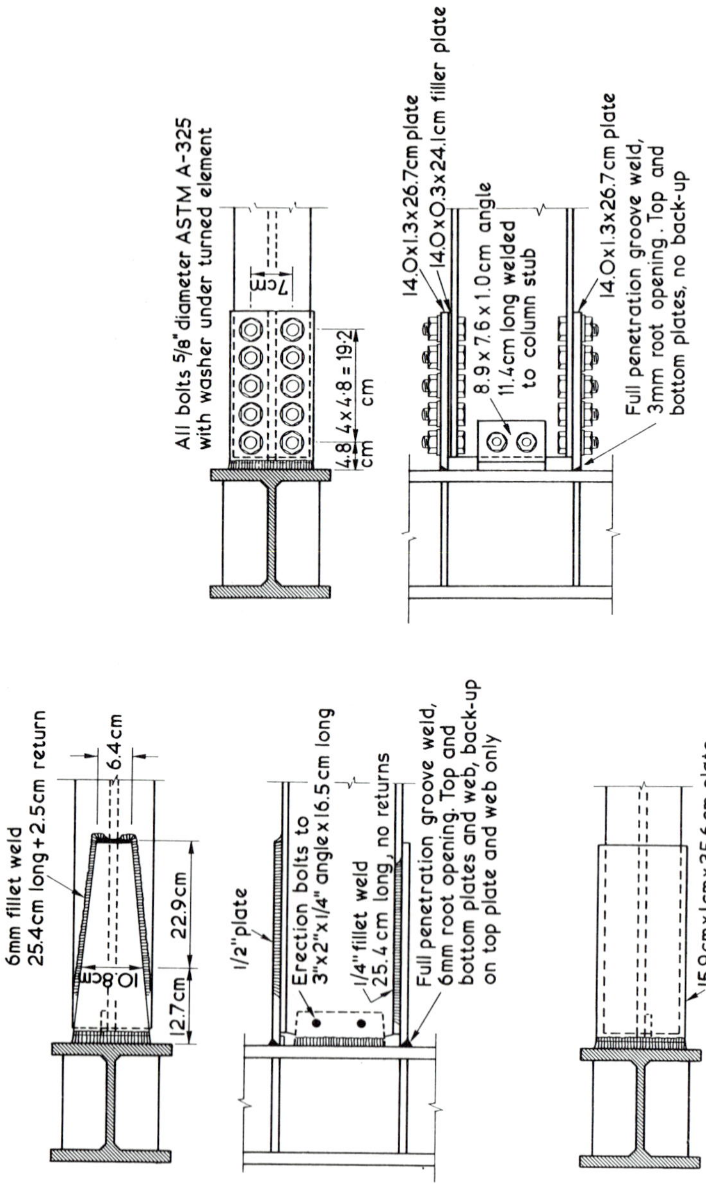

Fig. 4.19 Welded and bolted beam-to-column connection specimens (after Popov and Pinkney[4.9])

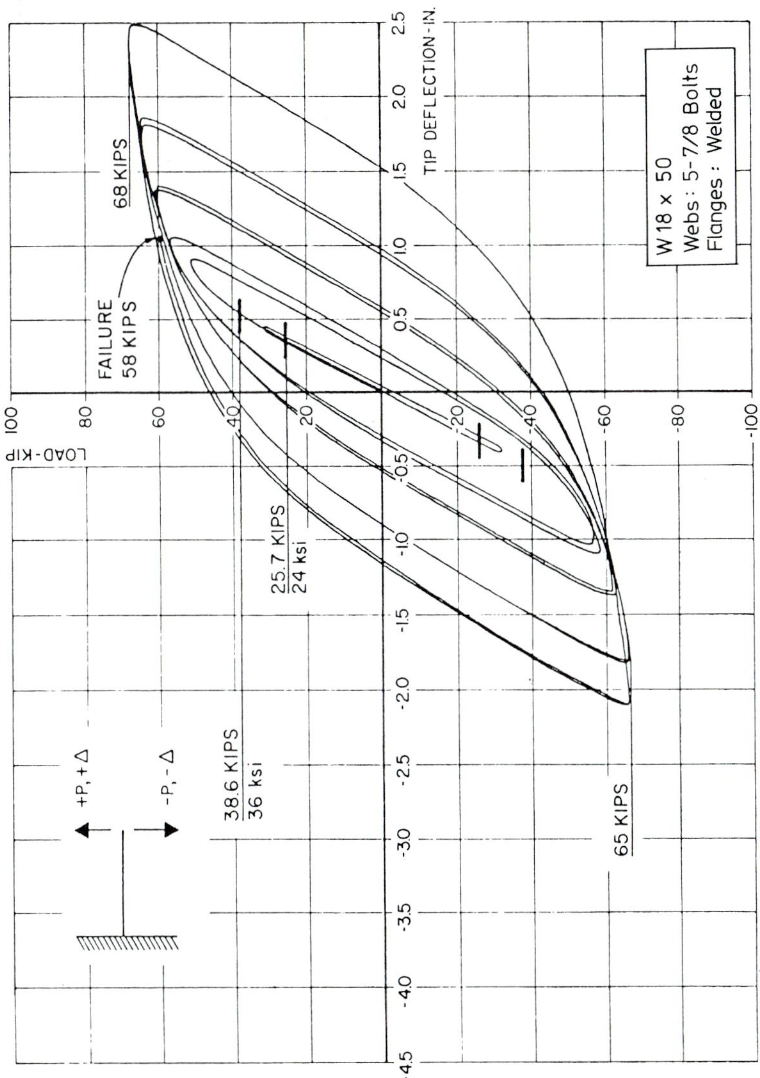

Fig. 4.20 Load-deflection hysteresis loops for a specimen with welded flanges and bolted web (after Popov and Stephen[4.40]*)*
(To convert loads from Kips to kN multiply by 4.448, deflections from in to mm multiply by 25.4, stresses from Ksi to kN/m^2 *multiply by 6.9 × 10*^{3}*)*

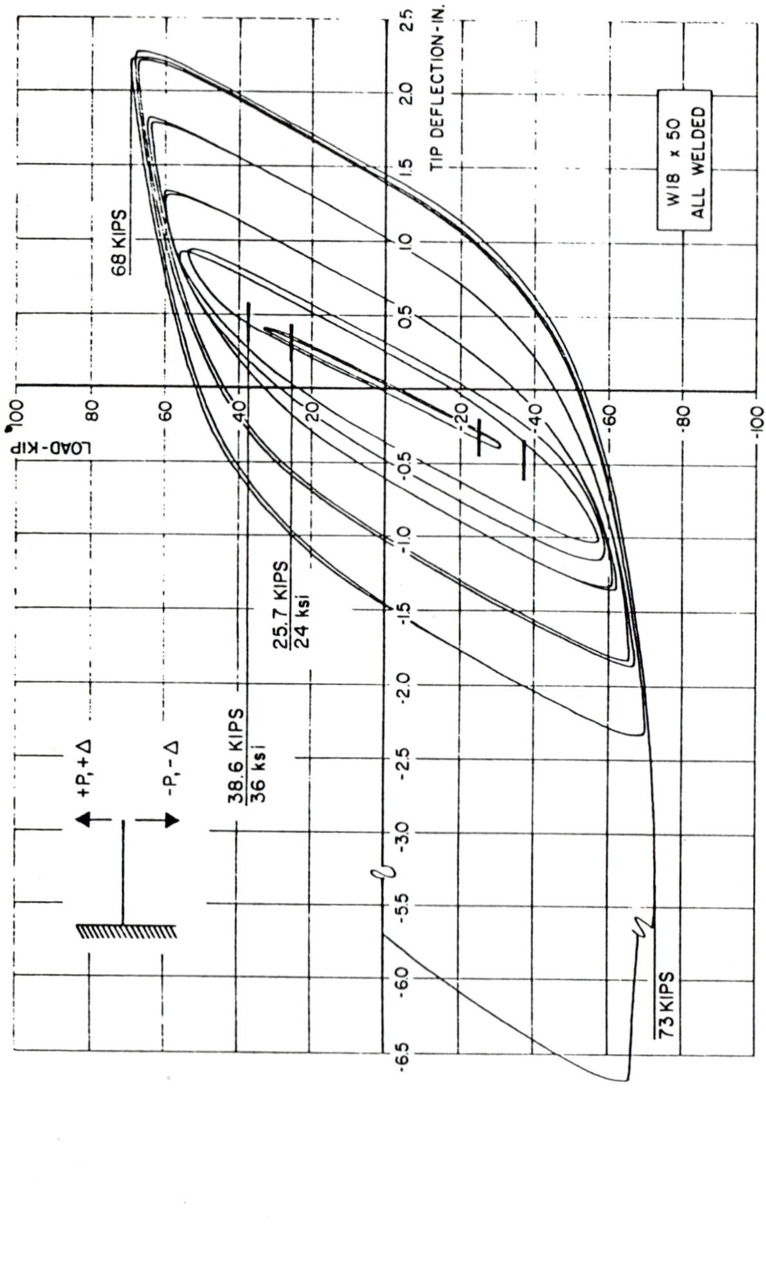

Fig. 4.21 Load-deflection hysteresis loops for an all-welded specimen (after Popov and Stephen[4.40]) (See Fig. 4.20 for conversions to SI units)

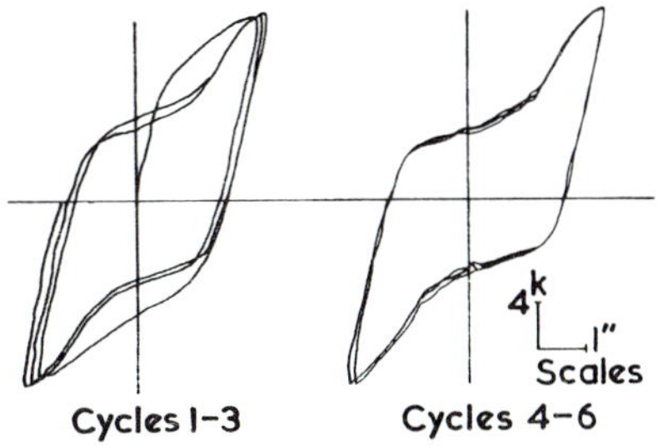

Fig. 4.22 Experimental load-deflection hysteresis loops for specimen connected with high-strength bolts (after Popov and Pinkney[4.9]*)*

those procedures behave satisfactorily under cyclic load. The last possible type of failure is studied below.

Response of a beam-to-column moment-resisting frame connection under vertical and horizontal loads is adequately represented by a $\Delta M - \gamma_p^{av}$ diagram; γ_p^{av} is the panel zone average shear distortion and ΔM is the sum of moments applied to the joint by the two beams, or by only one in exterior columns.

Because of the complexity of the problem, several simplifications are necessary to plot the $\Delta M - \gamma_p^{av}$ curve. The main parameters of the problem are panel zone shear strength, including reinforcing doubler plates or plates parallel to the column web, strength of elements bordering the panel zone (column-flange bending strength and beam-web stiffness play an important part in the joint's post-elastic behavior), beneficial effect of shear forces in columns, and column axial load.

Panel zone behavior is covered in detail in Refs. 4.11, 4.12 and 4.37. Column web crushing and flange distortion were avoided in all the specimens by designing them using the recommendations of Refs. 4.15 and 4.16, and employing horizontal stiffeners when necessary. Tests proved that, although developed for static loading, these recommendations are adequate to design cyclically loaded structures.

The panel zone of two upper-story subassemblages, revised by allowable stress methods, required stiffeners adjacent to both beam flanges; web reinforcement was not necessary. During the test the panel zone suffered significant distortions due to plastic shear, and plastic hinges could not develop in the beams. The lateral resistance of the subassemblages was low. Behavior of two similar subassemblages strengthened with vertical plates in the panel zone was satisfactory.

Four low-story subassemblages, with columns stronger and stiffer than the beams, did not require reinforcement. The beams developed their full plastic moment. Excessive panel-zone distortion in the first two subassemblages produced a rapid change in column curvature at the level of the horizontal stiffeners. This in turn caused kinks in the column flanges, high stress concentrations and cracks (Figure 4.23). Figure 4.24 shows the hysteresis loops of two specimens.

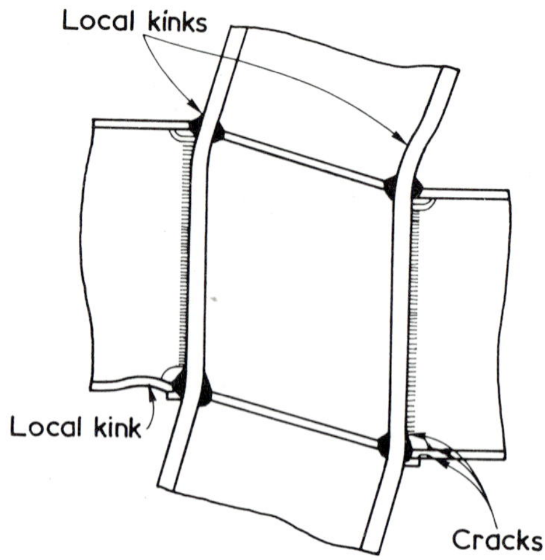

Fig. 4.23 Local effects caused by panel distortion (after Bertero et al.[4.37])

Connections can be very ductile structural elements, with significant strength above the yield level. Their rigidity decreases gradually in the plastic range and their hysteretic loops are stable. When carefully detailed and welded, and provided with adequate stiffeners to avoid web crippling and flange distorsion, connections have high capacity to dissipate energy.

Experiments show clearly that to develop the complete subassemblage load resistance, joint strength must be equal to or greater than the strength required to develop the full bending-moment capacity of the beams. Nevertheless, it is convenient to allow some plastic deformations to take place in connections as they reach their ultimate load capacity. If they are too rigid all plastic deformation takes place in the beams, and the latter's rotation capacity can become exhausted too early.

The average panel-zone shear stress can be computed from[4.11,4.42]

$$\tau_{av} = \frac{(\Delta M/d_b)(1-\rho)}{(d_c - t_c)w + 62.4 I_{cf}/d_b^2} \tag{4.19}$$

$\rho = V_{col} d_b / \Delta M$ pertains to the beneficial effect of the column shear force, and the terms in the denominator correspond to panel-zone shear strength and column-flange resistance to bending. The second term is usually much smaller than the first, but has some importance when the flanges are thick. I_{cf} is the moment of inertia of the column flange. It is advisable to take ΔM equal to the sum of the plastic moments of both beams to make sure that the joint will not fail before reaching the maximum resistance of individual members.

Axial load effects are taken into account, with good accuracy, computing the reduced shear yield stress

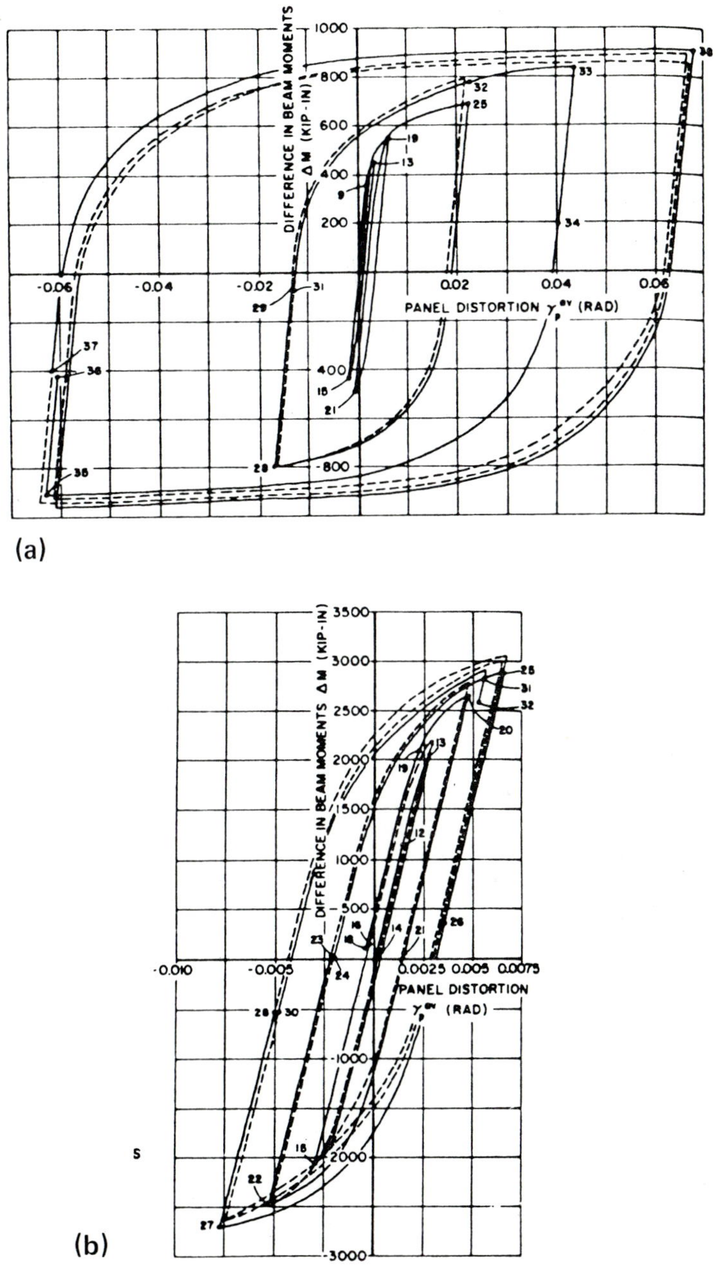

Fig. 4.24 *Moment-panel distortion hysteresis loops for two connections:* (*a*) *upper-story subassemblage,* (*b*) *low-story subassemblage* (*after Bertero* et al.[4.37]). *To convert moments from Kip-in to Kilonewton-metres, multiply by* 0.1130

$$\bar{\tau}_y = \frac{\sigma_y}{\sqrt{3}} \left\{ 1 - \left(\frac{P}{P_y} \right)^2 \right\}^{1/2} \tag{4.20}$$

and comparing it with τ_{av}; if $\tau_{av} \leqq \bar{\tau}_y$ there is no general plastic flow and the panel zone does not need reinforcement; if $\tau_{av} > \bar{\tau}_y$, doubler plates or plates parallel to the column web must be used to resist the excessive shear force. Equation 4.20 is valid if $P/P_y \leqq 0.5$.

4.5 DESIGN OF FULLY RIGID FRAMES

It is difficult to evaluate the energy that a structure must dissipate during an earthquake, the number of loading cycles that it will have to sustain, the required rotation capacity at plastic hinges, and the ductility factors of the members.

It is not easy to use directly in design the actual knowledge on ductility and energy absorption capacity of beams, columns and connections or subassemblages and complete frames. Nevertheless, structures can be designed providing them with their maximum possible strength and energy-absorption capacity, to obtain their best response under seismic loads.

Plastic hinges pertaining to the potential collapse mechanisms must generally form in the beams (except those at the column bases), which must have adequate strength and high rotation capacity, but measures must be taken to allow important plastic rotation to take place at column ends without loss of strength. Connections must not constitute weak links in the chain of resistance.

These conditions must be always satisfied in regions of high seismicity independently of the method, elastic or plastic, used to design the structure. Thus, slender plate girders should be avoided though they are efficient under static load.

Drift under working loads must always be computed and kept under adequate limits in order to avoid excessive damage to nonstructural elements. The structure's strength must also be evaluated to determine its safety factor against collapse.

A convenient way to study the behavior of a structure under working conditions and near collapse is using an elastic-plastic method to determine the complete load-displacement curves with the members selected in a preliminary design. Several computer programs, some highly sophisticated, have been developed to solve this problem. Up to the present though, most structures have been designed using simplified procedures, dividing the structures into simple subassemblages and obtaining their response by superposition of the subassemblage responses[4.43–4.45].

References 4.43 and 4.44 describe an approximate method that has been employed to obtain force-sway curves of isolated stories of rigid frames. To use it, a story is first isolated, and the upper columns are replaced with the forces and moments that they apply on the joints. The story is then divided in subassemblages consisting of a column and one or two beams. The force-sway curve of each subassemblage is determined. The story curve is obtained by superposition.

Joint deformation is often disregarded when computing lateral displacements. Thus $P-\Delta$ effects are underestimated. Joint deformations can be easily considered making an accurate study[4.46] or amplifying the displacements, computed in the usual way, in a percentage which is a function of the connections' shear stiffness.

References 4.10–4.12, 4.36 and 4.37 confirm that all-welded subassemblages and frames designed following AISC specification for plastic design are very ductile. They develop high plastic deformations under cyclic lateral displacement without loss of resistance. Their ability to resist constant vertical and cyclic horizontal loads considerably exceeds their strength under monotonically increasing lateral loads, because of strain hardening and P–Δ effects[4.47].

It has been suggested[4.11] that beams and columns that connect to a given joint must be designed to resist the total joint moment increased by 0.03 Ph; P = column load and h = story height. Too high P/P_y ratios should be avoided.

It is common practice to limit relative floor displacements under service conditions to between 0.0025 h and 0.003 h (allowable displacements can be increased by about 30% when joint deformations are considered). Actual displacements will usually be smaller because computations do not generally consider the stiffening effect of the floor system and nonstructural elements such as partitions and stairway shafts.

When the floor system has adequate strength and stiffness it is often advantageous to design and build it to work as a unit with the beams. It will thus increase beam strength and system stiffness and decrease lateral displacements. Results will ordinarily be more economical. Besides, description of structural behavior can be made more realistic since we are positive that floor systems will participate fully. We do not have the uncertainty stemming from possible participation when design does not force these systems to act with the beams, for plastic hinges can then change position relative to design assumptions and even appear in the columns, because of the beams increase in strength[4.48].

4.6 DRIFT CONTROL

Any reasonable drift control prevents failure by frame instability, even in the plastic range, by keeping second-order P–Δ moments below certain limits. Drift control is also necessary to guarantee joint integrity and minimize nonstructural damage. The way in which we control displacements is of the essence for economy and structural efficiency.

Lateral story displacements are due to five main factors:

(1) Column bending and shear deformation (Figure 4.25(a)).
(2) Joint rotation due to elastic and inelastic deformations of the beams

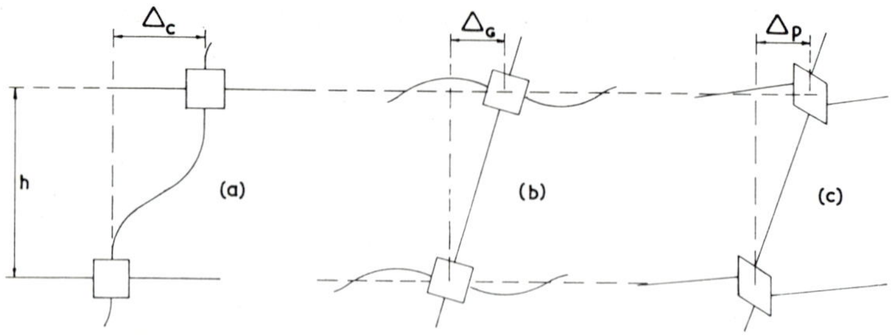

Fig. 4.25 Lateral story sway displacements

and the connecting media between beams and columns (Figure 4.25(b)).
(3) Panel-zone distortion (Figure 4.25(c)).
(4) Overall bending of the structure, due to changes in column lengths.
(5) Rotation of the building as a whole.

When significant, P–Δ effects must be taken into account to compute displacements, both under working loads and near collapse. Reference 4.12 shows that displacements due to joint deformation decrease pronouncedly when Equations 4.19 and 4.20 are satisfied.

To control lateral displacements economically we must know which sway component can be reduced the most with the smallest quantity of additional material and labor. As lateral displacements are usually almost proportional to the L/EI beam ratios, it is usually cheaper to reduce displacements by reducing beam deformations than column deformations. In so doing we must keep in mind that by increasing beam strength plastic hinges can shift from beams to columns.

Panel-zone distortions can be controlled using heavier columns, with thicker webs, or locally reinforcing the panel zone.

Lateral displacements depend strongly upon the number of columns that resist story shear; by increasing this number, the number of joints also increases, but every joint requires smaller welds, made in thinner plates, or a smaller number of fasteners. Beam lengths and the corresponding lateral-displacement component also decrease. This leads naturally to tubular structures, with many columns in the periphery, every 2 m or so, with short, stiff beams rigidly connected thereto, as an effective solution for tall buildings in strongly seismic areas.

Bracing and shear walls afford an efficient and economical way to control lateral displacements in medium-height buildings, maintaining ductilities at adequate levels.

4.7 BRACING

In ordinary framed structures it is difficult to simultaneously satisfy requirements of stiffness under working loads and strength and energy absorption capacity without over- or underdesigning for one of them. If structures are stiff enough to keep lateral displacements under prescribed limits their strength is generally much higher than required[4.48].

Lateral stiffness of moderately tall buildings is economically increased using a number of braced frames, compatible with architectural and functional requirements (end frames, service core periphery, facades); these frames restrict deformations of the remaining, unbraced, frames, with the floor systems acting as horizontal diaphragms. If possible, bracing must be continuous from bottom to top of the building. Several bracing configurations have been employed.

Braced frames are usually designed as two superimposed systems: an ordinary rigid frame, which supports vertical dead and live loads, plus a vertical bracing system, generally regarded as a pin-connected truss, which resists horizontal loads plus P–Δ effects, provides adequate lateral rigidity under working loads, and avoids overall frame buckling under factored vertical loads. Beams and columns of braced bays belong to both systems.

X- and K-bracing are discussed in Refs. 4.16 and 4.49, which include formulas to compute the area of the diagonals necessary to satisfy the foregoing conditions. Seismic or wind loads are replaced with equivalent static loads; dynamic effects are not considered.

Several experimental investigations on behavior of straight isolated bars under alternating tension and compression have lately been reported, as part of an effort to predict the seismic response of braced frames[4.50]

Figure 4.26 shows the theoretical behavior of a hinged bar of moderate slenderness under an axial tension that changes to compression and back to tension, assuming that the bar buckles when the smaller of the Euler critical load P_{cr} or the plastic load P_y is reached. Experimental behavior agrees fairly well with that predicted theoretically[4.51].

Tests on isolated members show that the maximum compression load (buckling load) decreases with the number of load cycles, that short members have greater energy absorption capacity than long ones, and that dynamic hysteretic deformation response is similar to the static response[4.52–4.54].

Analytical methods are fairly accurate during the first cycles. Their accuracy decreases in subsequent load applications. This is apparently due to difficulties found to take into account length changes in the specimen, and suggests the convenience of developing empirical load-deformation relationships[4.54]. Extensive experimental and analytical work in this field is still needed.

Experimental studies of isolated X-bracing have been carried-on[4.55]. One-floor, one-bay frames with different bracing arrangements have been tested under constant vertical load and horizontal cyclic forces to compare their behavior and to choose the most efficient arrangements[4.56,4.57,4.60].

Some analytical studies have also been done to determine the elasto-plastic response of tall braced buildings under strong recorded actual earthquakes[4.58,4.59]. They generally experience lesser displacements and plastic activity in beams and columns than similar unbraced frames. If

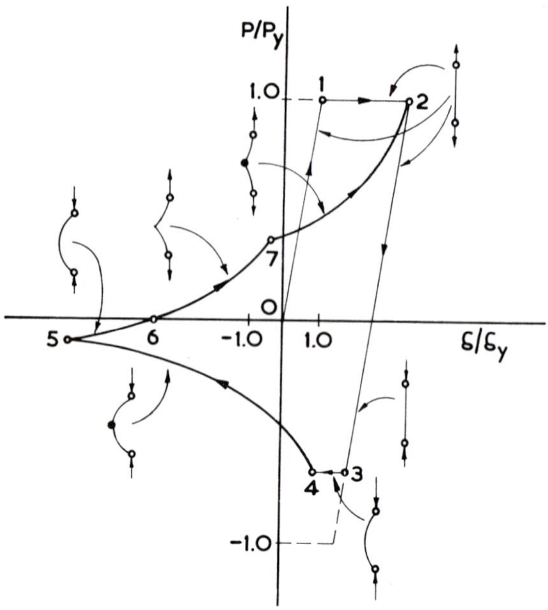

Fig. 4.26 Theoretical behaviour of a bar under cyclic tension-compression loading

designed according to building code recommendations, diagonals are generally too slender; they yield when the beams are still in the elastic range. As a result, ordinary braced frames resist equivalent static loads efficiently, but have a smaller energy absorption capacity than similar unbraced frames. To partially offset this disadvantage, diagonals are often designed to resist loads higher than those computed[4.63].

If diagonals and beams enter simultaneously into the plastic range, the structure gets the advantages of the bracing, greater rigidity and strength at smaller cost, without losing the fully rigid-frame high energy absorption capacity. If designed to fulfill this condition, bracing members turn out sturdier than obtained in accordance with Refs. 4.15, 4.16 and 4.49.

To increase the energy absorption capacity a method has been devised consisting in deliberately introducing computed eccentricities in the connections of the diagonals, leaving free beam segments under predominant bending moment to force the formation of plastic hinges in them. This has apparently not been used in actual structures, but extensive analytical and some experimental studies point out that the concept involved is very promising[4.61,4.62].

4.8 DIAPHRAGMS

Light-gauge steel infill panels are sometimes used as exterior curtain walls or interior partitions, as they provide an aesthetic solution with additional

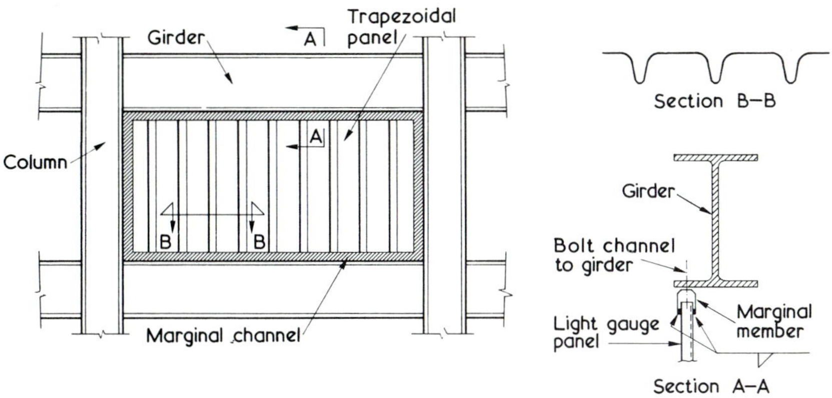

Fig. 4.27 Proposed construction for infilled frames (after Miller[4.68]*)*

advantages from the points of view of lightness, economy, and construction convenience. If no special precautions are taken to separate them, panel and structure deformation must be compatible, and the panel has to resist part of the seismic actions, in proportion to its stiffness. Because of their characteristics, light-gauge steel infill panels increase structural stiffness, ductility, and energy absorption capacity.

Panel connection with beams and columns is made with marginal elements designed to transmit shear forces in the panel plane and to reduce to a minimum vertical load transmission so as to avoid panel compression buckling. Figure 4.27 shows an infill panel and a possible connection detail. The infill panel is fillet welded to the marginal channel, and adjacent sheets are connected one to another through vertical seam welds.

Knowledge of infill panel stiffness and strength is necessary to evaluate their contribution to a building's seismic response, to compute the forces which will act upon them and to determine their safety factor against collapse. Those properties have been obtained experimentally and analytically, using finite element methods[4.64–4.67].

Diaphragm flexibility is the sum of flexibilities due to rib distortion, shear deformation and movement and deformation of the marginal channel and seam welds; failure can be due to panel tearing at the edges, fracture of seam welds, plastic flow, marginal channel instability or overall panel shear buckling.

The various factors that influence infill panel flexibility and strength are studied in Refs. 4.64, 4.66, and 4.67, while Ref. 4.65 deals with elastic shear buckling. Simplified formulas to evaluate strength are given.

Failure by elastic buckling corresponds to the maximum strength that can be provided by a given infill panel. It is often advisable to design seam welds and connections with the main structure to avoid premature failure, to develop the shear buckling strength.

It has been suggested[4.68] that light-gauge steel infill panels be used to reduce building drift to acceptable values, and that bare frames should resist the entire applied lateral loads. Structural safety does not depend then directly on the infill panels, as does serviceability. In the future, with more information on the behavior of the frame-panel system available, panel strength will doubtless also be taken into account.

A numerical example in Reference 4.68 shows that lateral deflections of a 26-story frame decrease markedly when using a small number of infill panels.

4.9 OTHER TYPES OF STRUCTURE

Conditions leading to satisfactory performance can be satisfied with no special difficulties in ordinary urban buildings, but not necessarily so in other types of steel structures, such as elevated water tanks or chimneys.

Behavior under earthquakes of elevated water tanks supported by four or more columns with struts and slender tension X-bracing has often been unsatisfactory. Failure of a diagonal or its end connection, generally in the upper part of the tower, causes a loss of symmetry in the structure. This induces significant torsional moments and collapse by column bending. Since it is difficult to achieve high ductility levels in these structures, they must resist high seismic loads remaining basically within the linear range. Their diagonals and connections must be designed using generous safety factors. This is discussed in Chapter 3.

In chimneys and other tall self-supported steel towers special attention must be paid to avoid wall buckling. As under strong earthquakes design loads will probably be exceeded, localized plastic flow must be expected; besides, as even moderate ductility factors produce dangerous local buckling when associated with inelastic action in a chimney stack, it is convenient to concentrate plastic deformations in elements that will not produce structural failure and can be easily replaced if necessary.

The most convenient of such elements are anchor bolts[4.14], whose length must be able to accommodate plastic deformations under maximum tension due to overturning moment. These bolts have to be easy to replace. This can be achieved by providing them with welded keys, to fasten them to special steel members buried in the concrete foundation, and allowing release of the bolts by turning.

4.10 BRITTLE FRACTURE AND LAMELLAR TEARING

Brittle fracture is more frequent in welded steel structures than in structures connected by mechanical fasteners. This is due to a combination of possible weld defects, high residual stresses, and continuity, which reduces the possibility of crack arrest. Adequate measures must be taken during fabrication and erection to minimize the phenomena that can induce brittle

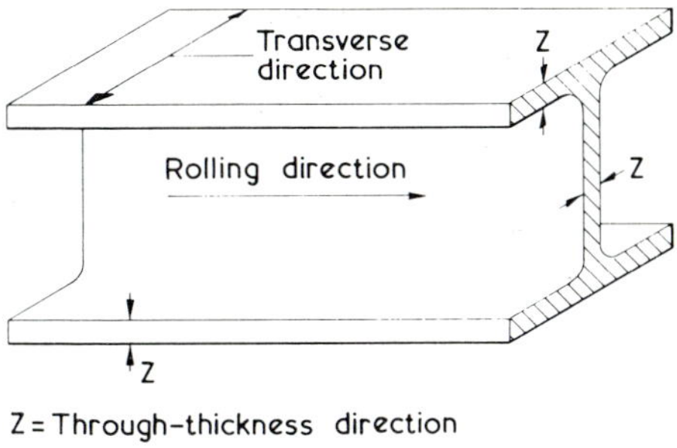

Fig. 4.28 Terminology related to rolling direction[4.69]

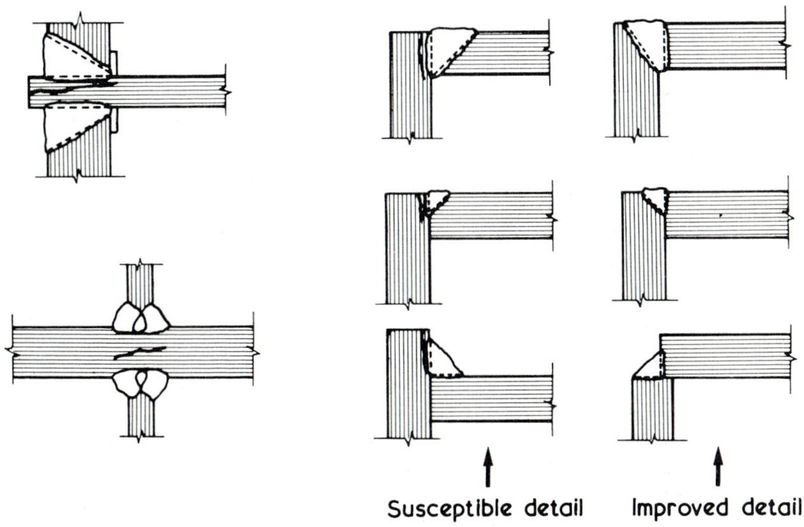

Fig. 4.29 Lamellar tearing due to contraction of large welded joints. Susceptible and improved details[4.69]

behavior. If conditions are especially severe, steels with adequate notch toughness at low temperature must be used.

Welding of rolled shapes or plates to transmit forces in the through-thickness direction (Figure 4.28) must be carefully done. Otherwise it can induce lamellar tearing. The through-thickness strength of steel is similar to the longitudinal or transverse strength, but capacity to deform in that direction is frequently only slightly in excess of the elastic-limit strain.

Lamellar tearing takes place in highly restrained plates when decohesion occurs at the interface between microscopic non-metallic inclusions and surrounding material, due to parent metal incapacity to accommodate strains imposed by weld shrinkage in the through-thickness direction. It can occur in plates of any thickness, but is much more frequent in thick plates because larger welds are deposited on them, and their cooling contraction is larger. Practical experience shows that lamellar tearing is almost non-existent in material less than 20 to 25 mm thick, but a few cases have occurred in thinner metal[4.69]. The most important design consideration is to minimize concentration of through-thickness strains in localized areas. Figure 4.29 shows details susceptible of lamellar tearing, and some ideas about how to improve them.

ACKNOWLEDGEMENTS

The valuable suggestions and criticisms from Emilio Rosenblueth and Luis Esteva are gratefully acknowledged.

REFERENCES

4.1 Degenkolb, H. J., Design of Earthquake-Resistant Structures – Steel Frame Structures, in *Earthquake Engineering*, (Edited by R. L. Wiegel), Prentice-Hall, Inc., Engelwood Cliffs, N.J. (1970)

4.2 Galambos, T. V., *Structural Members and Frames*, Prentice-Hall, Inc., Englewood Cliffs, N.J. (1968)

4.3 Lee, G. C., and Galambos, T. V., 'Post-Buckling Strength of Wide-Flange Beams', *J. Engng Mech. Div. Am. Soc. civ. Engrs*, Vol. 88, No. EM1 (1962)

4.4 Lee, G. C., Ferrara, A. T. and Galambos, T. V., 'Experiments on Braced Wide-Flange Beams', *Bulletin No.* 99, Welding Research Council, New York (1964)

4.5 Lukey, A. F., Smith, R. J., Hossan, M. U. and Adams, P. F., 'Experiments on Wide-Flange Beams under Moment Gradient', *Bulletin No.* 142, Welding Research Council, New York (1969)

4.6 Lay, M. G., 'Flame Local Buckling in Wide-Flange Shapes', *J. struct. Div. Am. Soc. civ. Engrs*, Vol. 91, No. ST6 (1965)

4.7 Lay, M. G. and Galambos, T. V., 'Inelastic Beams under Moment Gradient', *J. struct. Div. Am. Soc, civ. Engrs*, Vol. 93, No. ST1 (1967)

4.8 Bertero, V. V. and Popov, E. P., 'Effect of Large Alternating Strains of Steel Beams', *J. struct. Div. Am. Soc. civ. Engrs*, Vol. 91, No. ST1 (1965)

4.9 Popov, E. P. and Pinkney, R. B., 'Cyclic Yield Reversal in Steel Building Connections', *J. struct. Div. Am. Soc. civ. Engrs*, Vol. 95, No. ST3 (1969)

4.10 Carpenter, L. D. and Lu, L. W., 'Reversed and Repeated Load Tests of Full-Scale Steel Frames', *Bulletin No.* 24, American Iron and Steel Institute, New York (1973)

4.11 Krawinkler, H., Bertero, V. V. and Popov, E. P., 'Inelastic Behavior of Steel Beam-to-Column Subassemblages', *Report No. EERC* 71-7, Earthquake Engineering Research Center, College of Engineering, Univ. California, Berkeley, Cal. (1971)

4.12 Bertero, V. V., Krawinkler, H. and Popov, E. P., 'Further Studies on Seismic Behavior of Steel Beam-Column Subassemblages', *Report No. EERC* 73-27, Earthq. Eng. Res. Center, College of Eng., Univ. California, Berkeley, Cal. (1973)

4.13 Lay, M. G. and Galambos, T. V., 'Inelastic Steel Beams under Uniform Moment', *J. struct. Div. Am. Soc. civ. Engrs*, Vol. 91, No. ST6 (1965)

4.14 Newmark, N. M. and Rosenblueth, E., *Fundamentals of Earthquake Engineering*, Prentice-Hall Inc., Englewood Cliffs, N.J. (1971)

4.15 'Specification for the Design, Fabrication and Erection of Structural Steel for Buildings and Commentary to the Specification', American Institute of Steel Construction (1969), with Supplements Nos. 1, 2 and 3

4.16 'Plastic Design in Steel – A Guide and Commentary', ASCE Manual of Engineering Practice No. 41, 2nd edn. (1971)

4.17 Ketter, R. L., 'Stability of Beam-Columns above the Elastic Limit', *Proc. Am. Soc. civ. Engrs*, Vol. 81, Separate No. 692 (1955)

4.18 Galambos, T. V. and Ketter, R. L., 'Columns under Combined Bending and Thrust', *Trans. Am. Soc. civ. Engrs*, Vol. 126, Part 1 (1961)

4.19 Galambos, T. V., Lecture 10, 'Plastic Design of Multi-Story Frames Lecture Notes', Fritz Engineering Laboratory, *Report No.* 273.20, Lehigh University, Bethlehem, Pa (1965)

4.20 Ojalvo, M. and Fukumoto, Y., 'Nomographs for the Solution of Beam-Columns Problems', *Bulletin No.* 78, Welding Research Council, New York (1962)

4.21 Galambos, T. V. and Lay, M. G., 'Studies on the Ductility of Steel Structures', *J. Struct. Div. Am. Soc. civ. Engrs*, Vol. 91, No. ST4 (1965)

4.22 Galambos, T. V., Adams, P. F. and Fukumoto, Y., 'Further Studies on the Lateral-Torsional Buckling of Steel Beam-Columns', *Bulletin No.* 115, Welding Research Council, New York (1966)

4.23 Baker, J. F., Horne, M. R. and Heyman, J., *The Steel Skeleton*, Vol. 2, Cambridge University Press, Cambridge, England (1956)

4.24 'Joint Committee Report on Fully Rigid Multi-Storey Welded Steel Frames', Institution of Structural Engineers, London (1964)

4.25 'Joint Committee's Second Report on Fully Rigid Multi-Storey Welded Steel Frames', Institution of Structural Engineers, London (1971)

4.26 Santathadaporn, S. and Chen, W. F., 'Interaction Curves for Sections under Combined Biaxial Bending and Axial Forces', *Bulletin No.* 148, Welding Research Council, New York (1970)

4.27 Chen, W. F. and Atsuta, T., 'Interaction Equations for Biaxially Loaded Sections', *J. struct. Div. Am. Soc. civ. Engrs*, Vol. 98, No. ST5 (1972)

4.28 Santathadaporn, S. and Chen, W. F., 'Analysis of Biaxially Loaded Steel H-Columns', *J. struct. Div. Am. Soc. civ. Engrs*, Vol. 99, No. ST3 (1973)

4.29 Tebedge, N. and Chen, W. F., 'Design Criteria for H-Columns under Biaxial Loading', *J. struct. Div. Am. Soc. civ. Engrs*, Vol. 100, No. ST3 (1974)

4.30 Springfield, J., 'Design of Columns Subject to Biaxial Bending', *Engineering J.*, American Institute of Steel Construction, Vol. 12, No. 3 (1975)

4.31 Yamada, M., 'Low Cycle Fatigue Fracture Limits of Various Kinds of Structural Members Subjected to Alternately Repeated Plastic Bending under Axial Compression as an Evaluation Basis of Design Criteria for Aseismic Capacity', *Proc. Fourth World Conference on Earthquake Engineering*, Santiago, Chile (1969)

4.32 Yamada, M., 'Effect of Cyclic Loading on Buildings', State of Art Report No. 1, Technical Committee 18, International Conference on Planning and Design of Tall Buildings, Lehigh University, Bethlehem, Pa. (1972)

4.33 Suzuki, T. and Ono, T., 'An Experimental Study of Inelastic Behavior of Steel Members Subjected to Repeated Loading', *Preprints, Sixth World Conf. on Earthquake Engineering*, New Delhi, India (1977)

4.34 Mitani, T., Makino, M. and Matsui, C., 'Influence of Local Buckling on Cyclic Behavior of Steel Beam-Columns', *Preprints, Sixth World Conf. on Earthquake Engineering*, New Delhi, India (1977)

4.35 Kato, B. and Akiyama, H., 'Inelastic Bar Subjected to Thrust and Cyclic Bending', *J. struct. Div. Am. Soc. civ. Engrs*, Vol. 95, No. ST1 (1969)

4.36 Popov, E. P., Bertero, V. V. and Chandramouli, S., 'Hysteretic Behavior of Steel Columns', *Report No. EERC* 75-11, Earthq. Eng. Res. Center, College of Eng., Univ. California, Berkeley, Cal. (1975)

4.37 Bertero, V. V., Popov, E. P. and Krawinkler, H., 'Beam-Column Subassemblages under Repeated Loading', *J. struct. Div. Am. Soc. civ. Engrs*, Vol. 98, No. ST5 (1972)

4.38 Graham, J. D., Sherbourne, A. N., Khabbaz, R. N. and Jensen, C. D., 'Welded Interior Beam-to-Column Connections', American Institute of Steel Construction (1959)

4.39 Parfitt, J. Jr. and Chen, W. F., 'Tests of Welded Steel Beam-toColumn Moment Connections', *J. struct. Div. Am. Soc. civ. Engrs*, Vol. 102, No. ST1 (1976)

4.40 Popov, E. P. and Stephen, R. M., 'Cyclic Loading on Full-Size Steel Connections', *Report No. EERC* 70-3, Earth. Eng. Res. Center, College of Eng., Univ. California, Berkeley, Cal. (1970)

4.41 Popov, E. P. and Bertero, V. V., 'Cyclic Loading of Steel Beams and Connections', *J. struct. Div. Am. Soc. civ. Engrs*, Vol. 99, No. ST6 (1973)

4.42 Krawinkler, H., Bertero, V. V. and Popov, E. P., 'Shear behavior of Steel Frame Joints', *J. struct. Div. Am. Soc. civ. Engrs*, Vol. 101, No. ST11 (1975)

4.43 Daniels, J. H. and Lu, L. W., 'The Subassemblage Method of Designing Unbraced Multi-Story Frames', Fritz Eng. Lab., *Report* 273.37, Lehigh University, Bethlehem, Pa. (1966)

4.44 de Buen, O., 'A Modification to the Subassemblage Method of Designing Unbraced Multi-Story Frames', *Engineering J.*, American Institute of Steel Construction, Vol. 6, No. 4 (1969)

4.45 Powell, G. H. and El Hafez, M. B., 'Subassemblage Concept for Tall Frame Analysis', *J. struct. Div. Am. Soc. civ. Engrs*, Vol. 100, No. ST8 (1974)

4.46 Becker, R., 'Panel Zone Effect on the Strength and Stiffness of Steel Rigid Frames', *Engineering J.*, AISC, Vol. 12, No. 1 (1975)

4.47 Kato, B. and Lu, L. W., 'Instability Effects under Dynamic and Repeated Load', State of Art Report No. 3, Technical Committee 16, International Conference on Planning and Design of Tall Buildings, Lehigh University, Bethlehem, Pa. (1972)

4.48 de Buen, O., 'Antiseismic Design of Multi-Story Steel Frames by Plastic Methods', *Proc. Fourth World Conf. on Earthquake Engineering*, Santiago, Chile (1969)

4.49 'Plastic Design of Braced Multi-Story Steel Frames', American Iron and Steel Institute, New York (1968)

4.50 Wakabayashi, M., 'Frames under Strong Impulsive, Wind or Seismic Loading', State of Art Report No. 6, Technical Committee No. 15, International Conference on Planning and Design of Tall Buildings, Lehigh University, Bethlehem, Pa. (1972)

4.51 Shibata, M., Nakamura, T., Yoshida, N., Morino, S., Nonaka, T., and Wakabayashi, M., 'Elastic-Plastic Behavior of Steel Braces under Repeated Axial Loading', *Proc. Fifth World Conference on Earthquake Engineering*, Rome, Italy (1974)

4.52 Kahn, L. F. and Hanson, R. D., 'Inelastic Cycles of Axially Loaded Steel Members', *J. struct. Div. Am. Soc. civ. Engrs*, Vol. 102, No. ST5 (1976)

4.53 Wakayashi, M., Matsui, C. and Mitani, I., 'Cyclic Behavior of a Restrained Steel Brace under Axial Loading', *Preprints, Sixth World Conference on Earthquake Engineering* New Delhi, India (1977)

4.54 Higginbotham, A. B. and Hanson, R. D., 'Axial Hysteretic Behavior of Steel Members', *J. struct. Div. Am. Soc. civ. Engrs*, Vol. 102, No. ST7 (1976)

4.55 Wakabayashi, M., Nakamura, T., Shibata, M., Yoshida, N. and Masuda, H., 'Hysteretic Behavior of Steel Braces Subjected to Horizontal Load due to Earthquake', *Preprints, Sixth World Conf. on Earthquake Engineering*, New Delhi, India (1977)

4.56 Wakabayashi, M., Matsui, C., Minami, K. and Mitani, I., 'Inelastic Behavior of Steel Frames Subjected to Constant Vertical and Alternating Horizontal Loads', *Proc. Fifth World Conference on Earthquake Engineering*, Rome, Italy (1974)

4.57 Tanabashi, R., Kaneta, K. and Ishida, T., 'On the Rigidity and Ductility of Steel Bracing Assemblage', *Proc. Fifth World Conference on Earthquake Engineering*, Rome, Italy (1974)

4.58 Goel, S. C. and Hanson, R. D., 'Seismic Behavior of Multistory Braced Steel Frames', *J. struct. Div. Am. Soc. civ. Engrs*, Vol. 100, No. ST1 (1974)

4.59 Anderson, J. C., 'Seismic Behavior of K-Braced Framing Systems', *J. struct. Div. Am. Soc. civ. Engrs*, Vol. 101, No. ST10 (1975)

4.60 Yamada, M., Tsuji, B. and Nakanishi, S., 'Elasto-Plastic Behavior of Braced Frames under Cyclic Horizontal Loading', *Preprints, Sixth World Conference on Earthquake Engineering*, New Delhi, India (1977)

4.61 Wyllie, L. A. Jr. and Degenkolb, H. J., 'Improving the Seismic Response of Braced Frames', *Preprints, Sixth World Conference on Earth Engineering*, New Delhi, India (1977)

4.62 Kaldjian, M. J., 'Inelastic Cyclic Response of Split K-Braced Frames', *Preprints, Sixth*

World Conference on Earthquake Engineering, New Delhi, India (1977)
4.63 'Uniform Building Code', 1976 Edition, International Conference of Building Officials, Whittier, Cal. (1976)
4.64 Bryan, E. R., *The Stressed Skin Design of Steel Buildings*, Constrado Monographs, Crosby Lockwood Staples, London (1972)
4.65 Easley, J. T., 'Buckling Formulas for Corrugated Metal Shear Diaphragms', *J. struct. Div. Am. Soc. civ. Engrs*, Vol. 101, No. ST7 (1975)
4.66 Davies, J. M., 'Calculation of Steel Diaphragm Behavior', *J. struct. Div. Am. Soc. civ. Engrs*, Vol. 102, No. ST7 (1976)
4.67 El-Dakhakhni, 'Shear of Light-Gage Partitions in Tall Buildings', *J. struct. Div. Am. Soc. civ. Engrs*, Vol. 102, No. ST7 (1976)
4.68 Miller, C. J., 'Ligut Gage Steel Infill Panels in Multistory Steel Frames', *Engineering J.*, American Institute of Steel Constructions, Vol. 11, No. 2 (1974)
4.69 American Institute of Steel Construction, 'Commentary on Highly Restrained Welded Connections', *Engineering J.*, American Institute of Steel Constructions, Vol. 10, No. 3 (1973) (See also discussion by N. B. Jones and W. A. Milek, *Engineering J.*, Vol. 12, No. 1 (1975)

5 Concrete structures

Robert Park and Thomas Paulay

5.1 INTRODUCTION

Experience of earthquakes and laboratory tests has shown that well designed and detailed reinforced concrete is suitable for earthquake resistant structures. Earlier chapters have emphasized that the static design horizontal loads recommended by most codes are such that structures will be able to resist only moderate earthquakes without structural damage. The most severe likely earthquake can only be survived if the members are sufficiently ductile to absorb and dissipate seismic energy by inelastic deformations. This requires a designer to realistically assess the possible levels of strength in flexure and shear of elements and their connections and to ensure adequate ductility. All forms of brittle failure should be avoided.

This chapter will commence by discussing the strength and ductility of reinforced concrete elements. Then the design of earthquake resistant reinforced concrete frame and shear wall structures is described. Some attention is also given to reinforced concrete bridge piers and to prestressed concrete frames. Extensive reference will be made to the building code of the American Concrete Institute (ACI 318-77)[5.1] since it is one of the most widely used reinforced concrete codes with seismic provisions. Also, the strength method of design is used because, as mentioned earlier, comprehensive seismic design requires an assessment of the behaviour of the structure in the inelastic range. The levels of strength available for various possible failure modes cannot be assessed accurately using the working stress method.

5.2 STRENGTH AND DUCTILITY PROVISIONS

5.2.1 Design loads

In strength design the design (ultimate) loads are found from the service loads using load factors which are intended to ensure adequate safety. For example, the ACI Code[5.1] recommends that the required strength U provided to resist dead load D, live load L and earthquake load E shall be at least equal to that required by the worst of the following load cases:

$$U = 1.4D + 1.7L \quad (5.1)$$

$$U = 0.75(1.4D + 1.7L + 1.87E) \quad (5.2)$$

where L may have its full value or zero, and

$$U = 0.9D + 1.43E \quad (5.3)$$

5.2.2 Section strength

The ideal strength of a section of a member S_i is obtained from theory predicting failure behaviour of the section and on given member dimensions and specified material strengths. Other strength levels can be conveniently related to the ideal strength.

The dependable strength S_d is given by

$$S_d = \varphi S_i \quad (5.4)$$

where φ is the strength reduction factor to allow for approximations in calculations and variations in the material strengths, workmanship and dimensions. Values for φ recommended by the ACI Code depend on the importance of the variable quantities: for flexure of beams $\varphi = 0.9$; for flexure and axial compression of columns if spirally reinforced $\varphi = 0.75$, otherwise $\varphi = 0.7$ (for sections with axial compression approaching zero, $\varphi \rightarrow 0.9$); and for shear and torsion $\varphi = 0.85$.

The overstrength S_o is given by

$$S_o = \varphi_o S_i \quad (5.5)$$

where φ_o is the overstrength factor allowing for all sources of strength increase and is greater than unity. For example, the actual yield strength of the steel and the actual strength of the concrete may be considerably higher than the specified strengths. Other factors include the increase in steel strength due to strain hardening at large inelastic deformation of members, more steel placed at sections than required by the calculations, and concrete section sizes larger than assumed.

The various levels of strength described by Equations 5.4 and 5.5 may be used in design to ensure that the desired sequence of strength attainment in the members of a structure loaded by a severe earthquake will occur.

5.2.3 Section ductility

Ductile behaviour means the ability to undergo large inelastic deformations with little decrease in strength. The available ductility of reinforced and prestressed concrete sections is mainly dependent on the contents of longitudinal tension and compression steel, the content of transverse steel for concrete confinement and restraint against buckling of bars, the concrete and steel strengths, and the level of axial load. The possibility of shear and bond failure should be avoided. Codes at present do not indicate specifically the level of ductility plastic hinge sections should be capable of achieving but recommend detailing practices with the aim of ensuring adequate ductility.

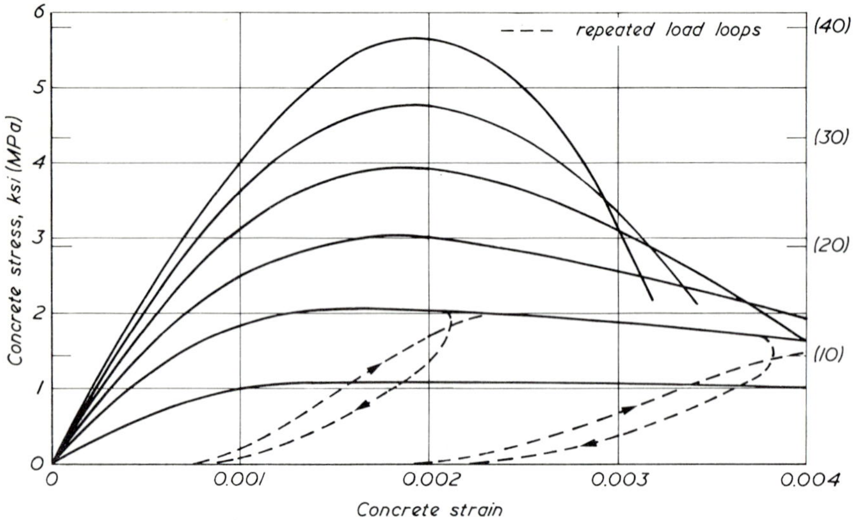

Fig. 5.1 Stress–strain curves for concrete cylinders loaded in uniaxial compression

5.3 STRESS–STRAIN BEHAVIOR OF CONCRETE AND STEEL

5.3.1 Concrete

Figure 5.1 shows typical stress–strain curves obtained from normal weight concrete cylinders loaded in uniaxial compression in a test conducted over several minutes. The curves are almost linear up to about one-half of the compressive strength, and in this linear region the modulus of elasticity for normal weight concrete may be considered to be $57{,}000\sqrt{f'_c}$ psi ($4730\sqrt{f'_c}$ MPa), where f'_c is the concrete compressive cylinder strength. The strain at maximum stress is approximately 0.002. When the load is applied at a fast strain rate both the modulus of elasticity and the strength of the concrete increase. For example, for a strain rate of 0.01/s the concrete strength may increase by as much as 17%. Repeated high intensity compressive loading produces a pronounced hysteresis effect in the stress–strain curve, as illustrated in Figure 5.1. Such tests indicate that the envelope curve is almost identical to the curve obtained in a single load application. For normal weight concrete the modulus of rupture can be taken as approximately $7.5\sqrt{f'_c}$ psi ($0.63\sqrt{f'_c}$ MPa).

When concrete is compressed to stresses approaching the uniaxial compressive strength the transverse strains become very high because of progressive internal cracking. If transverse reinforcement is present it will apply a confining reaction to the concrete at this stage and will considerably improve the stress–strain characteristics of the concrete at high strains[5.2]. Figure 5.2 shows stress–strain curves for concrete confined by spirals or

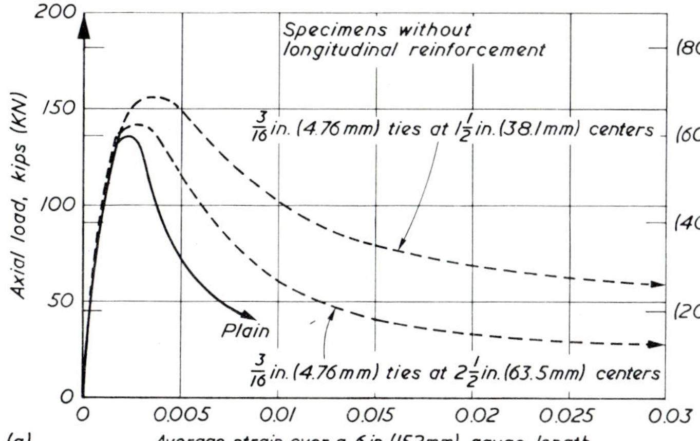

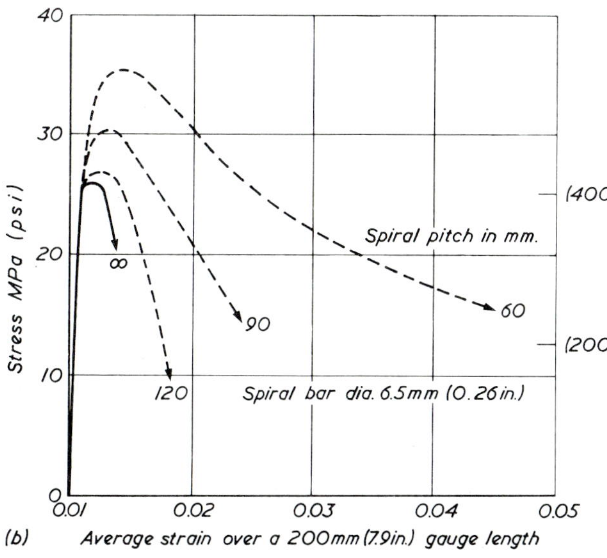

Fig. 5.2 Axial deformation of confined concrete specimens. (a) Axial load-strain curves for $4\frac{1}{4}$ *in* (108 *mm*) *square concrete prisms with square ties*[5.3]. (*b*) *Stress–strain curves for* 150 *mm* (5.9 *in*) *diameter by* 300 *mm* (11.8 *in*) *long concrete cylinders with circular spirals*[5.4]

square ties. Spirals confine the concrete more effectively because the circular shape enables it to provide a continuous confining pressure around the whole circumference. A square tie can generally only apply confining pressure near the corners because the pressure of the concrete tends to bow the sides of the tie outwards. Empirical stress–strain relationships for concrete confined by spirals have been proposed by Iyengar, *et al*[5.4], and for concrete confined by rectangular hoops by Kent and Park and others[5.2].

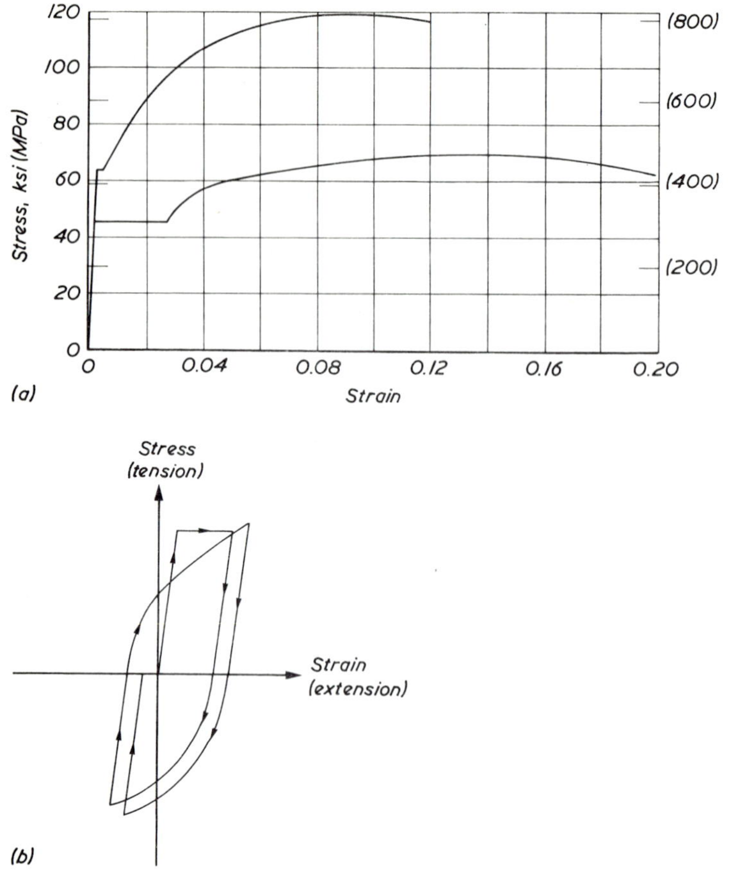

Fig. 5.3 Stress–strain curves for steel reinforcement; (a) monotonic loading, (b) reversed loading

5.3.2 Reinforcing steel

Figure 5.3(a) shows typical stress–strain curves for steel bars, used in reinforced concrete construction, loaded monotonically in tension. The modulus of elasticity of steel for the elastic linear portion of the curve is usually taken as 29×10^6 psi (0.2×10^6 MPa). High strength steels generally have a shorter length of horizontal yield plateau than lower strength steels and thus strain-harden at lower strains. High strength steels also have a smaller elongation before fracture than lower strength steels. Generally stress–strain curves in tension and compression are assumed to be identical. The effect of a fast rate of loading is to increase the yield strength of steel. For example, for a strain rate of 0.01/s the yield strength may increase by up to 14%. For repeated loading of the same sign the specimen recovers along a

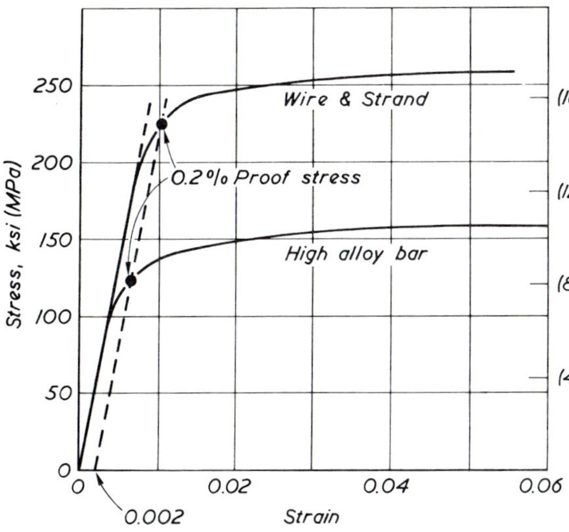

Fig. 5.4 Stress–strain curves for high tensile steel

path parallel to the initial elastic portion and on reloading follows the same path up to the original curve with perhaps a small hysteresis and/or strain hardening effect. The monotonic loading curve gives a good idealisation for the envelope curve for repeated loading of the same sign.

If reversed (tension-compression) axial loading is applied to a specimen in the yield range a stress–strain curve of the type shown in Figure 5.3(b) is obtained. The figure shows the Bauschinger effect in which the stress–strain curve becomes nonlinear at a stress much lower than the initial yield stress. This steel behaviour is strongly influenced by previous strain history. Reversed loading curves are important when considering the effect of high intensity seismic loading on members. Idealisations for such curves are discussed elsewhere[5.2].

It is recommended that deformed bars be used for main reinforcement in seismic design because of the better bond characteristics of such reinforcement.

5.3.3 Prestressing steel

Figure 5.4 shows typical stress–strain curves for various types of high tensile steel, used in prestressed concrete construction, loaded monotonically in tension. The most notable features of the curves are the lack of a horizontal yield plateau before strain hardening and a lower ultimate elongation than reinforcing steel. It is important for adequate ductility in seismic design that the elongation at fracture be not too small; an ultimate elongation of at least 4% would appear to be necessary. For repeated tensile loading in the inelastic range some hysteresis occurs in the unloading–loading loop, the width of the loop becoming greater at large strains.

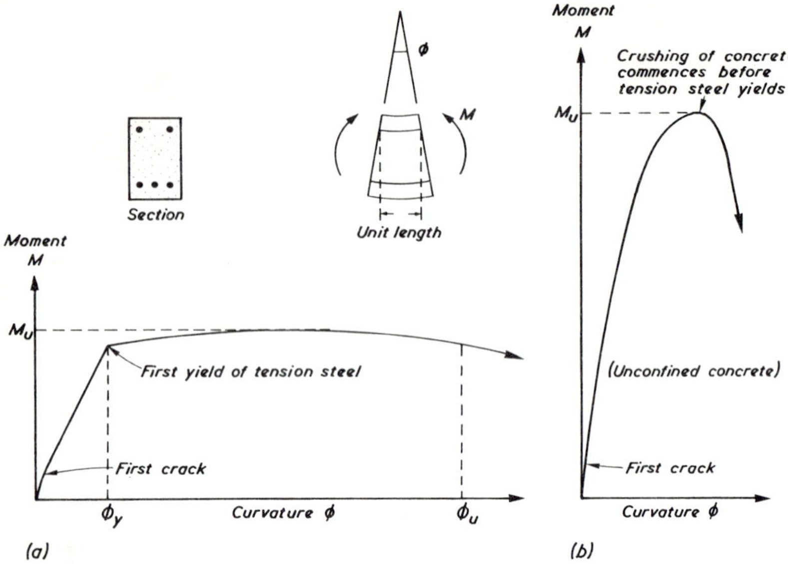

Fig. 5.5 Moment-curvature relationships for reinforced concrete beam sections; (a) tension failure, $\rho < \rho_b$, (b) compression failure, $\rho > \rho_b$

5.4 REINFORCED CONCRETE MEMBERS

5.4.1 Strength and ductility of sections with flexure

Moment-curvature behaviour The load-deflection characteristics of earthquake resistant structures are mainly dependent on the moment–curvature relationships of the sections, since for members of normal proportions most of the deformations arise from strains associated with flexure providing other types of failure are suppressed and deformations due to shear and axial force are not significant. Figure 5.5 shows moment–curvature relationships obtained from measurements on beams at the critical sections of a structure. The curvature φ is the rotation per unit length of member. When the tension steel content is low and/or the compression steel content is high the tension steel reaches the yield strength and then a large increase in curvature can occur at near constant bending moment (Figure 5.5(a)). This type of failure is known as a "tension failure" although eventually crushing of concrete will occur. Alternatively, when the tension steel content is high and the compression steel content is low the tension steel does not yield and the failure is brittle if the concrete is unconfined (Figure 5.5(b)). This is known as a "compression failure". In design beams are always proportioned so as to be capable of exhibiting the ductile characteristics of a tension failure. The aspects of section behaviour of interest are the flexural strength M_u and the curvature ductility factor expressed by the ratio φ_u/φ_y where φ_u is the ultimate curvature and φ_y is the curvature at first yield of the tension steel.

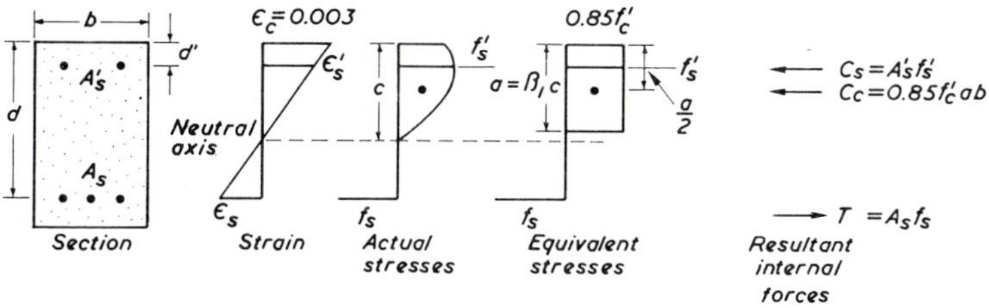

Fig. 5.6 Doubly reinforced concrete section when the flexural strength is reached

Flexural strength A doubly reinforced rectangular concrete beam section when the flexural strength is reached is shown in Figure 5.6. The tensile strength of the concrete is ignored. The ACI Code[5.1] assumes that at the flexural strength of the section the extreme fibre concrete compressive strain is 0.003. This is a reasonably conservative maximum strain[5.2] as most sections will maintain near maximum strength over a considerable range of strains greater than 0.002 (see Figure 5.5(a)). The ACI Code replaces the actual shape of the concrete compressive stress block by an equivalent rectangle as a means of simplification. The factor β_1 in Figure 5.6 is taken as 0.85 for f'_c up to and including 4000 psi (27.6 MPa), and is reduced continuously at the rate of 0.05 for each 1000 psi (6.9 MPa) of strength in excess of 4000 psi, but is not taken as less than 0.65. In design it is always ensured that at the flexural strength of the section the tension steel has reached the yield stress and hence that $f_s = f_y$. By taking moments of the internal forces about the tension steel the flexural strength is given by

$$M_u = 0.85 f'_c ab(d - 0.5a) + A'_s f'_s (d - d') \tag{5.6}$$

where from equilibrium of forces

$$a = (A_s f_y - A'_s f'_s)/(0.85 f'_c b) \tag{5.7}$$

and from the strain diagram

$$f'_s = \varepsilon'_s E_s = 0.003 E_s (a - \beta_1 d')/a \leqslant f_y \tag{5.8}$$

A balanced flexural failure occurs when the tension steel reaches yield strength f_y just as the extreme compression fibre of the concrete reaches a strain of 0.003. From the strain diagram of Figure 5.6, at a balanced failure

$$a_b = \frac{0.003 \beta_1 d}{0.003 + (f_y/E_s)} \tag{5.9}$$

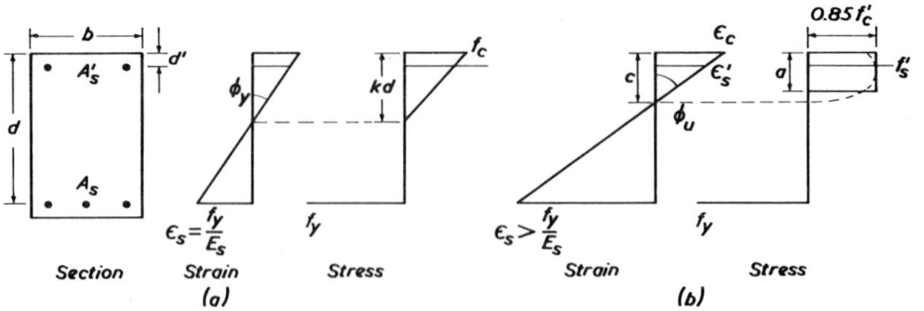

Fig. 5.7 Doubly reinforced concrete beam section; (a) at first yield of tension steel, (b) at ultimate strain of concrete

Equating Equations 5.9 and 5.7, with $\rho_b = A_s/bd$ and $\rho' = A'_s/bd$ gives

$$\rho_b = \frac{0.85 f'_c \beta_1}{f_y} \left(\frac{0.003 E_s}{0.003 E_s + f_y} \right) + \frac{\rho' f'_s}{f_y} \tag{5.10}$$

where f'_s is given by Equation 5.8 with a_b from Equation 5.9 substituted for a. A tension failure will occur (i.e. tension steel yielding) if a section contains a tension steel content $\rho = A_s/bd$ which is equal to or less than ρ_b from Equation 5.10.

For a T section, when the flexural strength is reached, the neutral axis depth is generally small because of the large flange width. In most cases the neutral axis lies in the flange and hence the section may be analysed as a rectangular section of width b. If the calculation indicates that the neutral axis lies in the web the equations need to be rewritten to take into account the T-shaped compressed area[5.2].

Comprehensive design aids have been published by the American Concrete Institute[5.5] which are of considerable assistance in design.

Curvature ductility The available ductility of the section can be expressed by the ratio of ultimate curvature φ_u to curvature at first yield φ_y. Figure 5.7 represents the general case of a doubly reinforced section at first yield of tension steel and at ultimate concrete strain. When the tension steel first reaches the yield strength the concrete stress distribution may be still linear, due to maximum concrete stress being significantly less than the concrete strength, and the neutral axis depth kd may be calculated using elastic (straight line) theory[5.2] as

$$k = \sqrt{(\rho + \rho')^2 n^2 + 2\left(\rho + \frac{\rho' d'}{d}\right) n} - (\rho + \rho') n \tag{5.11}$$

where $n = E_s/E_c$, E_s is the modulus of elasticity of steel, and E_c is the modulus of elasticity of concrete. The curvature is given by the extension of the tension steel at first yield per unit length (that is, the yield strain) divided by the distance from the tension steel to the neutral axis.

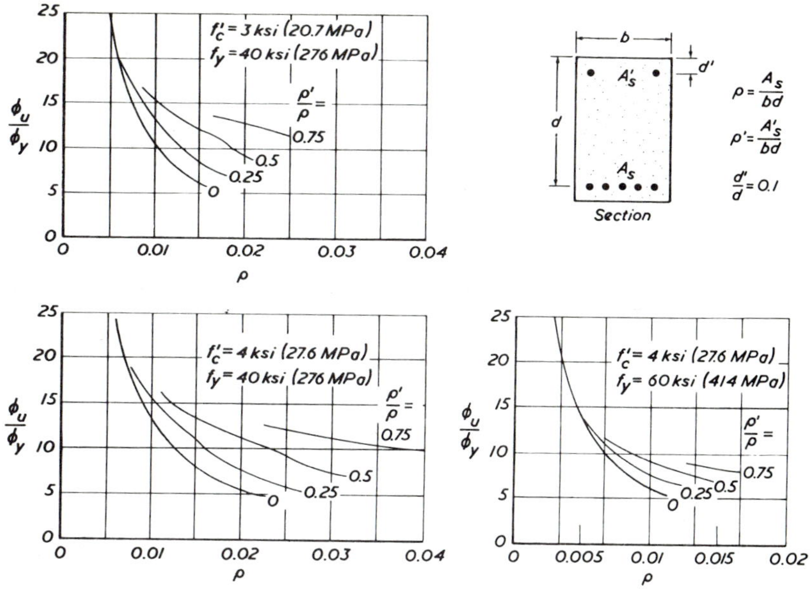

Fig. 5.8 Variation of φ_u/φ_y for beam sections with unconfined concrete and $\varepsilon_c = 0.004$[5.2]

$$\therefore\ \varphi_y = \frac{f_y/E_s}{d(1-k)} \tag{5.12}$$

Similarly, the ultimate curvature is given by

$$\varphi_u = \varepsilon_c \beta_1 / a \tag{5.13}$$

where a is given by Equations 5.7 and 5.8 and ε_c may be assumed to be 0.004 for unconfined concrete[5.2]. It is to be noted that $\varepsilon_c = 0.003$ is too conservative for ultimate curvature calculations. The curvature ductility factor of the section is given by

$$\frac{\varphi_u}{\varphi_y} = \frac{\varepsilon_c}{f_y/E_s}\ \frac{d(1-k)}{a/\beta_1} \tag{5.14}$$

Figure 5.8 shows φ_u/φ_y from Equation 5.14 plotted for $\varepsilon_c = 0.004$ and a range of steel ratios and practical combinations of steel and concrete strengths. It is evident that, with other variables held constant, the available curvature ductility factor increases with decrease in tension steel content, with increase in compression steel content, with decrease in steel yield strength and with increase in concrete strength.

If the compression zone of a member is confined by closely spaced closed stirrups or spirals the ductility of the concrete is greatly improved and a more ductile performance of the member will result. Experimental results have given an indication of the maximum usable concrete compressive strain in confined members[5.2]. For example, from the results of tests on simply supported beams performed by Corley, Mattock[5.6] has proposed that the maximum concrete strain be taken as

$$\varepsilon_c = 0.003 + 0.02\frac{b}{z} + 0.2\rho_s \tag{5.15}$$

where b is the beam width, z is the distance from critical section to point of contraflexure or zero moment, and ρ_s is the ratio of volume of confining steel (including compression steel) to volume of concrete core. The inclusion of the b/z ratio in Equation 5.15 indicates that concrete confinement is assisted by a high strain gradient along the member (that is a small z value) and by a large beam width. It is obvious that Equation 5.15 cannot be valid for the large b/z values found in slabs. Equation 5.15 was derived from beam tests with $b/z \leqslant 0.33$. It is evident from Equation 5.15 that with some confining steel it is not difficult to increase the available concrete strain in the extreme compression fibre to 0.01 or higher.

5.4.2 Strength and ductility of sections with flexure and axial load

Axial load strength The ultimate load of an axially loaded reinforced concrete column may be written as

$$P_0 = 0.85f'_c(A_g - A_{st}) + f_y A_{st} \tag{5.16}$$

where A_g is the gross area of the section and A_{st} is the total area of longitudinal steel in the section. This equation is the sum of the concrete and steel strengths because the steel will be at its yield strength when the concrete reaches its strength at a strain of about 0.002. The concrete strength is taken as $0.85f'_c$ mainly because the vertical casting of the column leads to water gain and sedimentation at the top of the column resulting in a concrete strength lower than the cylinder strength f'_c.

A poorly tied column will immediately collapse once the load P_0 is reached. However if the column has closely spaced ties or spiral reinforcement around the longitudinal steel it will continue to carry significant load at high strains since, although the cover concrete may be lost, the confined concrete core will still be effective and the longitudinal bars will not buckle. The behaviour of poorly confined and well confined columns after a severe earthquake is contrasted in Figure 5.9.

For spiral columns the ACI Code[5.1] requires that the ratio of the volume of spiral steel to the volume of concrete core ρ_s be not less than the value given by

$$\rho_s = 0.45\frac{f'_c}{f_y}\left(\frac{A_g}{A_c} - 1\right) \tag{5.17}$$

Fig. 5.9 Some columns from Olive View Hospital after the 1971 *San Fernando earthquake.* (*a*) *Column with nominal ties.* (*b*) *Spiral column*

where A_c is the concrete core area measured to the outside of the spiral. This quantity of spiral steel is intended to ensure that the load carried after crushing of the cover concrete (when the spiral reinforcement reaches yield and the confined concrete core reaches its enhanced strength) is at least equal to P_0 from Equation 5.16.

Strength in flexure and axial load Figure 5.10(a) shows a rectangular concrete section symmetrically reinforced with bars at two faces loaded eccentrically at the ultimate load. The total steel area is A_{st}. The same extreme fibre compressive concrete strain and equivalent rectangular concrete stress block parameters as used for beams is assumed by the ACI Code[5.1]. The equilibrium equations when the neutral axis lies within the section are

$$P_u = 0.85 f'_c ab + 0.5 f'_s A_{st} - 0.5 f_s A_{st} \tag{5.18}$$

$$P_u e = 0.85 f'_c ab(0.5h - 0.5a) + 0.5 f'_s A_{st}(0.5h - d') + 0.5 f_s A_{st}(0.5h - d') \tag{5.19}$$

where the compression steel stress f'_s is given by Equation 5.8. The axial load and moment at a balanced failure, P_b and $P_b e_b$, may be found by substituting $f_s = f_y$ and $a = a_b$ from Equation 5.9 into Equations 5.18 and 5.19. A tension failure occurs if the applied load P_u is less than P_b, since the smaller column load means that $a < a_b$ and reference to the strain diagram shows that therefore ε_s is greater than the steel yield strain; then Equation 5.18 and 5.19 apply with $f_s = f_y$. A compression failure occurs if the applied load P_u is greater

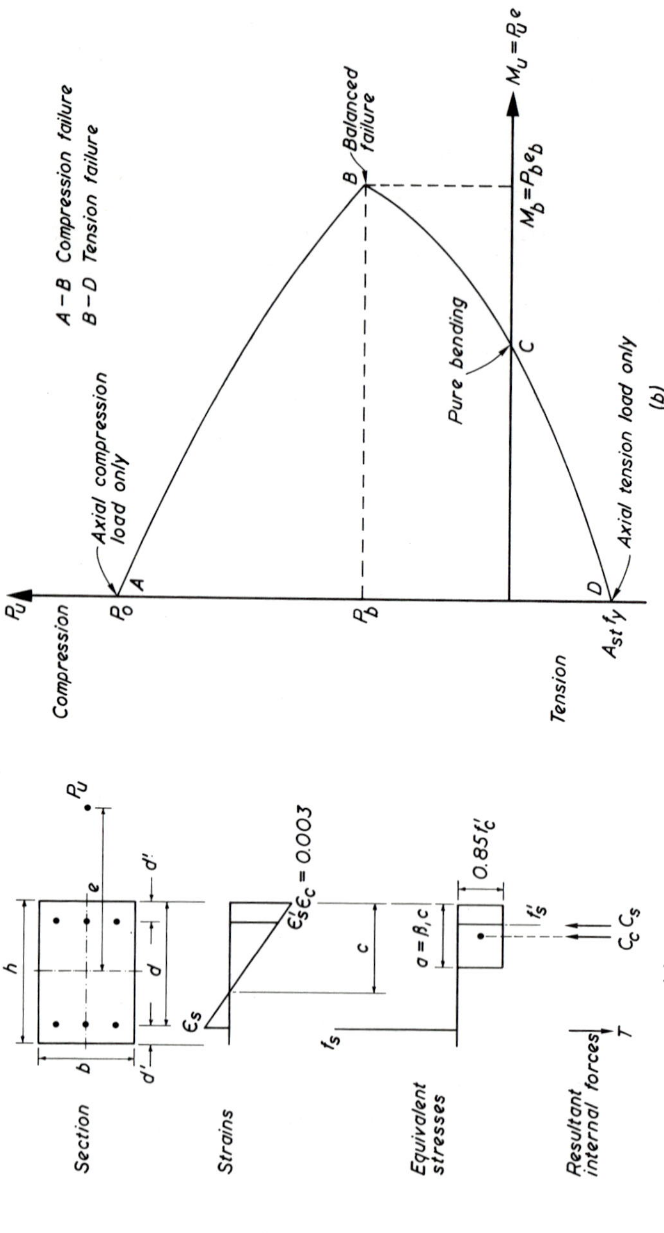

Fig. 5.10 Symmetrically reinforced concrete column (*a*) at ultimate eccentrically applied load (*b*) Interaction diagram showing combinations of load and moment that cause failure

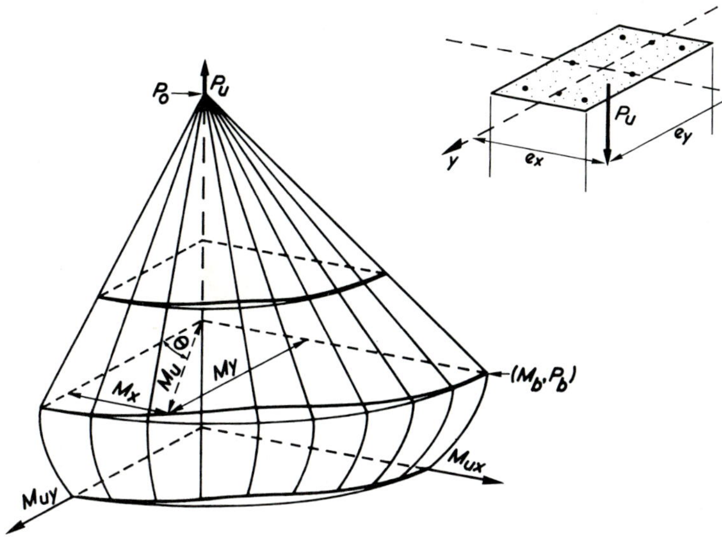

Fig. 5.11 Interaction diagram for reinforced concrete column with biaxial bending at failure

than P_b, since the greater compression load means that $a > a_b$ and reference to the strain diagram shows that therefore ε_s is less than the steel yield strain; then Equations 5.18 and 5.19 apply with f_s determined from the strain diagram as

$$f_s = \varepsilon_s E_s = 0.003 E_s(\beta_1 d - a)/a \tag{5.19}$$

The combinations of P_u and $P_u e$ which cause failure of a given column section are best illustrated by an interaction diagram (see Figure 5.10(b)). Any combination of load and moment that can be plotted within the area of the interaction diagram can be supported without failure.

When a column has bars distributed at all faces the derivation of the equations for design and analysis becomes difficult because the bars may be at various stress levels throughout the section. In practice the design and analysis of column sections can be carried out conveniently using design tables and charts. A comprehensive set of tables and charts for a wide range of columns has been published by the American Concrete Institute[5.5,5.7].

Many columns are subjected to bending about both major axes simultaneously, that is, biaxial bending. The strength of columns with biaxial bending can be represented by interaction surfaces. By analysing the column section for various angles of inclination of the neutral axis a range of interaction curves may be obtained which describe the interaction surface as illustrated in Figure 5.11. Each point on this surface represents a particular set of load and moments about the principal axes which will cause failure of the section. A horizontal section drawn through the interaction surface

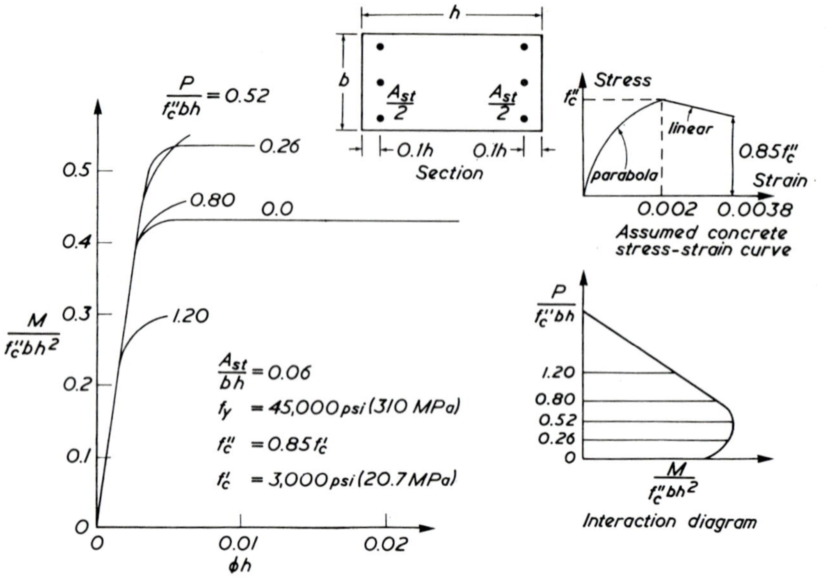

Fig. 5.12 Theoretical moment–curvature relationships for column sections at various levels of axial load[5.8]

shows that at constant load the shape of the curve relating the moment strengths in the two directions is not elliptical. The number of variables involved means that the preparation of design charts for the general case of columns with biaxial bending is difficult. However, there are available design approaches which use simplifying approximations[5.2].

The slenderness of a column may cause the ultimate load to be reduced by additional eccentricity caused by lateral deflection of the column. This so-called P–Δ effect may be of significance, particularly for columns of frames not braced sideways. The ACI Code[5.1] gives a moment magnifier method for use in design to take into account effects of slenderness, but that method does not properly apply to a frame with seismic loading. Slenderness is often not an important aspect of the seismic design of reinforced concrete frames, because the columns may have a relatively large cross section.

Curvature ductility Axial load on a column section influences the neutral axis depth and therefore the curvature. Theoretical moment–curvature curves for a column section derived by Pfrang, *et al*,[5.8] are presented in Figure 5.12. Each curve is for an axial load held constant at a particular level while the column is bent to failure. The ultimate curvature was assumed to be reached when the maximum concrete compressive strain was 0.0038. The curves illustrate that at axial load levels greater than the balanced failure load the ductility is negligible, being only due to the inelastic deformation of the concrete. At levels less than the balanced failure load the tension steel

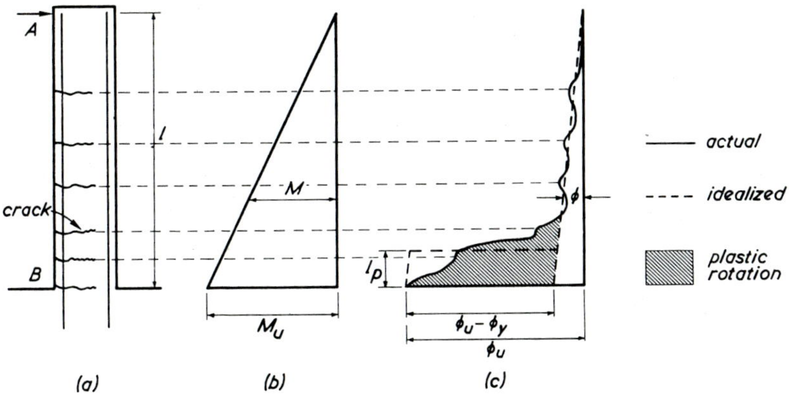

Fig. 5.13 Member with ultimate curvature reached; (a) member, (b) bending moment diagram, (c) curvature diagram

yields and the ductility increases as the load level is reduced. As with beams, the ductility of column sections can be considerably increased by the presence of transverse confining steel.

5.4.3 Rotations and deflections of members

Figure 5.13 shows a reinforced concrete member that has reached ultimate curvature and moment at the critical section. The region of inelastic curvature is spread over a length of member, this region being at least that over which the bending moment exceeds the yield moment of the section. Local peaks of curvature occur at the cracks because of the increased rigidity of the member between the cracks due to the tension carried by the concrete between the cracks. The actual curvature distribution can be idealised into elastic and inelastic regions as in Figure 5.13(c). The shaded area gives the inelastic rotation that can occur at the 'plastic hinge' in the vicinity of the critical section. The shaded area can be replaced by an equal equivalent area of height $\varphi_u - \varphi_y$ and length l_p. The length l_p is the equivalent length of the plastic hinge. When diagonal tension cracks are present in plastic hinge zones the inelastic curvature is spread further along the member than the bending moment implies[5.2], and therefore values for l_p proposed have been based on experimental evidence. For example, Mattock[5.6] has proposed that

$$l_p = 0.5d + 0.05z \tag{5.20}$$

where d is the distance from the tension steel to the extreme compressed fibre and z is the distance from the critical section to the point of contraflexure or zero moment.

The rotation between two points A and B may be found from the area of the curvature diagram between A and B. The transverse deflection of point A from the tangent to the axis of the member at point B is given by the first moment of area of the curvature diagram between A and B about A. For example for the cantilever column or tall shear wall shown in Figure 5.13 the transverse deflection at A at the ultimate load is given by

$$\Delta_u = \left(\frac{\varphi_y l}{2}\frac{2l}{3}\right) + (\varphi_u - \varphi_y) l_p (l - 0.5 l_p) \tag{5.21}$$

Also the transverse deflection at A at the first yield load is given by

$$\Delta_y = \frac{\varphi_y l}{2}\frac{2l}{3} \tag{5.22}$$

Therefore

$$\mu = \frac{\Delta_u}{\Delta_y} = 1 + \left(\frac{\varphi_u - \varphi_y}{\varphi_y}\right)\frac{3 l_p (l - 0.5 l_p)}{l^2}$$

and hence

$$\frac{\varphi_u}{\varphi_y} = 1 + \frac{l^2(\mu - 1)}{3 l_p (l - 0.5 l_p)} \tag{5.23}$$

Equation 5.23 can be used to demonstrate the difference between the curvature ductility factor φ_u/φ_y required to achieve a particular value for the displacement ductility factor $\mu = \Delta_u/\Delta_y$. For example, if $\mu = 4$ the required φ_u/φ_y values are:

l_p/l	0.05	0.1	0.15	0.2	0.25	0.3	0.35	0.4
φ_u/φ_y	21.5	11.5	8.2	6.6	5.6	4.9	4.5	4.1

Hence if the equivalent plastic hinge length is a small proportion of the length of the member the curvature ductility factor demand will be much higher than the imposed displacement ductility factor since the yielding is concentrated over a small region of the member.

5.4.4 Cyclic loading of flexural members

Moment reversals in the inelastic range cause a significant change in the moment–curvature relationship for a section. Figure 5.14 illustrates the behaviour of a doubly reinforced beam section with unequal top and bottom steel. When the beam is loaded well into the inelastic range in one direction (say with positive bending moment), the large cracks will not close completely on unloading but will remain open because of the residual plastic

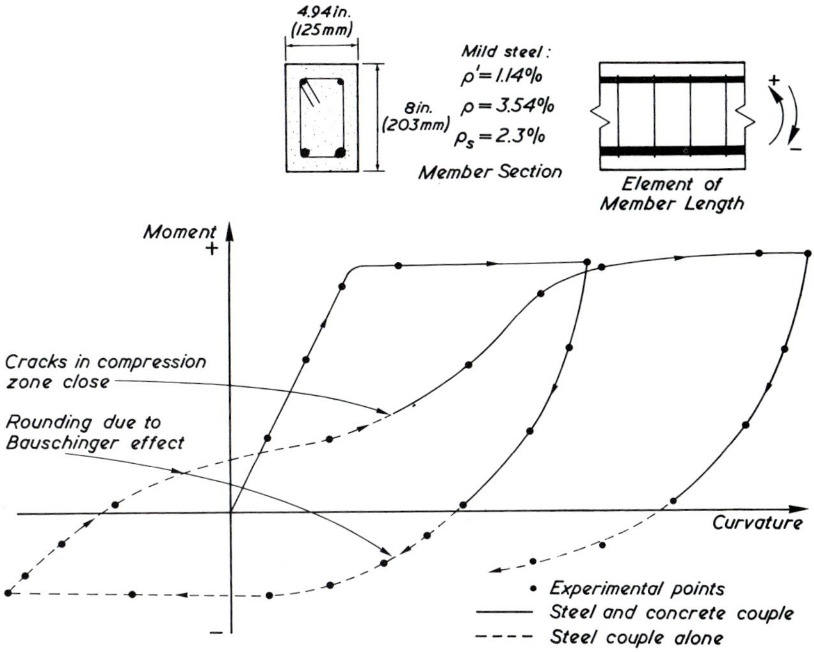

Fig. 5.14 Moment curvature relationship for doubly reinforced beam section with reversed flexure[5.2]

strain in the steel. If the beam section is then loaded in the opposite direction (negative bending moment) the presence of open cracks in the compression zone means that the whole of the compressive force is carried by the compression steel and open cracks exist down the whole beam depth. The flexural stiffness is then due only to the steel and is further reduced when the compression and tension steel reaches the stress level at which the Bauschinger effect commences, and behaves inelastically (see Figure 5.3(b)). The open cracks in the compression zone will not close since the area of bottom steel is greater than the area of top steel. On unloading, the full depth open cracks remain. On reloading in the direction of positive bending moment again the flexural stiffness is initially due only to the steel, and reduces when the Bauschinger effect commences. However for this direction of moment the cracks in the compression zone will eventually close since the area of top steel is less than the area of bottom steel, resulting in an increase in flexural stiffness.

The reduction in stiffness due to open cracks in the compression zone and the Bauschinger effect of the steel will not prevent a properly detailed member from reaching its flexural strength, but the deformation at which the flexural strength is reached is increased. The behaviour shown in Figure 5.14, as well as being observed experimentally, can be determined theoretically[5.2].

The opening and closing of cracks may eventually lead to deterioration in the compressive strength of the concrete, because the faces of the crack may not come into even contact owing to slight lateral movement or debris in the crack. Also, buckling of compression steel becomes more likely due to the reduced tangent modulus of the steel at low stress levels caused by the Bauschinger effect.

It was mentioned previously that high strain rates, such as can be associated with earthquake loading, cause significant increases in the strength of steel and concrete as measured in monotonic load tests. Tests conducted at Berkeley[5.9] on reinforced concrete beams under high strain rates have indicated an increase in the first yield moment of about 20%, but a reduction in the effect of high strain rate occurred at larger deformations, and after the first cycle of loading in which the beam is yielded the hysteresis loops are little affected by the strain rate. Thus there is good justification for ignoring the effects of high strain rates on the material strengths in seismic design.

5.4.5 Strength of members with shear

The combination of flexure and significant shear force in reinforced concrete members causes inclined cracks, referred to as diagonal tension cracks. Figure 5.15(a) shows part of a simply supported beam without shear reinforcement to one side of a diagonal tension crack. The shear force V is resisted by a shear force V_{cz} carried across the compression zone, a dowel force V_d transmitted across the crack by the flexural reinforcement, and the vertical components of inclined shear stress v_a transmitted across the inclined crack by means of interlocking of the aggregate particles. If the resultant aggregate interlock force is assumed to pass through the intersection of C and V_{cz} the external bending moment at section 1 of Figure 5.15(a) is given by

$$M = Tjd + V_d f \tag{5.24}$$

If V_d is neglected in design, then $M = Tjd$ and it is evident that the tensile force at section 2 is due to bending moment at section 1. This shift in tensile force causes spreading of flexural yielding along the member and influences the plastic hinge length as discussed in Section 5.4.3. It also needs to be taken into account when the curtailment of flexural bars is being considered.

In reinforced concrete beams of normal proportions without shear reinforcement shear failure can occur by breakdown of the concrete cantilevers between the diagonal tension cracks due to bond forces arising from beam action. The ACI Code[5.1] recommends that the shear force carried by the concrete be taken as

$$V_c = v_c b_w d \tag{5.25}$$

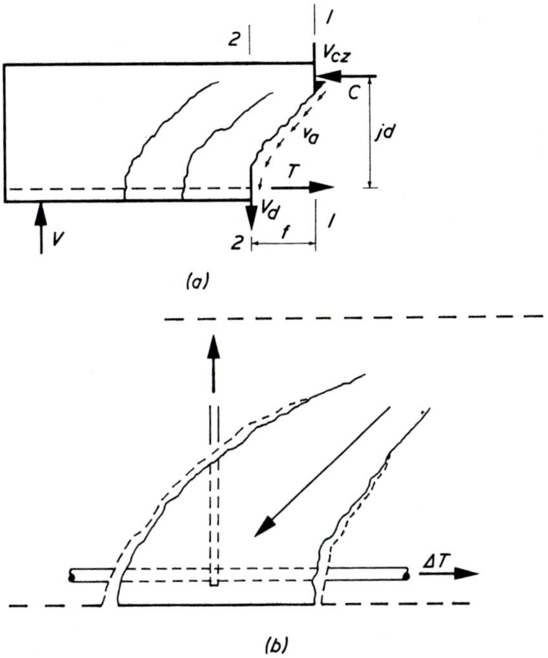

Fig. 5.15 Shear resisting mechanism in beam; (a) internal forces resisting shear without web reinforcement present, (b) concrete cantilever acting as strut with web reinforcement present

where b_w is the width of the web, d is the effective depth, and for a beam

$$v_c = 2\sqrt{f'_c} \tag{5.26}$$

where v_c and f'_c are in psi ($0.17\sqrt{f'_c}$ where v_c and f'_c are in MPa).

Axial compression increases the shear carried by the concrete because of better crack control and larger neutral axis depth; conversely, axial tension has the opposite effect. The ACI Code[5.1] recommends for members carrying axial compressive force N_u on a section of gross area A_g that

$$v_c = 2\left(1 + 0.0005\frac{N_u}{A_g}\right)\sqrt{f'_c} \tag{5.27}$$

where v_c and f'_c are in psi, or

$$v_c = 0.17\left(1 + 0.073\frac{N_u}{A_g}\right)\sqrt{f'_c}$$

where v_c and f'_c are in MPa. For members carrying axial tension N_u

$$v_c = 2\left(1 + 0.002\frac{N_u}{A_g}\right)\sqrt{f'_c} \tag{5.28}$$

where v_c and f'_c are in psi, or

$$v_c = 0.17\left(1 + 0.29\frac{N_u}{A_g}\right)\sqrt{f'_c}$$

where v_c and f'_c are in MPa, where N_u is taken as negative for tension and positive for compression.

Shear reinforcement, commonly in the form of stirrups, can carry shear force by truss action in which the concrete cantilevers form diagonal compression struts (see Figure 5.15(b)). The number of stirrups perpendicular to the axis of the member crossing an assumed 45° crack is d/s where s is the stirrup spacing, and therefore the shear force carried by perpendicular stirrups is

$$V_s = f_y A_v d/s \tag{5.29}$$

where f_y is the steel yield strength and A_v is the area of stirrups within spacing s. If the shear reinforcement is inclined at angle α to the longitudinal axis of the member the shear force carried is given by Equation 5.29 divided by $(\sin\alpha + \cos\alpha)$.

The shear strength of a member with shear reinforcement is found summing the contribution from concrete and steel as

$$V_u = v_c b_w d + V_s \tag{5.30}$$

The ACI Code also limits the shear force carried by the truss action to $8\sqrt{f'_c}b_w d$ in lb and in. units $(0.67\sqrt{f'_c}b_w d$ in N and mm units).

Cyclic loading into the yield range of the longitudinal reinforcement, such as due to severe earthquake motions, causes a deterioration of the shear carried by aggregate interlock, dowel action and across the compression zone. This is because when full depth cracks exist (see Section 5.4.4) the shear is carried mainly by dowel action which tends to cause splitting along the longitudinal bars, and the alternating opening and closing of diagonal tension cracks can cause deterioration of the concrete at the cracks. Tests have shown that when the nominal shear stress $V/b_w d$ exceeds $3\sqrt{f'_c}$ psi where f'_c is in psi $(0.25\sqrt{f'_c}$ MPa where f'_c is in MPa) reduction of stiffness may occur due to shear and shear failure may occur after cycles of loading by sliding along the vertical crack at the critical section at the plastic hinge zone. Use of inclined shear reinforcement is an effective way of preventing such a shear failure. The design approach to shear strength in potential plastic hinges is discussed in Section 5.5.2.

When a column with a rectangular section is subjected to biaxial flexure, discussed in Section 5.4.2, the corresponding shear force will have to be resisted at some angle to the principal axes of the section. The forces in the stirrup tie legs will therefore be at some angle to the direction of the applied shear force. For a symmetrically reinforced square column section it may be shown that the shear resistance of stirrup ties symmetrically arranged and perpendicular to the axes of the member is the same for all directions of shear force application. This comes about because, although the component of force in the stirrup tie legs in the direction of shear force is smaller when the shear force is at a general angle to the principal axes of the section, the diagonal tension crack has a greater projected length and therefore intercepts more stirrup tie legs. Some recent tests have confirmed this prediction. For circular sections the shear force resisted by the spiral reinforcement may be found by summing the components of the force in the spiral bars intercepted by a 45° diagonal tension crack in the direction of the shear force.

In rectilinear space frames, rotations of beam–column joints will inevitably introduce torsion into some beams. In the evaluation of strength no reliance is normally placed on the torsional resistance of individual members, particularly because the torsional stiffness of members reduces significantly after diagonal tension cracking[5.2]. Provided that stirrup ties are present in the critical areas, secondary torsion should have no detrimental effect because beams with stirrup ties are reasonably ductile in torsion.

5.4.6 Bond and anchorage

Bond failure of deformed bars generally occurs by concrete splitting. For No. 11 (36 mm diameter) or smaller deformed bars with $f_y \leqslant 60$ ksi (414 MPa) the ACI Code[5.1] recommends a minimum development length

$$l_d = m_f \, 0.04 A_b f_y / \sqrt{f'_c} \tag{5.31}$$

but not less than $m_f\ 0.0004 d_b f_y$ or 12 in where all units are in lb and in., or

$$l_d = m_f 0.019 A_b f_y / \sqrt{f'_c}$$

but not less than $m_f 0.058 d_b f_y$ or 305 mm, where all units are in N and mm, and where A_b is the bar area and d_b is the bar diameter. The factor m_f allows for adverse performance of top cast bars ($m_f = 1.0$ for bars with less than 12 in (305 mm) of concrete cast below the bar, otherwise $m_f = 1.4$), lightweight concrete, cover and lateral spacing between bars, excess reinforcement, and the effect of a spiral around bars. For bars in compression, the contribution from end bearing allows a smaller l_d value than above to be specified.

Splices should ideally be staggered and located away from sections of maximum tension in seismic design. The ACI Code[5.1] recommends that where the bar tension is less than $0.5f_y$, the lap length shall be $1.3l_d$ if more

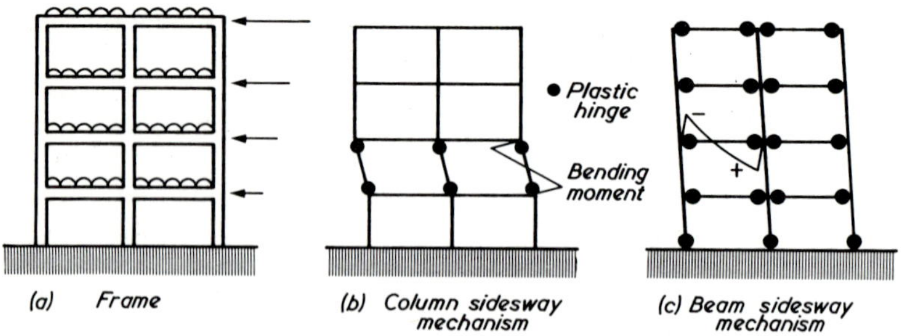

Fig. 5.16 Frame and sidesway mechanisms

than three quarters of the bars are lap spliced within a lap length otherwise the lap length shall be $1.0l_d$. Lap lengths for bars in compression are smaller. Lapped splices should never be located in potential plastic hinge zones of beams. In columns of buildings they should be positioned in the mid-height region between floors.

5.5 REINFORCED CONCRETE BUILDING FRAMES

5.5.1 Basic concepts

Nonlinear dynamic analyses of structures designed to code static loading, responding to severe earthquakes, have given an indication of the order of the inelastic deformations required. However it is difficult to reach general conclusions because of the large number of variables involved (see earlier chapters). Californian[5.10] and New Zealand[5.11] codes have indicated that a displacement ductility factor Δ_u/Δ_y in the order of 3 to 5 is required for ductile earthquake resistant structures. The designer should recognise that the required curvature ductility factor φ_u/φ_y at the plastic hinge sections will be significantly greater than the Δ_u/Δ_y value, because once yielding commences further displacement occurs mainly by rotation of the plastic hinges, as is illustrated by Equation 5.23 for a cantilever column. Dynamic analyses have indicated that the ductility demand concentrates in the weak part of frames. This can also be illustrated by examination of static collapse mechanisms[5.2]. Figure 5.16 shows a frame under seismic (code type) loading and possible basic mechanisms which could form due to flexural yielding. If yielding commences in the columns before in the beams, a column sidesway mechanism can form and in the worst case the plastic hinges will form in the columns of only one storey since the columns of the other storeys may be stronger. Such a mechanism will make very large curvature ductility demands at the plastic hinges of the critical storey and in tall buildings it is unlikely that sufficient curvature ductility would be available for this mechanism to survive a severe earthquake. On the other hand, if the yielding commences in the beams first, a beam sidesway mechanism can develop which makes much more moderate demands on the curvature ductility required at the plastic hinges of the beams and column bases. Also, curvature ductility can be provided more easily in beams, and columns are more difficult to repair. The mechanisms of Figure 5.16 are somewhat idealised but nevertheless they indicate the desirability of a strong column–weak beam approach in seismic design.

Thus the capacity design procedure for earthquake resistant frames is to detail the beams and column bases so that a ductile energy dissipating mechanism can develop by plastic hinging there during a severe earthquake, while sufficient reserve strength is provided so that the probability of column yielding, or shear or bond failure, is low. Design codes are not specific about the level of curvature ductility factor to be available at the critical sections but recommend detailing procedures which are aimed at achieving adequate ductility. The provisions for seismic design in the following sections are based on Appendix A of the ACI Code[5.1].

5.5.2 Special provisions for flexural members

The sections should be designed for flexure using a strength reduction factor φ of 0.9 (see Equation 5.4). An upper limit to the tension steel ratio ρ of $0.5\rho_b$, where ρ_b is given by Equation 5.10, is recommended to ensure ductile sections. Also, at column connections the code requires the beam positive moment capacity to be at least 50% of the beam negative moment capacity. Both top and bottom steel should consist of not less than two bars and ρ should be at least $200/f_y$ with f_y in psi ($1.37/f_y$ with f_y in MPa) throughout the length of the member. Therefore in a beam at a column face ρ' will be approximately 0.5ρ or more and $\rho \leqslant 0.5\rho_b$ and this will ensure that $\varphi_u/\varphi_y > 6$ for $\varepsilon_c = 0.004$ (see Figure 5.8). Hence if the curvature ductility demand is two or three times this value, as is likely in a severe earthquake, the concrete will need to be confined effectively. For example, reference to Mattock's Equation 5.15 shows that if $z = 4d$ and if No. 3 (9.5 mm diameter) closed stirrups at 4 in (102 mm) centres enclose a 10 in (254 mm) by 20 in (508 mm) beam core giving $\rho_s = 0.00825$ excluding compression steel, then $\varepsilon_c = 0.0097$ which results in the ductility level required but with crushing of the cover concrete and possibly some reduction in moment capacity as a result. In the plastic hinge zones of members the spacing of closed stirrups should not exceed 4 in (102 mm) or $0.25d$, where d is the effective depth of the member. Also, as discussed in Section 5.4.4, reversed loading to yield of reinforcing bars lowers the tangent modulus of that steel at low stress levels and this, and the presence of open cracks in the compression zone, may lead to bar buckling. Therefore it is recommended that in plastic hinge zones the spacing of closed stirrups should not exceed six compression steel bar diameters, a spacing which is smaller than the ACI Code requirements. Also, the stirrup ties should be arranged so that each longitudinal bar or bundle of bars is restrained against buckling by a 90° bend in the stirrup, except that when bars at not more than 8 in (204 mm) centres are so restrained by the same

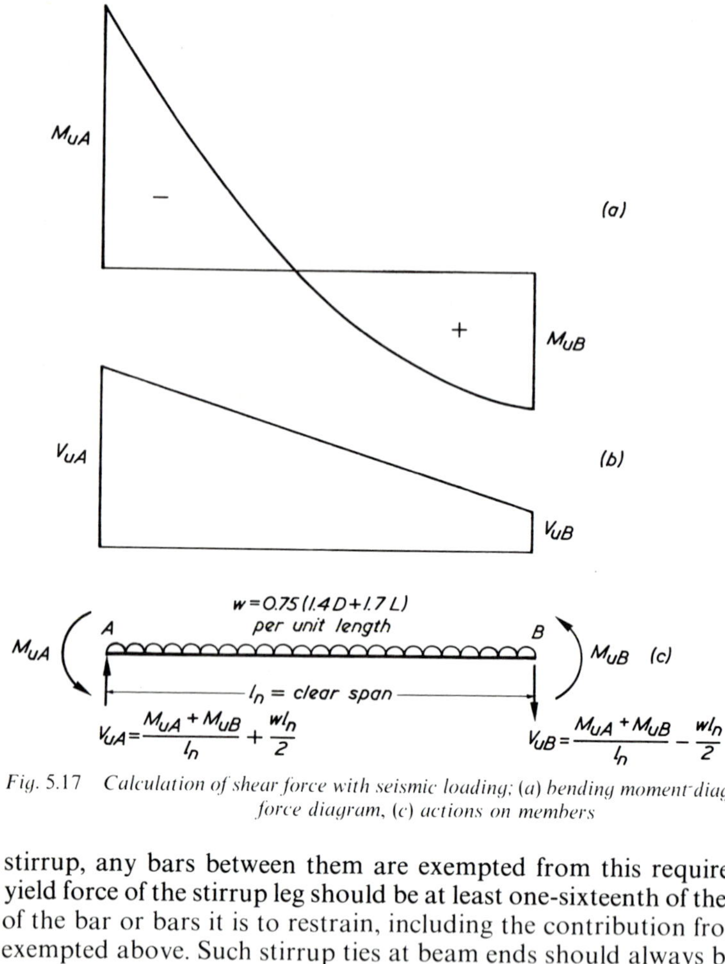

Fig. 5.17 Calculation of shear force with seismic loading; (a) bending moment diagram, (b) shear force diagram, (c) actions on members

stirrup, any bars between them are exempted from this requirement. The yield force of the stirrup leg should be at least one-sixteenth of the yield force of the bar or bars it is to restrain, including the contribution from the bars exempted above. Such stirrup ties at beam ends should always be provided for a distance of twice the effective beam depth from the column face, and wherever else in the beam inelastic deformations of the frame may cause the flexural strength to be reached.

The dependable shear capacity, using a strength reduction factor φ of 0.85, should at least equal the design shear force calculated using the design gravity loads and the overstrength plastic hinge moments (see Equation 5.5), since only then will the probability of a shear failure be sufficiently low. The determination of the design shear force is illustrated in Figure 5.17 for the case where plastic hinges form at the ends of the beam. The beam steel for negative moment should include the slab steel in an effective width of slab of say four slab thicknesses each side of an interior column. The steel strength could be taken as $1.25f_y$ for steel with a specified yield strength f_y of 40 ksi (276 MPa) and $1.4f_y$ for steel with a specified yield strength of 60 ksi (414 MPa), these values reflecting the actual steel stress (including strain hardening, which is particularly significant for high strength steel) when the required curvature ductility is reached. In plastic hinge zones reversals of moment cause a reduction in the shear carried by the concrete by aggregate interlock, dowel action and across the compression zone, as discussed in Section 5.4.5. Thus in plastic hinge zones the contribution of these concrete shear resisting mechanisms should be ignored ($v_c = 0$) and the whole of the shear force carried by the shear reinforcement at a spacing not exceeding $0.25d$, which is more stringent than the ACI Code requirement. Outside plastic hinge zones v_c may be taken as the value recommended by Equation 5.26. Where the nominal reversed shear stress exceeds $3\sqrt{f'_c}$ psi where f'_c is in psi ($0.25\sqrt{f'_c}$ MPa where f'_c is in MPa) diagonal bars could well be used as shear reinforcement in plastic hinge zones to help prevent a sliding shear failure along an open vertical crack.

5.5.3 Special provisions for columns

The vertical reinforcement ratio should be between 0.01 and 0.06. The ACI Code provisions aim at having plastic hinges form in the beams rather than in the columns by requiring that the sum of the moment strengths of the columns exceeds the sum of the moment strengths of the beams along each principal plane at the connection, unless special provisions are made. However this requirement will not prevent plastic hinges forming in the columns for two reasons[5.2]. Nonlinear dynamic analyses have shown that in frames, due to higher mode effects, points of contraflexure may occur well away from the mid-height of columns. Thus bending moment distributions in columns such as in Figure 5.18 are possible at times during the earthquake[5.2]. The greatest column moment is given by

$$M_{c1} = M_{b1} + M_{b2} - M_{c2} \tag{5.32}$$

Hence the beam input moment $M_{b1} + M_{b2}$ may have to be resisted almost entirely by one column section if M_{c2} is small. Also, in design it is customary to consider seismic loading to act in the direction of the principal axes of a structure and in one direction at a time. However a general direction of seismic loading may cause yielding to occur in the beams in both directions simultaneously (see Reference 5.2 and Chapters 1 to 3). This results in an increased beam input moment which is to be resisted by a decreased column section strength. For example for a symmetrical structure with equal strengths in the two principal directions the resultant beam input moment applied diagonally to the column section is $\sqrt{2}$ times the uniaxial beam input moment, and the diagonal bending strength of the column section may be some 10% less than the uniaxial bending strength. To prevent higher mode effects and general seismic loading causing plastic hinges to form in columns requires column strengths of up to three times the ACI Code requirement. The difficulty of preventing plastic hinges from forming in columns is such

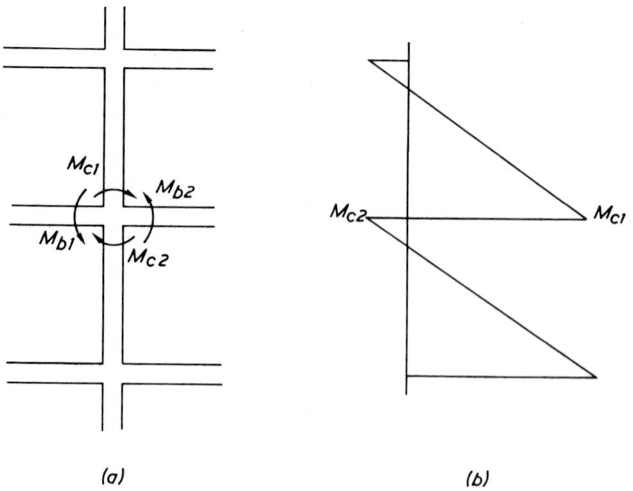

Fig. 5.18 Column moments at beam–column joint; (a) part of frame, (b) column bending moment diagram

that some column yielding must be considered to be inevitable. Note that yielding due to shift of the points of contraflexure will only occur at one end of the columns at a time and therefore will not lead to a column sidesway mechanism in that storey. Methods for determining column actions are given in Chapters 1 to 3.

The possibility of plastic hinges forming at the ends of columns due to the effects discussed above makes it important to ensure that the columns are capable of behaving in a ductile manner. The ACI Code recommends that if the maximum design axial load P_e is less than $0.4P_b$, where P_b is the dependable balanced load capacity of the column, the columns can be detailed as flexural members. This is because at low load levels reasonable ductility is available from column sections as shown in Figure 5.12. When the axial load level is increased the available curvature ductility decreases significantly and the ACI Code requires special transverse steel for confinement over the end regions of columns when $P_e > 0.4P_b$. Each end region is at least equal to the overall column depth, or 18 in (457 mm) or $\frac{1}{6}$th of the clear height of the column. Where a spiral is used the volumetric ratio ρ_s is recommended to be not less than required by Equation 5.17, but not less than $0.12f'_c/f_y$.

When rectangular loops are used the area of the hoop bar may be calculated from

$$A_{sh} = l_h \rho_s s_h / 2 \tag{5.33}$$

where l_h is the maximum unsupported length of the hoop side measured between perpendicular legs of the hoop or supplementary cross ties, ρ_s is as for a spiral with area of rectangular core to outside of hoops substituted for A_c, and s_h is the centre to centre spacing between hoops, not to exceed 4 in (102 mm). Supplementary cross ties, if used to reduce l_h should be of the same bar diameter as the hoop and should engage the hoop and a longitudinal bar. Equation 5.33 for rectangular hoops was devised to provide the same confinement for a rectangular core as would exist in the core of the equivalent spiral column, assuming that the efficiency of rectangular hoops as confining reinforcement is 50% of that of a spiral. Also, the amount of special transverse spiral steel specified is based on preserving the axial load strength of the column after the concrete cover has spalled rather than aiming at a particular curvature ductility factor under eccentric loading. However theoretical moment–curvature analyses of columns with rectangular hoops have shown that the ACI provisions are conservative except for heavily loaded columns having a small cross section and small longitudinal steel contents[5.2].

As for beams it is recommended that in the potential plastic hinge zones the spacing of transverse steel surrounding the longitudinal bars should not exceed six longitudinal bar diameters if buckling under reversed load conditions is to be restrained. Splices should preferably be made in the mid-height regions of columns, and the lap length should not be less than 30 d_b or 16 in (406 mm).

The design shear force can be found from the likely maximum moment gradient in the column when plastic hinges develop in the frame under the design seismic and gravity loads. The ACI provisions allow shear to be carried by the concrete and shear reinforcement without allowance for the possible deterioration of the concrete shear resisting mechanisms due to the opening and closing of cracks under reversed loading. This is a reasonable assumption only if plastic hinges in columns occur infrequently and if substantial axial compression acts on the column.

5.5.4 Special provisions for beam–column joints

Special transverse reinforcement as for the end regions of columns should be present in joint cores. Also, the joint core should have sufficient shear strength to at least equal the design shear force calculated from the column shear and the shears induced by the yield forces in the beam reinforcement. For example, for the interior beam-column joint shown in Figure 5.19(c) the maximum horizontal shear force is given by the horizontal forces above a typical horizontal plane as

$$V_j = C'_s + C'_c + T - V = T' + T - V \tag{5.34}$$

For joints with a beam entering only one side of the column in the direction considered (see Figure 5.19(a)) only one steel force term will appear in Equation 5.34. The design shear force should be calculated as above using the maximum likely steel strength. For example, for steel with f_y of 40 ksi (276 MPa) the steel strength could be taken as $1.25f_y$ and for steel with f_y of 60 ksi

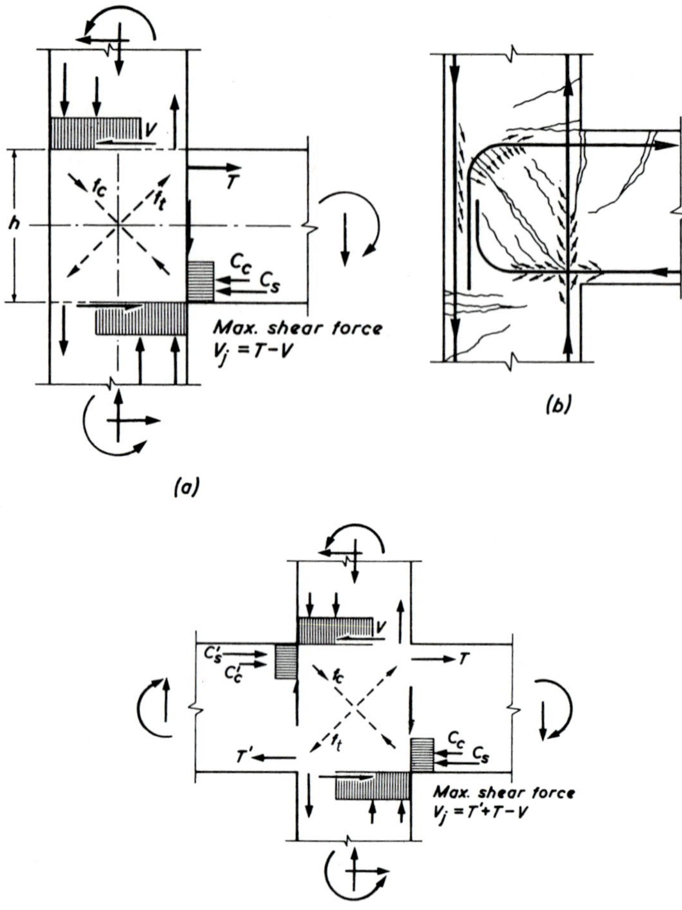

Fig. 5.19 Column–beam joints of frame with seismic loading. (a) Actions and stress resultants at exterior column–beam joint. (b) Crack pattern and bond forces at exterior column–beam joint. (c) Actions and stress resultants at interior column–beam joint

(414 MPa) the steel strength could be taken as $1.4f_y$. The design horizontal shear force is then allocated to the concrete and steel shear resisting mechanisms as follows

$$V_j = V_c + V_s \tag{5.35}$$

A strength reduction factor φ of 0.85 should be used when calculating V_c and V_s.

Recent experimental investigations[5.2] have shown that when reversible plastic hinges form in the beams adjacent to the connection, the joint core may break down due to alternating diagonal tension cracks which open and close, and high alternating bond stresses. A particularly severe condition arises after the first inelastic load cycle, when a full depth crack exists in the beam at the column face and all the beam forces are transferred to the joint core by the steel through bond forces. Hence it is considered that when the axial load level in the column is small, V_c should be ignored since shear transfer by a wide diagonal compression strut across the joint core will be less. Thus it is recommended that the shear carried by the concrete mechanism V_c be given as by Equation 5.25 and 5.27 except that V_c should be taken as zero when the minimum average compressive stress on the gross concrete area of column above the joint is less than $0.1f'_c$. This limitation is more stringent than the ACI Code provisions, as are the recommendations to follow.

Tests have also shown[5.2] that the critical diagonal tension crack runs between diagonally opposite corners of the joint core (i.e. parallel to the diagonal compression strut as in Fig. 5.19(b)), and not at 45° to the axes of the members. Therefore V_s is the shear carried by the horizontal ties in the joint core crossing this crack and is given by

$$V_s = nf_yA_v \qquad (5.36)$$

where A_v is the area of each set of ties and n is the number of sets of ties placed between the top and bottom reinforcement in the beam. It should also be realised that vertical shear forces need to be transferred across the joint core as well. However, column bars do not normally yield during the earthquake, and the column is normally carrying some compressive load, and hence the vertical shear forces are not so critical. The best method is to carry such shear on existing column bars which cross the corner to corner crack, and it is important therefore that at least one column bar should exist between corner bars on each face. It is suggested that column bar spacing in the relevant faces should not exceed 6 in (152 mm). The capacity of intermediate column bars crossing the joint should not be less than the tension developed in one outer layer of bars in the column section.

The horizontal joint core hoops placed should not be less than the special transverse steel placed in the column ends for confining reinforcement. It is considered that the presence of beams on four faces of the column should not allow a reduction in this requirement since extensive beam yielding in the two directions during an earthquake is likely to reduce the confining effect of the beams. An exception is when the designer purposely enforces the beam plastic hinges to form away from the column faces. Then the minimum amount of joint steel could be one half of the quantity of special transverse steel. For beam hinges forming away from the column face a less conservative approach to the design of joint core shear reinforcement could also be adopted, since the joint core remains elastic and no degradation of shear strength will occur.

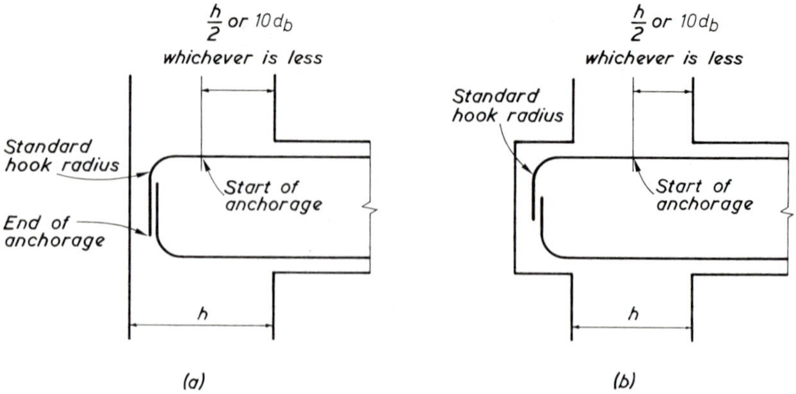

Fig. 5.20 Anchorage of beam bars at exterior column when plastic hinge can form in beam at column face. (a) Anchorage in columns, (b) anchorage in beam stub

In order to prevent bond failure due to yield penetration into columns during cyclic loading, where beams enter opposite sides of a column, the maximum diameter of the beam bars passing through the column should not exceed 0.04 of the column dimension in the direction of the bar, except that when no yielding can occur at the end of the beams in which case bar diameters up to 0.05 of the appropriate column dimension can be used. This recommendation is a result of observation of the performance of steel with f_y of 40 ksi (276 MPa). For higher grade steel smaller bar diameters than above should be adopted.

Where beam bars are to be anchored in a column it is considered that the anchorage length should be deemed to commence at either the midpoint of the column overall depth or 10 bar diameters from the face at which the steel enters the column, whichever is less. An exception is when no yielding can occur at the end of the beam in which case anchorage can be considered to commence at the column face. Notwithstanding the adequacy of the anchorage no bar shall be anchored within a column core without terminating with at least a standard hook near the far side of the core as is practically possible. It may be necessary to have a beam stub at the far face of the column to give adequate anchorage length. Such beam stubs have been shown to be helpful in confining the joint core and resulting in a general improvement in joint core behaviour. Figure 5.20 shows some anchorage details at exterior columns.

For two-way frames, when a general direction of seismic loading causes yielding in the beams in both directions simultaneously, the joint core will be subjected to biaxial shear forces. The resultant joint core shear force will act approximately along the diagonal of the section and the contribution of the horizontal shear reinforcement to the shear strength can be assessed by summing the components of the shear reinforcement bar forces intersected by the diagonal tension crack in the direction of the resultant shear force.

Fig. 5.21 Slab-column specimen with stirrup ties as shear reinforcement[5.12]

5.5.5 Special provisions for slab-column joints

Multistorey flat plate or flat slab buildings should not normally be used as seismic resistant structures without the presence of walls or frames to stiffen the building against excessive horizontal deflections caused by seismic forces which could result in serious non-structural damage during a severe earthquake. However even when such stiffening elements are present substantial unbalanced moments may need to be transferred at slab-column connections and such connections without shear reinforcement can fail by diagonal tension cracking and splitting of the concrete along the bars in the top of the slab with little ductility. A structural steel shearhead can be used to increase the strength and ductility of the connection. However tests have shown[5.12], that the use of stirrup ties in the slab around those slab bars that pass through the column results in an increase in strength and a substantial increase in ductility of the connection when subjected to cyclic unbalanced moments well into the inelastic range. Two-legged or four-legged stirrup ties, at a spacing of one-half the effective depth of the slab, were placed as shown in Figure 5.21 and were found to result in more ductile behaviour than structural steel shearheads[5.12].

5.6 REINFORCED CONCRETE SHEAR WALLS IN BUILDINGS

5.6.1 Introduction

The protection against damage of non-structural components, that reinforced concrete shear walls can offer in earthquake resistant buildings, is now well appreciated. The good performance of well designed shear walls has been demonstrated in several recent earthquakes. Some codes[5.10] require that shear wall structures be designed for greater strength against lateral

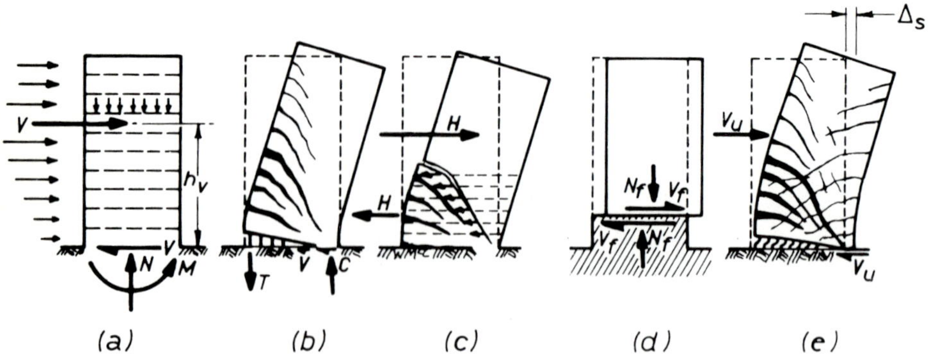

Fig. 5.22 Failure modes in a laterally loaded reinforced concrete cantilever shear wall

loads than ductile reinforced concrete frames with similar dynamic characteristics. This caution stems from the belief that shear walls are inherently less ductile and that perhaps the dominant mode of failure is shear. A shear wall should perhaps more properly be referred to as a 'structural wall'. In addition to lateral forces shear walls also have to carry some gravity load. In the following sections principles are discussed which may be used to ensure that shear walls will possess all the desirable characteristics of ductile earthquake resistant structures.

5.6.2 Cantilever shear walls without openings

In spite of their large scale most cantilever shear walls in multistorey buildings are slender enough to be treated as ordinary structural members. There is no reason to suggest that in their behaviour such walls disobey the fundamental principles of reinforced concrete, outlined in the previous sections. A cast-in-place cantilever shear wall (Figure 5.22(a)) will tend to act as an integral unit, irrespective of the configuration of the cross section. The subdivision of such shear walls into boundary elements and web panels, sometimes used in Japanese practice, is not recommended. The designer should be encouraged to think in terms of the actual overall behaviour of the whole wall.

The seismic response of a cantilever shear wall (Figure 5.22(a)) should be dominated by flexure. The considerations to be given to the potential plastic hinge at the wall base (Figure 5.22(b)) are the same as for ordinary structural members. Because of load reversals, shear wall sections necessarily contain substantial quantities of compression reinforcement. Gravity loads commonly produce axial compression that is considerably less than that required at balanced failure of the section. Consequently the available curvature ductility at the critical wall section is usually ample. However, the curvature ductility factor available is affected by the distribution of the reinforcement. For rectangular sections it is best to distribute the vertical reinforcement so that the reinforcement ratio in the web is the minimum allowable value of

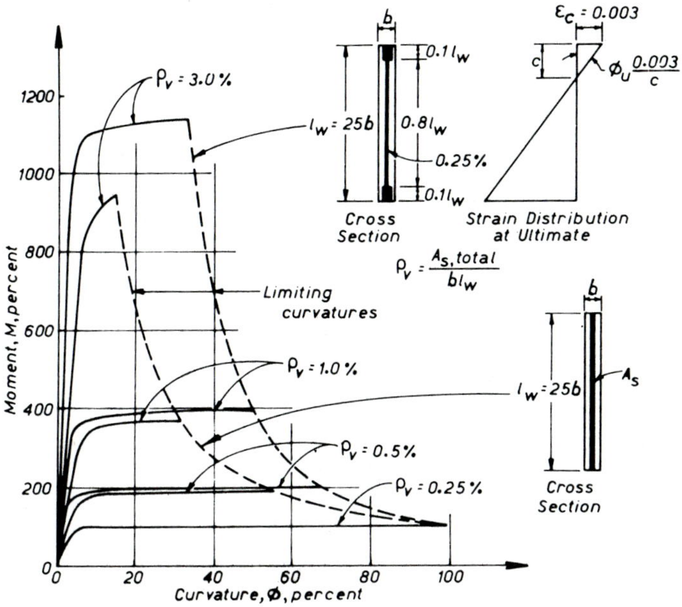

Fig. 5.23 The effect of the amount and distribution of vertical reinforcement on the ultimate curvature of wall sections[5.13]

0.25% and all additional reinforcement is concentrated near the edges of the wall section. As shown by the theoretical moment–curvature relationships in Figure 5.23, this distribution of steel results in an increase in the available strength and curvature ductility. Flanges will also increase the available curvature ductility of walls. Transverse ties should be placed at a spacing not exceeding six longitudinal bar diameters at the extremities of the wall section in the potential plastic hinge zone to prevent buckling of the longitudinal reinforcement. Some experimentally measured load–rotation relationships for cantilever shear walls are given elsewhere[5.2] which illustrate that well designed walls do possess considerable curvature ductility.

Shear wall sections are often thin and therefore under reversed cyclic yielding there is a danger of section instability. It is recommended that the thickness b of rectangular multistorey shear walls should be at least $l_u/10$, where l_u is the unsupported height of the walls between floors or other effective lateral supports, wherever the computed compression strain in the cross section at the ultimate load exceeds 0.0015, unless this fibre occurs closer than $2b$ or $0.2l_w$ horizontally from the compression edge. Figure 5.24(a), which also defines the symbols used, shows that the wall thickness beyond sections 1 and 2 would need to be checked in the case of strain patterns (a) and (b), respectively. If necessary the wall thickness can be increased as shown in Figure 5.24(c). Alternatively, a small flange with a minimum width of $l_u/5$ can be provided (see Figure 5.24(b)). In flanged

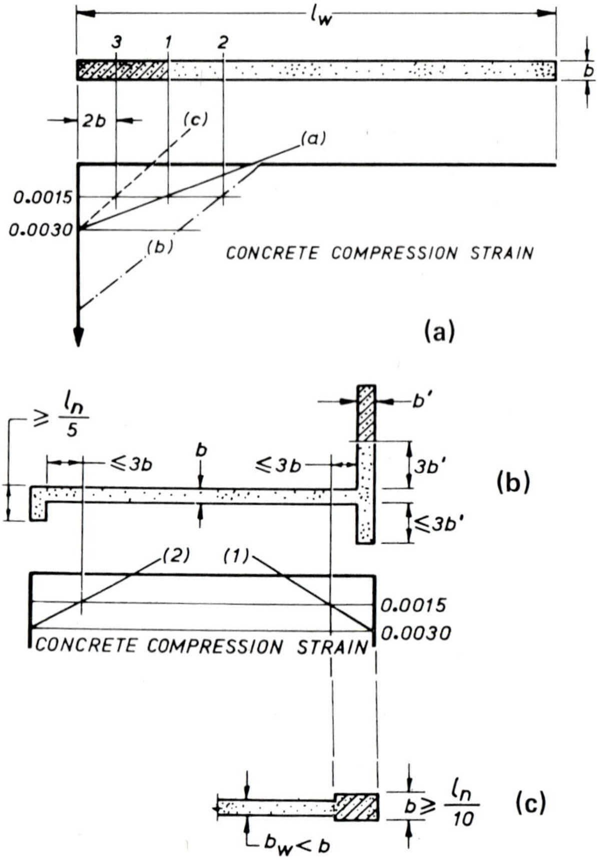

Fig. 5.24 Strain profiles and critical dimensions that affect the stability of thin wall sections; (a) rectangular wall section, (b) wall with flanges, (c) wall without flanges

sections, adjacent parts of the wall will provide lateral support to each other. Therefore it is recommended that the slenderness limitation of $l_u/10$ should apply to those parts which may be subjected to compression strains in excess of 0.0015 and which are located farther than $3b$ from a transverse wall. For example, the thickness of the shaded area of the flange shown in Figure 5.24(b) would need to be checked. The vertical extent of the potential plastic hinge, near the base of a cantilever shear wall, will indicate the region of wall where these slenderness limitations should be complied with.

It is now more generally recognised that the shear strength of multistorey shear walls should be assessed the same way as that of beams. In the potential plastic hinge zone wide flexural cracks will combine with diagonal tension cracks, due to shear (Figure 5.22(c)), and after a few load reversals the contribution of shear resisting mechanisms, other than the horizontal web reinforcement, will diminish. Therefore web reinforcement in the potential plastic hinge zone, the height of which should not be taken as less than $1.5l_w$, should be provided for the entire expected horizontal shear force.

For cantilever shear walls the code specified equivalent lateral static force generally does not give satisfactory protection against a shear failure during severe excitation when the maximum flexural strength at the base is being developed. During certain combinations of the modes of vibration the line of action of the resultant of the lateral inertia forces, located at h_v from the wall base in Figure 5.22(a),may be lower than that obtained from customary code[5.10,5.11] prescribed load patterns. That is, h_v may be significantly less than the two-thirds of the wall height given by the customary inverted triangular load. Consequently considerably larger shear forces may be generated when the moment capacity at the base is fully developed[5.14].

There are two potential locations in cantilever shear walls where failure by sliding shear could occur. One is a horizontal construction joint which is sensitive to the quality of surface preparation. The other is the plastic hinge zone, usually immediately above foundation level. The phenomenon of sliding along wide full depth flexural cracks (Figure 5.22(e)), especially when the axial compression from gravity loads on the wall is small, is similar to that encountered in the plastic hinge zone of beams, discussed in Section 5.4.5. The sliding along a construction joint, shown in Figure 5.22(d), often observed in shear walls damaged by earthquakes, can be suppressed if, in accordance with the concept of shear friction[5.2], adequate vertical reinforcement is provided over the length of the wall to supply, together with the minimum available gravity load, the necessary clamping force N_f. The dependable shear strength of a horizontal construction joint can be assured if the vertical reinforcement across the joint is provided in accordance with[5.2]

$$A_{vf}=\left(v_u-0.85\frac{N}{A_g}\right)\frac{A_g}{f_y}\geqslant 0.0025A_g \tag{5.37}$$

where v_u is the nominal computed horizontal shear stress, N is the minimum axial compressive load on the wall and A_g is the gross area of the section. This reinforcement should be placed in the web area of the section in between the main flexural reinforcement that is normally situated at the extreme edges or flanges of the wall section. The inelastic response of mechanisms associated with sliding shear indicate drastic loss of stiffness and strength with reversed cyclic loading. Therefore sliding shear should be considered as being an unsuitable energy dissipating mechanism in earthquake resistant structures.

In detailing of shear wall sections, attention must be given to transverse reinforcement in areas where the calculated concrete compression strain could exceed 0.0015, to ensure that the longitudinal wall reinforcement does not buckle. The diameter of stirrup ties used for this purpose should not be smaller than one quarter the diameter of the main longitudinal bars they restrain, and they should be spaced no further than 6 times the main bar

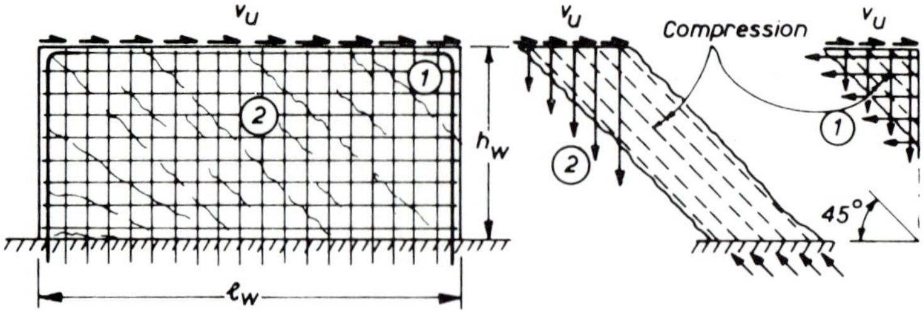

Fig. 5.25 The shear resistance in low-rise shear walls

diameter. When the neutral axis depth in the section of a wall that is expected to sustain a displacement of ductility of 4 exceeds $0.15l_w$, the required curvature ductility will be reached when the concrete compression strain is in excess of 0.004. Therefore in such cases special confining reinforcement, as is required for columns (see Section 5.5.3), should be placed in the area of the wall where the computed compression strain at the development of full flexural strength exceeds 0.0015.

5.6.3 Squat shear walls

In low-rise buildings the height of a shear wall may be less than its length. Such walls cannot be designed with the customary techniques of reinforced concrete theory. However, because the earthquake load for squat walls is seldom critical, an approximate design that ensures at least limited ductility will often suffice. The strength of many low rise shear walls will be limited by the capacity of the foundations to resist the overturning moments. In such cases a rocking structure results and thus ductility becomes irrelevant.

As Figure 5.25 indicates, after diagonal cracking the horizontal shear introduced at the top of a low-rise cantilever will need to be resolved into diagonal compression and vertical tensile forces. Thus, distributed vertical flexural reinforcement will also enable the shear to be transmitted to the foundations. The equilibrium condition of the free body marked 2 shows this in Figure 5.25. Where the diagonal compression field does not find a support at foundation level, as is the case with the triangular free body marked 1, an equal amount of horizontal shear reinforcement will be required. Figure 5.25 thus shows that for a squat shear wall a steel mesh with equal area in both directions will be required if a compression field acting at 45° is conservatively assumed. The flexural strength at the base must be carefully evaluated, taking the contribution of all vertical bars into account, to ensure that the required shear strength can be provided. This way most squat shear walls can be made ductile and a brittle failure will be avoided.

5.6.4 Coupled shear walls

In a rectangular cantilever wall the maximum shear stress is induced along the fibre at the neutral axis up the height of the wall. If this or a nearby fibre is

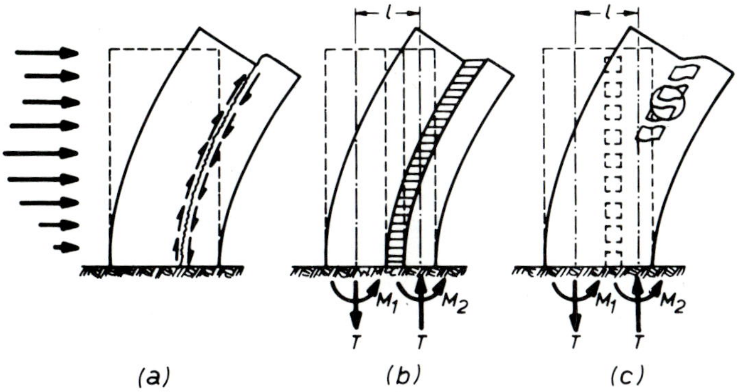

Fig. 5.26 Laterally loaded coupled shear walls

potentially weak, as may be the case in precast panel construction, a sliding shear failure, such as shown in Figure 5.26(a), may occur. However, if this failure mechanism can be made ductile, and also made to give stable hysteresis loops under reversed cyclic loading, then it can be used as a viable energy dissipating mechanism in an earthquake resistant structure. An equivalent of such a system, shown in Figure 5.26(b), illustrates that the total overturning moment M_0 is resisted in the following manner

$$M_0 = M_1 + M_2 + Tl \tag{5.38}$$

where T is the accumulation of shear forces in the coupling system between the two walls. The potential of this structure, as an efficient earthquake resistant construction, would stem from its ability to dissipate energy, when required, within the shear transfer system over the full height of the structure. This would be in addition to the energy dissipated in the plastic hinge at the base of each wall. With a skillful selection of relative stiffness and strength properties it is possible to reinforce the various components of the walls so that under monotonic static loading the strength of the coupling system is developed before the onset of yielding at the base of the walls. Therefore in comparison with cantilever walls, coupled shear walls offer more than one line of defence when energy dissipation is required. The wider dispersal of energy dissipating devices is likely to result in better structural damage control. Studies have indicated that in walls so designed extensive yielding during a typical severe seismic excitation is largely confined to the coupling system, which is more easily repaired than the wall bases.

In practice the shear transfer from one wall to another is afforded by discrete beams, formed between openings and arranged in one or more vertical rows, as shown in Figure 5.26(c). For an efficient coupled shear wall structure the beam stiffnesses should be chosen so that

$$Tl > \frac{2}{3} M_0 \tag{5.39}$$

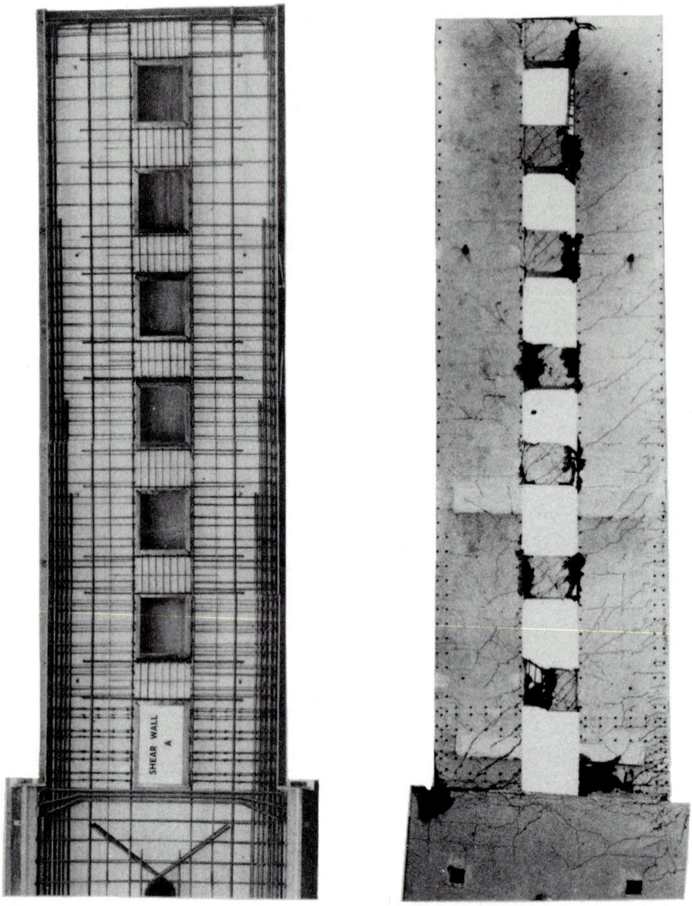

Fig. 5.27 A seven-storey $\frac{1}{4}$-scale reinforced concrete shear wall with conventionally reinforced coupling beams that failed by sliding shear[5.15]

This usually involves the use of short and relatively deep coupling beams in which the development of the flexural strength is associated with shear forces that may be large enough to destroy the beams in diagonal tension. This type of failure was observed in Anchorage during the 1964 Alaskan earthquake. If the entire shear is resisted by vertical stirrups, a sliding shear failure may occur along a vertical plane (instead of a diagonal tension failure) after a few cyclic excursions into the inelastic range of response. Figure 5.27 shows such beams in a model of a seven-storey coupled shear wall[5.15]. The best way to overcome a sliding shear failure in coupling beams is to use diagonal reinforcement instead of the conventional horizontal flexural and vertical shear reinforcement. A typical example of such a suggested arrangement is

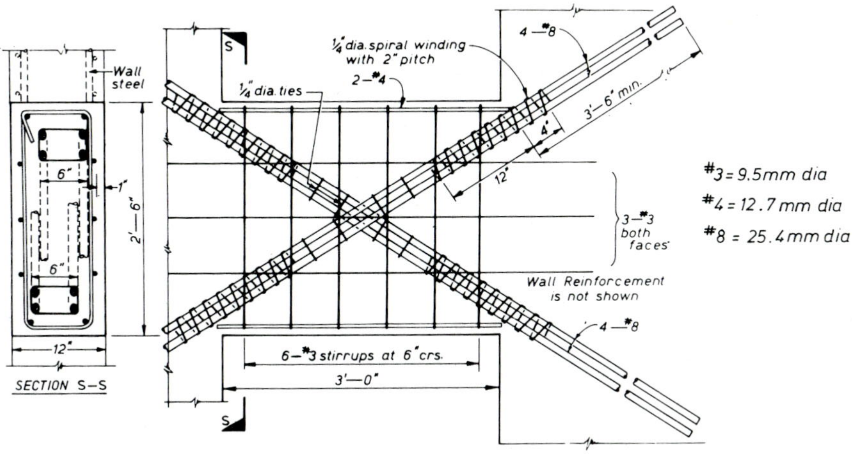

Fig. 5.28 Suggested steel arrangement in a diagonally reinforced coupling beam[5.2]

shown in Figure 5.28. Such beams, widely used now in New Zealand, have been found to be extremely ductile[5.15] provided that the buckling of the diagonal compression bars is prevented. After one excursion beyond yield in each of the two directions of loading the flexural and shear forces are transferred by the diagonal steel, while the concrete provides flexural stiffness and hence ensures lateral stability. As may be expected the hysteresis loops for such a member bear the characteristics of a steel member. Tests have shown[5.15] that a coupled shear wall with diagonally reinforced coupling beams, proportioned so that yielding of the beams commences before that of the wall bases, has an energy dissipation capacity well in excess of a wall containing conventionally reinforced coupling beams. Moreover, such coupled shear walls show less stiffness degradation and less visible damage for the same lateral displacement. Therefore coupled shear walls can be considered to possess all the desirable features of an efficient earthquake resistant structure.

In cores, containing stairs and elevator shafts, of multistorey buildings, walls of different configurations may be coupled. A typical example is a box section pierced by two rows of openings, resulting in the coupling of two channel shaped walls. The assessment of the behaviour of such walls may be carried out with the aid of interaction diagrams, such as shown in Figure 5.29. The positive moment refers to a load which produces compression in the tips of the wall flanges. Such a diagram showing the position of the neutral axis (c/l_w) is also useful when the regions of the wall section are to be determined where confining reinforcement might be required. It will be noted from Figure 5.29 that for positive moment with only a small axial compression the c/l_w ratio is relatively high and therefore the available

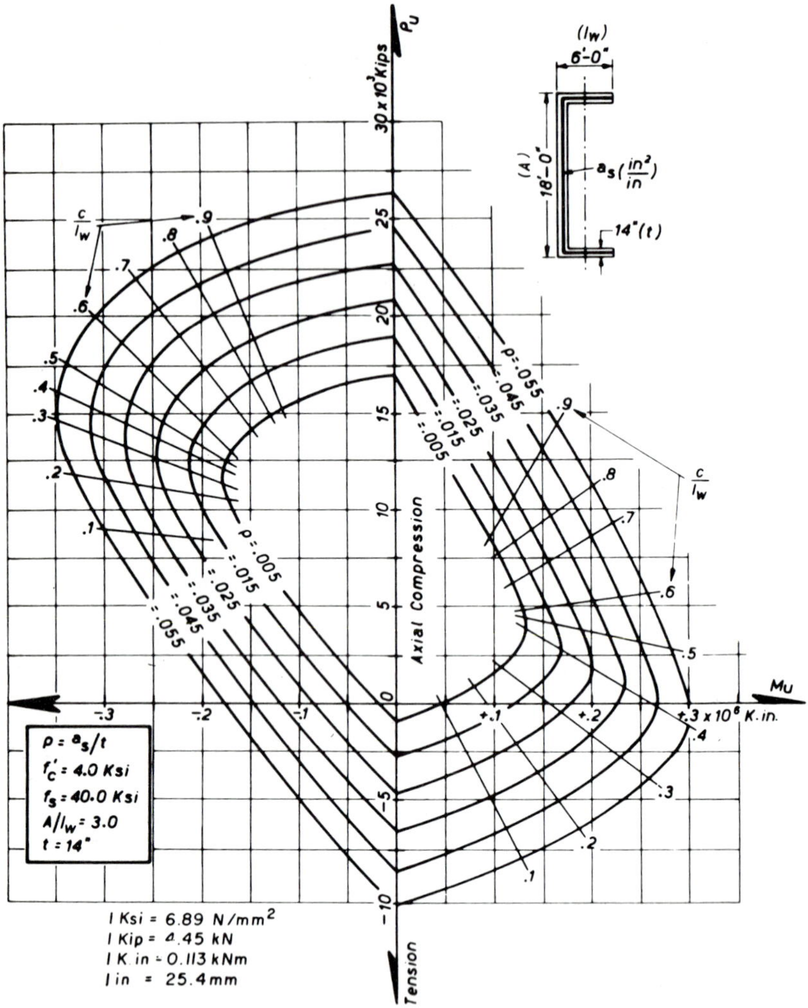

Fig. 5.29 Typical axial load-moment interaction relationships for a channel-shaped shear wall section with uniformly distributed vertical reinforcement[5.2]

curvature ductility of the wall section is small. However, for this moment to exist the coupling system will transmit large axial tension and this will increase the curvature ductility of the critical section. Conversely it may be seen that for negative moments the wall can accommodate a considerable axial compression force while possessing ample curvature ductility.

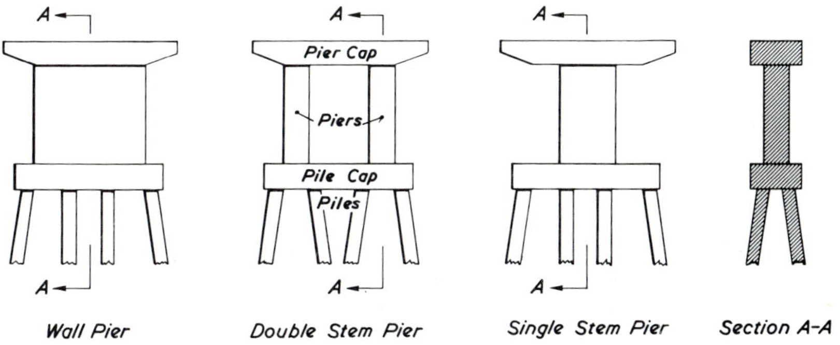

Fig. 5.30 *Some bridge pier arrangements*

5.7 REINFORCED CONCRETE BRIDGE PIERS

Many of the principles of seismic design discussed for buildings apply to other types of reinforced concrete structures such as bridge piers. If a code static design load is used the designer should ensure that an energy dissipating mechanism is available in the structure. For bridges the greater part of the mass exists in the superstructure consisting of girders, slabs, diaphragms, etc., and therefore this is where the main inertia forces are generated during earthquake ground shaking. However the superstructure is not significantly stressed by seismic loading and the inertia forces only impose significant stresses on the bridge piers and foundations. Special energy dissipating devices, such as piston dampers utilizing oil or lead, can be encorporated between the piers and the superstructure (see Chapter 3) thus minimizing structural damage to the piers and foundations. However most bridge structures will not have such special devices and will rely on energy dissipation below the tops of the piers. Yielding of the foundation structure should always be avoided since damage there is difficult to inspect and repair. Therefore, the bridge piers are the most logical elements to detail as energy dissipating regions. Generally the bridge superstructure transmits horizontal shears and axial or eccentric vertical loads to the piers. The vertical loading on most piers is relatively small and hence ductility can be provided relatively easily there. Figure 5.30 shows some typical bridge piers. Design procedures already discussed for potential plastic hinge zones in columns and walls of buildings can be followed for detailing the piers. The most important consideration is in providing sufficient transverse reinforcement in the form of spirals or hoops to confine the concrete, prevent buckling of longitudinal bars, and act as shear reinforcement. For wall piers or single-stem piers the plastic hinge regions will normally be only at the pier base. For multi-stem piers plastic hinging could develop top and bottom in the piers and both ends need to be detailed for ductility. For important or complex bridge structures, special dynamic studies may be required to determine the level of displacement ductility factor to be used in the design of bridge piers (see Chapter 3).

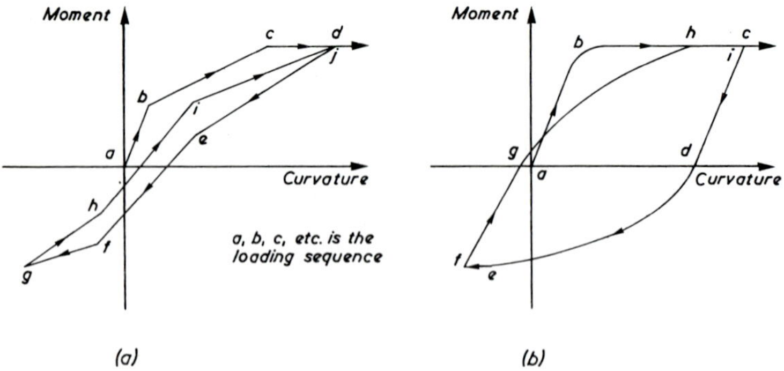

Fig. 5.31 Moment–curvature idealizations for prestressed concrete and reinforced concrete; (a) prestressed concrete section, (b) reinforced concrete section

The approximate Equation 5.23, or an adjusted form of it to take different bending moment distributions into account, could be used to relate the curvature ductility demand of the critical sections to the required displacement ductility factor.

5.8 PRESTRESSED CONCRETE BUILDING FRAMES

5.8.1 Introduction

Prestressed concrete has been widely used for structures carrying gravity loads but has not had the same acceptance for structural systems which resist seismic loads. This has been mainly due to a paucity of research and design information and experience of the behaviour of prestressed structures during severe earthquakes. The ACI Code[5.1] contains no special provisions for the seismic design of prestressed concrete structures. However design recommendations have been prepared recently by the New Zealand Prestressed Concrete Institute[5.16] and the Commission on Seismic Structures of the FIP[5.17]. These recommendations and other recent studies give clear evidence that properly detailed prestressed concrete frames will give satisfactory seismic load resistance.

Idealized moment–curvature hysteresis loops for prestressed concrete and reinforced concrete sections are illustrated in Figure 5.31. For prestressed concrete, the initial elastic tensile strain in the tendons due to prestress causes a large deflection recovery even after large deflections. Hence the hysteretic energy dissipation of a prestressed concrete system will be lower than that of a reinforced concrete system of similar strength and initial stiffness, generally resulting in a greater deflection response of the prestressed concrete system to a severe earthquake. This difference in deflection response for code designed single degree of freedom structures of similar strength, initial stiffness and viscous damping, responding nonlinearly to the 1940 El Centro and synthetic earthquakes, has been found on average to be approximately 30% [5.18]. This feature has led the New Zealand loadings code[5.11] to specify a design horizontal seismic load for prestressed concrete structures which is 20% higher than that for equivalent reinforced concrete structures.

An appealing application for prestressing is in precast concrete construction. Precast concrete beams and columns can be post-tensioned together on the site to form continuous seismic resistant frames. Insitu prestressed concrete construction can also be used with or without precast concrete members. The ductility and energy dissipation of the critical sections can be improved by the presence of nonprestressed longitudinal steel. An excellent design concept is to balance gravity loads (dead load plus some proportion of live load) with draped prestressed tendons and then place sufficient nonprestressed steel in the section to provide the additional moment capacity required when seismic loading acts on the frame.

The use of unbonded tendons in seismic design has caused considerable controversy. It is considered that tendons in ductile frames should be grouted because for unbonded tendons the steel strains in the event of a severe earthquake remain elastic and thus total reliance is placed on the concrete for energy dissipation. Also, fluctuation of tendon forces could cause failure of the anchorages of unbonded tendons with the catastrophic result of release of prestressing forces. A better case could be made for unbonded tendons if nonprestressed bonded steel were present in the sections as well. However there would appear to be no real objections to the use of unbonded tendons in floor or roof systems not contributing to the strength of the frame.

The general philosophy for the seismic design of prestressed concrete structures is similar to that for reinforced concrete. That is, structures designed to withstand the static design loads of codes should be capable of inelastic deformations by the formation of plastic hinges having adequate ductility.

5.8.2 Strength and ductility of flexural members

The flexural strength of rectangular sections, or flanged sections in which the neutral axis lies within the flange, where the tendons are concentrated near the extreme tension fibre, may be calculated from

$$M_u = A_{ps} f_{ps}\left(d - \frac{a}{2}\right) \tag{5.40}$$

where

$$a = A_{ps} f_{ps}/(0.85 f'_c b) \tag{5.41}$$

where A_{ps} is the area of prestressing steel, f_{ps} is the stress in the prestressing steel at the flexural strength, d is the depth from the centroid of the tendon to the extreme compression fibre, and b is the width of the compression face of the member. The same assumptions for the concrete compressive stress block are made as for reinforced concrete. For bonded members the ACI Code[5.1]

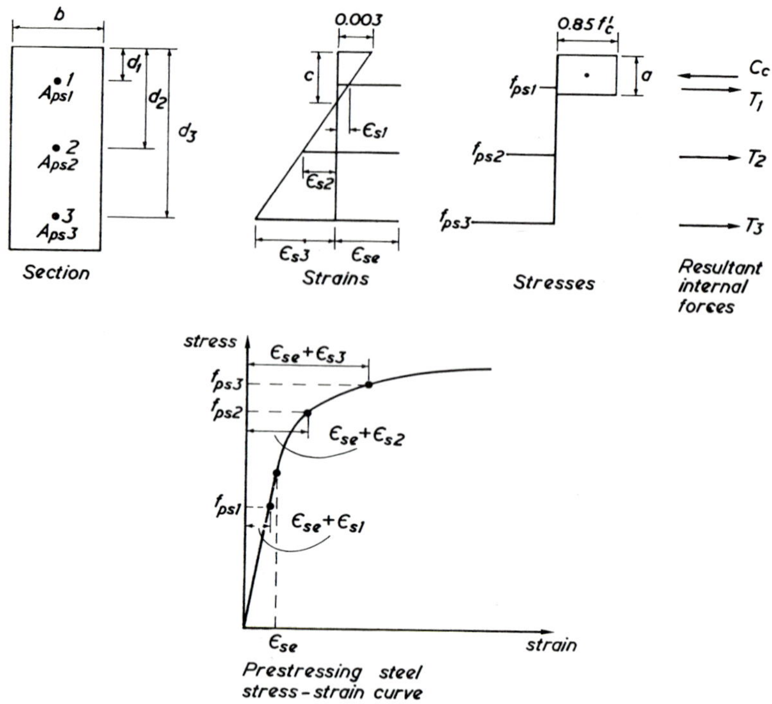

Fig. 5.32 Prestressed concrete section when flexural strength is reached

recommends that provided f_{se} is not less than $0.5f_{pu}$ the following approximate value for f_{ps} may be used:

$$f_{ps}=f_{pu}\left(1-0.5\rho_p\frac{f_{pu}}{f'_c}\right) \tag{5.42}$$

where f_{pu} is the ultimate strength of the prestressing steel, f_{se} is the stress in the prestressing steel due to prestress alone, and $\rho_p=A_{ps}/bd$. This empirical expression reflects the effect of the variables on the neutral axis depth and hence the level of steel strain reached.

A more general approach using strain compatibility and equilibrium is necessary to accurately assess the flexural strength when bonded prestressing tendons exist at various depths throughout the section. The stress–strain curve for the tendons is required for such an analysis. For example, Figure 5.32 shows a section with three tendons, each with an initial strain ε_{se} due to prestress alone. Strictly, ε_{se} is the strain in the prestressing steel when the concrete strain there is zero. However, taking ε_{se} as the steel strain due to effective prestress is sufficiently accurate. The additional strains in the prestressing steel due to flexure caused by external loading are

$$\varepsilon_{si}=0.003(d_i-c)/c \tag{5.43}$$

where $i=1, 2, 3$. Also, for equilibrium

$$0.85f'_c ab = A_{ps1}f_{ps1} + A_{ps2}f_{ps2} + A_{ps3}f_{ps3} \tag{5.44}$$

The flexural strength may be calculated by an iterative procedure as follows. A value for the neutral axis depth c is chosen, the strains ε_{s1}, ε_{s2} and ε_{s3} calculated from Equation 5.43 and the steel stresses f_{ps1}, f_{ps2} and f_{ps3} are determined from the total steel strains $\varepsilon_{se}+\varepsilon_{s1}$, $\varepsilon_{se}+\varepsilon_{s2}$ and $\varepsilon_{se}+\varepsilon_{s3}$ using the steel stress-strain curve (see the stress-strain curve in Figure 3.32). Finally, the equilibrium Equation 5.44 is checked. These steps are repeated until a value for the neutral axis depth c is found which allows Equation 5.44 to be satisfied. Then $a=\beta_1 c$ is calculated and the ultimate moment is given by

$$M_u = A_{ps1}f_{ps1}(d_1 - 0.5a) + A_{ps2}f_{ps2}(d_2 - 0.5a) + A_{ps3}f_{ps3}(d_3 - 0.5a) \tag{5.45}$$

The contribution of any nonprestressed steel which may be present in the section can be taken into account in the same manner as for the prestressed steel, except that $\varepsilon_{se}=0$ may be assumed and the appropriate steel stress–strain curve used.

Figure 5.33 shows theoretical moment–curvature relationships derived for a prestressed concrete section with tendons concentrated near the extreme tension fibre. The concrete stress–strain curves assumed in the analysis for core concrete and cover concrete outside the transverse steel are shown in the figure, the core concrete being confined by typical plastic hinge zone quantities of closed stirrups [for example, No. 3 (9.5 mm diameter) stirrups at $3\frac{1}{2}$ in (89 mm) centres confining a 15 in (381 mm) deep by 6 in (152 mm) wide concrete core]. The moment–curvature curves give a good indication of the reduction of available curvature ductility with increasing content of prestressing steel. For seismic design it is recommended that the following requirement should be observed

$$\rho_p f_{ps}/f'_c \leqslant 0.2 \tag{5.46}$$

For the section shown in Figure 5.33, Equation 5.46 requires $A_{ps}/bh \leqslant 0.0046$ and it is evident from the figure that this requirement results in adequate curvature ductility.

Figure 5.34 shows that with prestressing steel present in the compression zone the theoretical curvature ductility of the section is not so affected by increase in prestressing steel content. This is a consequence of the prestressing steel acting as compression steel at large curvatures, providing that steel is restrained against buckling by the surrounding concrete and transverse steel. Successive cycles of reversed flexure may cause concrete damage leading to buckling. It would appear to be reasonable to require that all beam sections are capable of reaching a specified curvature at a given extreme fibre

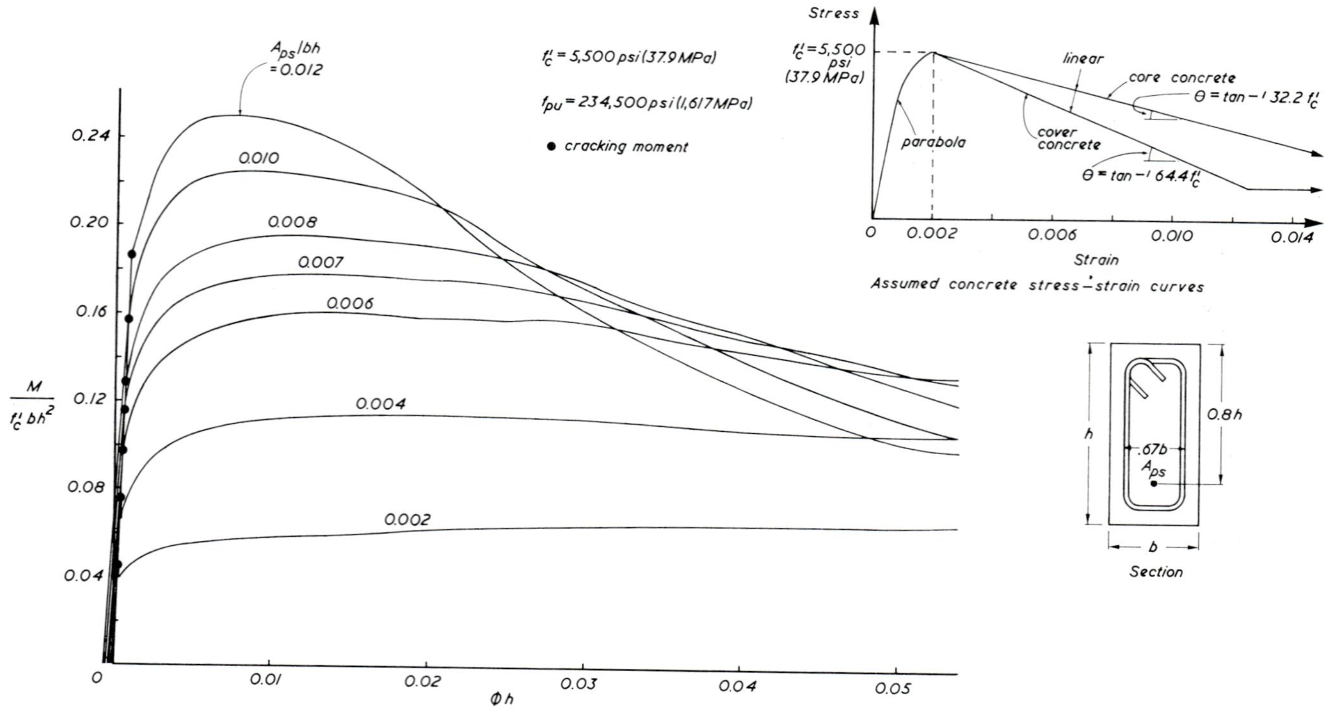

Fig. 5.33 Moment–curvature relationships for sections with various contents of eccentrically placed prestressing steel[5.18]

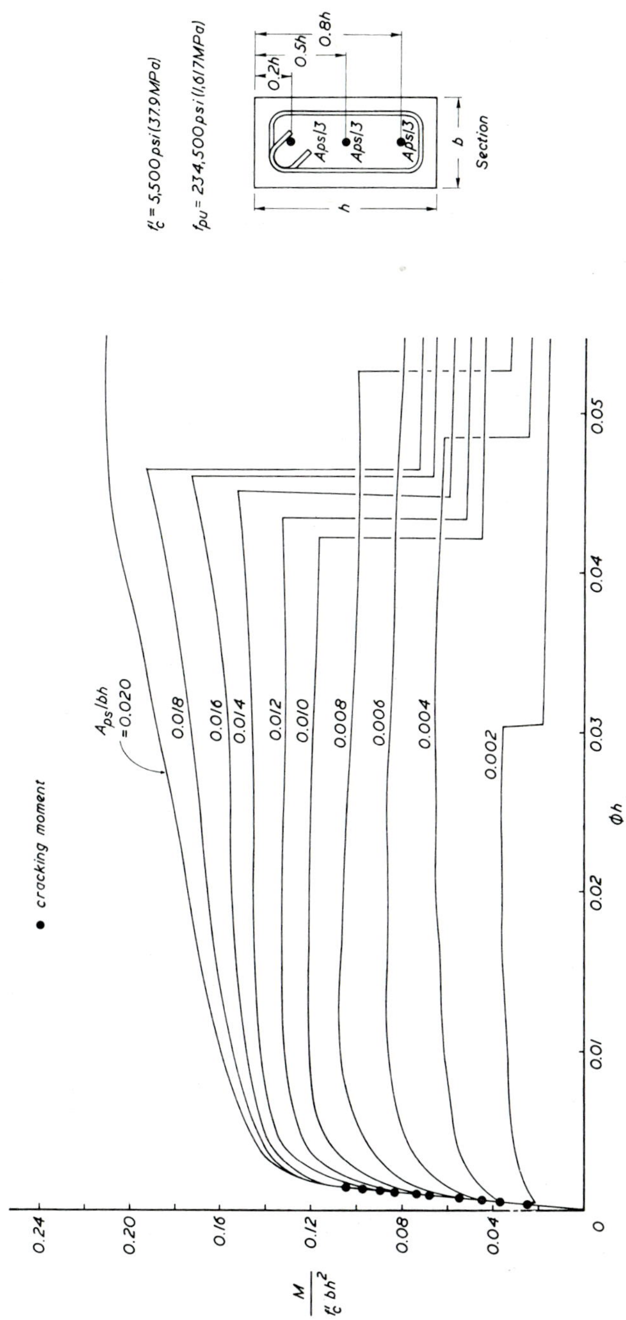

Fig. 5.34 Moment-curvature relationship for section with three symmetrically placed prestressing tendons[5.18]

concrete strain. This means that at the ultimate moment the neutral axis depth should not exceed a limiting value. Equation 5.46 for tendons concentrated near the extreme tension fibre requires $A_{ps}f_{ps} \leqslant 0.2f'_c bd$, which means that at the flexural strength the maximum concrete force is $0.2f'_c bd$ and hence that the maximum depth of the concrete rectangular stress block is $a = 0.2/0.85 = 0.235d$. But since d is approximately 0.8 of the overall depth h, and a is approximately 0.75 of the neutral axis depth c, the requirement becomes $c \leqslant 0.25h$. For sections with tendons at various levels down the depth it is difficult to set a limiting $\rho_p f_{ps}/f'_c$ value because tendons at different levels lead to different moment–curvature relationships. However for any tendon arrangement the curvature at ultimate will at least equal that with all tendons placed near the extreme tension fibre if $c \leqslant 0.25h$. This requirement has been adopted by the FIP recommendations[5.17].

In seismic design moment reversals in beams near column faces will require the sections to have both negative and positive moment strength and hence tendons will exist near both extreme fibres of the section and near mid-depth. Figure 5.35 shows theoretical moment–curvature curves for a section with up to five tendons symmetrically distributed down the depth. The total prestressing steel content is the same for each of the five cases, being 0.00969 of the gross concrete section. For the case of all steel concentrated in a single central tendon, $N=1$, the moment capacity reduces significantly at high curvatures. However there is little difference in the moment capacity for two or more tendons and such sections are able to maintain near maximum moment capacity at high curvatures. Therefore two or more tendons are to be preferred.

5.8.3 Special provisions for seismic design

Flexural members The sections should be designed for flexure using a strength reduction factor φ of 0.9. In order to ensure adequate ductility the content of flexural steel (prestressed plus nonprestressed) should be such that the neutral axis depth at potential hinge regions is not greater than 0.25 of the overall depth of the section. At plastic hinge positions, where moment reversal occurs, tendons should be present in the top and bottom of the section, and not exist at the mid-depth region alone unless a significant quantity of nonprestressed steel is present top and bottom. This is because prestressed or nonprestressed steel in the compression zone aids the ductility of the section at large deformations. In order to avoid failure the dependable flexural strength should be greater than the maximum likely cracking moment. In the absence of special studies it may be assumed that the tensile stress in the concrete prior to cracking is $12\sqrt{f'_c}$ psi where f'_c is in psi ($\sqrt{f'_c}$ MPa where f'_c is in MPa) to take the possibility of a high modulus of rupture into account, and that the prestressing force is 10% more than calculated. The maximum design shear force should be calculated using the design gravity loads, and the ultimate moment capacities of the beam should be taken as 1.15 times the ideal flexural strength to include the effect of possible

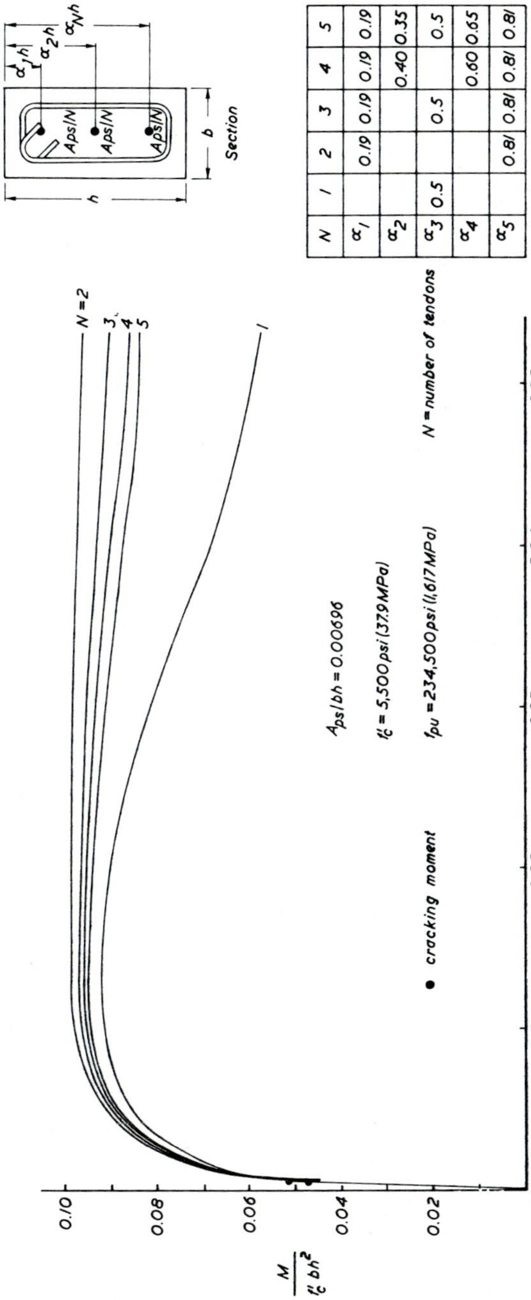

N	1	2	3	4	5
α_1		0.19	0.19	0.19	0.19
α_2				0.40	0.35
α_3	0.5		0.5		0.5
α_4				0.60	0.65
α_5		0.81	0.81	0.81	0.81

Fig. 5.35 *Moment-curvature relationships for sections with various numbers of symmetrically placed prestressing tendons*[5.18]

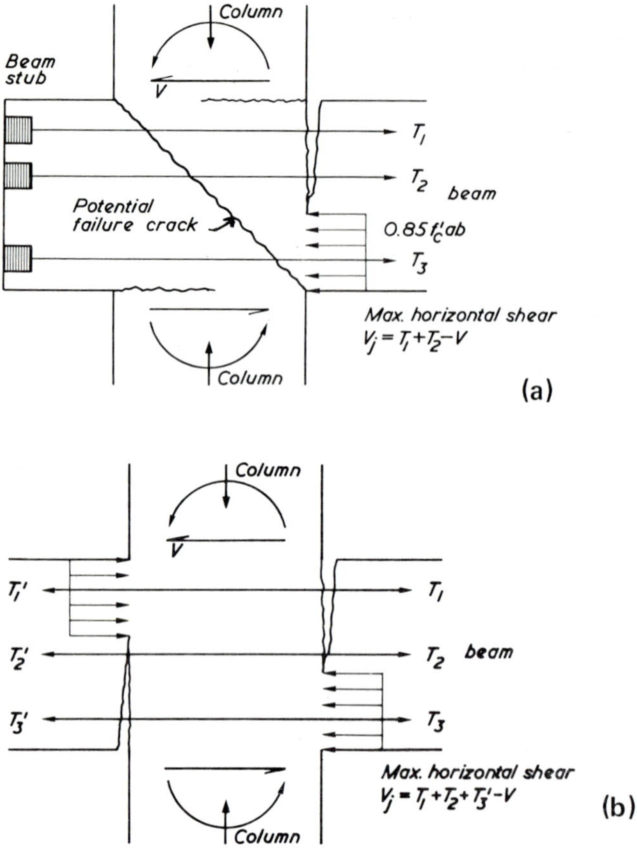

Fig. 5.36 Prestressed concrete beam-column joints; (a) exterior column, (b) interior column

over-strength of the materials. In plastic hinge zones the contribution of the concrete to the shear capacity should be ignored and all the shear carried by truss action of closed stirrups at spacing not exceeding 4in (102 mm) or $0.25d$, where d is the effective depth of the member which need not be taken as less than 0.8 of the overall depth. Elsewhere the stirrup spacing should not exceed $0.5d$. Longitudinal steel should be present at each corner of the stirrups. The spacing of closed stirrups around longitudinal steel in compression zones of plastic hinge regions should not exceed 6 longitudinal bar diameters.

Columns The dependable strength of column sections should be greater than the maximum likely cracking moment including the effect of axial load. Assumptions may be made concerning the possible increase in modulus of rupture of the concrete and the level of prestress similar to those made for beams. Special transverse steel should be provided at column ends as for reinforced concrete columns (using either Equation 5.17 but not less than $0.12f'_c/f_y$, or Equation 5.33) when the neutral axis depth exceeds 0.25 of the overall depth of the section or the maximum design axial load exceeds $0.1P_0$, where P_0 is the axial load strength of the column without flexure. Otherwise the member should be treated as a flexural member. The determination of column design moments and shears should be as for reinforced concrete frames.

Beam–column connections In order to avoid tensile bursting stresses in a region already subjected to severe diagonal tension from beam and column forces, anchorages for post-tensioned tendons should not be placed within beam–column joint cores. At exterior joints anchorages can be placed in stubs outside the joint core region (see Figure 5.36). Connections between faces of precast members at beam–column joints are acceptable, providing that the jointing material is strong enough and that the interfaces are roughened to ensure good shear transfer and retention of the jointing material after cracking. Where possible it is preferable to locate the jointing faces away from plastic hinge zones. Transverse reinforcement for beam–column joint cores should be designed as for reinforced concrete frames. It has been shown[5.18] that the presence of a tendon near the mid-depth of members helps carry the joint core shear. Thus in calculating the shear force to be carried by the transverse hoops in the joint core using Equations 5.35 and 5.36, V_c may be increased by $0.7P_s$ where P_s is the prestressing force in the tendon or tendons in the middle third of the beam depth. The calculation of the applied horizontal shear force is illustrated in Figure 5.36. Ideal beam forces should be increased by 15% to take possible overstrength into account when calculating the design horizontal shear force.

REFERENCES

5.1 ACI Committee 318, *Building Code Requirements for Reinforced Concrete* (*ACI* 318-77), American Concrete Institute, Detroit (1977)

5.2 Park, R. and Paulay, T., *Reinforced Concrete Structures*, John Wiley, New York (1975)

5.3 Bertero, V. V. and Felippa, C., Discussion of 'Ductility of Concrete', by H. E. H. Roy and M. A. Sozen, Proc. International Symposium on Flexural Mechanics of Reinforced Concrete, ASCE–American Concrete Institute, Miami, pp. 227–234 (1964)

5.4 Iyengar, K. T. R. J., Desayi, P. and Reddy, K. N., 'Stress–Strain Characteristics of Concrete Confined in Steel Binders', *Magazine of Concrete Research*, Vol. 22, No. 72, pp. 173–184 (1970

5.5 ACI Committee 340, *Design Handbook in Accordance with Strength Design Method of ACI* 318-71, Vol. 1, SP-17(73), American Concrete Institute, Detroit (1973)

5.6 Mattock, A. H., Discussion of 'Rotational Capacity of Reinforced Concrete Beams' by W. G. Corley, *J. struct. Div. Am. Soc. civ. Engrs*, Vol. 93, ST2, pp. 519–522 (1967)

5.7 ACI Committee 340, *Design Handbook in Accordance with the Strength Design Method of ACI* 318-77, Vol. 2, Columns, SP-17A(78), American Concrete Institute, Detroit (1978)

5.8 Pfrang, E. O., Siess, C. P. and Sozen, M. A., 'Load–Moment–Curvature Characteristics of Reinforced Concrete Cross Sections', *J. Am. Concr. Inst.*, Vol. 61, No. 7, pp. 763–778 (1964)

5.9 Mahin, S. A. and Bertero, V. V. 'Rate of Loading Effects on Uncracked and Repaired Reinforced Concrete Members', Report No. EERC 72-9, Univ. California, Berkeley, December (1972)

5.10 SEAOC, 'Recommended Lateral Force Requirements and Commentary', Seismology Committee, Structural Engineers' Association of California, San Francisco (1975)

5.11 NZS 'Code of Practice for General Structural Design and Design Loadings for Buildings (NZS 4203: 1976)', Standards Association of New Zealand, Wellington (1976)

5.12 Islam, S. and Park, R. 'Tests on Slab–Column Connections With Shear and Unbalanced Flexure', *J. struct. Div. Am. Soc. civ. Engrs*, Vol. 102, No. ST3, pp. 549–568 (1976)

5.13 Cardenas, A. E. and Magura, D. D., 'Strength of High-Rise Shear Walls–Rectangular Cross Sections', ACI Special Publication 36, Response of Multistorey Concrete Structures to Lateral Forces, pp. 119–150 (1973)

5.14 Blakeley, R. W. G., Cooney, R. C. and Megget, L. M., 'Seismic Shear Loading at Flexural Capacity in Cantilever Wall Structures', *Bull. N. Z. Natn. Soc. Earthq. Engng.*, Vol. 8, No. 4, pp. 278–290 (1975)

5.15 Paulay, T. and Santhakumar, A. R., 'Ductile Behaviour of Coupled Shear Walls', *J. struct. Div. Am. Soc. civ. Engrs*, Vol. 102, No. ST1, pp. 93–108 (1976)

5.16 NZPCI Seismic Committee, 'Recommendations for the Design and Detailing of Ductile Prestressed Concrete Frames for Seismic Loading', *Bull. N. Z. natn. Soc. Earthq. Engng*, Vol. 9, No. 2, pp. 89–96 (1976)

5.17 FIP Commission on Seismic Structures, 'Recommendations for the Design of Aseismic Prestressed Concrete Structures', Federation Internationale de la Precontrainte, London (1977)

5.18 Thompson, K. J., 'Ductility of Concrete Frames Under Seismic Loading', Ph.D. Thesis, Univ. Canterbury, N.Z. (1975)

6 Masonry

M. J. N. Priestley

6.1 INTRODUCTION

Structural masonry does not enjoy a reputation for sound performance under seismic loading. Although the reason for a certain lack of confidence in the material by structural engineers lies largely in the unsatisfactory performance of unreinforced, or inadequately reinforced masonry structures during the early part of this century, behaviour of reinforced masonry in recent earthquakes has been less than exemplary.

Glogau[6.1], assessing masonry performance in earthquakes, has noted many common forms of failure, including

(1) Vertical shear failures in stack-bonded walls with inadequate horizontal reinforcement (Figure 6.1(a)).
(2) Early structural degradation under high shear forces, resulting from horizontal reinforcement placed in weak, horizontal joint planes.
(3) Toe failure of walls due to the combined action of flexural compression and shear (Figure 6.1(b)).
(4) Modification of seismic forces by supposedly non-structural partitions and in-fill panels. An example of this is to be found in the behaviour of the Macuto Sheraton Hotel in the 1967 Caracas earthquake, where very weak hollow masonry partitions stiffened the upper floors to the extent that energy dissipation was concentrated in undesirable inelastic deformation in lower storey columns.

Despite the comparatively poor record of masonry in recent earthquakes, masonry structures of substantial size can be designed to perform adequately under major earthquakes, provided careful design and detailing requirements are followed. Earthquakes have a tendency to detect weak links in structural systems, and concentrate damage there. By virtue of the form of construction, masonry has a larger number of potential weak links than other materials. As masonry is a comparatively brittle material, it is generally necessary to design for higher seismic forces than appropriate for other materials.

Masonry includes a wide range of material types:– shale, clay, slate, stone, and concrete, and various construction forms, from unmortared stacked stone blocks, where resistance to lateral forces is provided solely by gravity stability and friction, to carefully detailed reinforced masonry walls designed

Fig. 6.1 Failure of masonry walls under seismic forces. (a) Failure of stack-bonded wall (Courtesy of O. A. Glogau). (b) Compression failure of shear-wall toe (Courtesy of O. A. Glogau)

for ductile response under seismic attack. This chapter will concentrate on the design of reinforced masonry structures. Unreinforced masonry is only suitable for small structures in seismically active areas, and the principles involved in design are sufficiently straightforward to warrant their exclusion from this text. It should, however, be realised when designing unreinforced

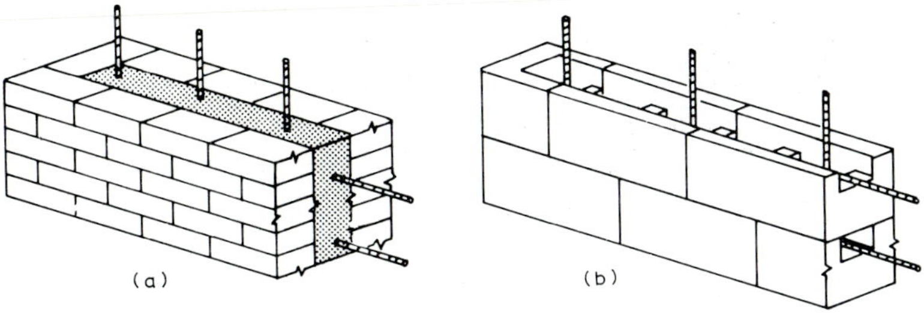

Fig. 6.2 Common forms of masonry construction: (a) reinforced grouted masonry (RGM), (b) reinforced hollow unit masonry (RHUM) using open-end bond-beam units

masonry structures for seismic forces that codified lateral load levels, which generally imply a degree of ductility in response, may substantially underestimate the true response of a material unsuited for energy dissipation. Thus reduction factors based on ductility, which are applicable to the design of reinforced masonry elements do not apply to unreinforced masonry.

6.2 METHODS OF CONSTRUCTION

In many of its characteristics, reinforced masonry may be treated, in design, in the same fashion as reinforced concrete. The fundamental behaviour of masonry and concrete under axial load is similar, with both being relatively strong in compression and weak in tension, and of course reinforcement for flexure and shear plays the same role for each material. There are, however, sufficient differences in behaviour to necessitate careful examination of masonry material properties. These differences are partially due to the different component materials, and partially due to the different methods of construction. To fully understand the former, it is necessary to be familiar with the different forms of reinforced masonry common in practice.

Figure 6.2 illustrates the two main forms of reinforced masonry construction. Reinforced grouted masonry (Figure 6.2(a)) consists of two skins, or wythes of brick or concrete block, normally solid, separated by a cavity in which the vertical and horizontal reinforcement is placed, and which is grouted as the wythes are built. The wythes are normally 2–4 in (50–100 mm) thick, and the grout gap $2\frac{1}{2}$–4 in (60–100 mm) wide. Reinforced hollow unit masonry (Figure 6.2(b)) is constructed from concrete or brick units with internal vertical voids or cells, some or all of which will contain vertical reinforcement, and will be grouted. The size and design of the units varies considerably, and widths from 4–12 in (100–300 mm) are common. Other common structural forms include masonry infill panels in reinforced concrete or structural steel frame structures, and a modified infill construction method consisting of peripheral reinforced masonry columns and bond beam elements with unreinforced masonry infill.

For seismic resistant structures, the blocks or bricks in both forms of construction illustrated in Figure 6.2 should be laid in running, or stretcher bond. Stack bonding, which may be architecturally attractive, creates vertical planes of weakness, particularly in reinforced hollow unit masonry constructed from units with both ends closed. Shear capacity can be reduced virtually to nil, with the stack-bonded wall behaving under earthquakes as a series of independent vertical cantilevers, one unit in section. Several examples of this behaviour were evident in the 1971 San Fernando earthquake[6.2] (see Figure 6.1(a)). Consequently, except at wall ends, or where movement control joints are required, or where seismic shears are low, stack-bonding should be avoided.

Reinforced grouted masonry has the advantage of providing the designer with greater flexibility in selecting wall widths, and in locating vertical and transverse reinforcement. However, the grout core generally forms a smaller portion of the total wall section area than for reinforced hollow unit masonry, thus reducing its ultimate shear strength, and is a more labour intensive construction form.

In reinforced hollow unit masonry, open-end units (units with one end-shell missing, or knocked-out) are to be preferred to standard units because of the poor characteristics, from both strength and waterproofing considerations, of vertical mortared header joints passing through the wall thickness. Horizontal reinforcement is placed in bond-beam units (see Figure 6.2(b)) which consist of standard or open-end units with webs knocked-out to approximately half-depth, allowing the horizontal rebar to be located away from the weak horizontal joint plane. Bond beam units may be laid inverted at the bottom course to facilitate clean-out of mortar droppings prior to grouting. In general, walls subjected to in-plane lateral forces under seismic loading should have all cells grout-filled, whether or not they contain vertical reinforcement. This is necessary because of the poor and unpredictable shear strength of ungrouted portions of the wall, and desirable also because of the construction difficulties in placing horizontal reinforcement when some cells are left ungrouted. Exceptions may be made when walls are subjected to face-loading only (veneers) and shear walls subjected to very low shear stresses.

6.3 BEHAVIOUR OF CONSTITUENT MATERIALS

6.3.1 Compression stress–strain characteristics

Reinforced masonry consists of four separate materials: brick or concrete masonry units, grout, mortar and reinforcement, which are required to act in composite fashion. It is difficult to predict the compression characteristics of the assemblage on the basis of the constituent properties, in the same way that concrete properties cannot easily be predicted from the individual strengths and modulii of elasticity of the aggregate and cement paste. Figure 6.3 shows typical stress–strain curves for prisms consisting of at least three courses of solid brick or concrete masonry units subjected to uniaxial

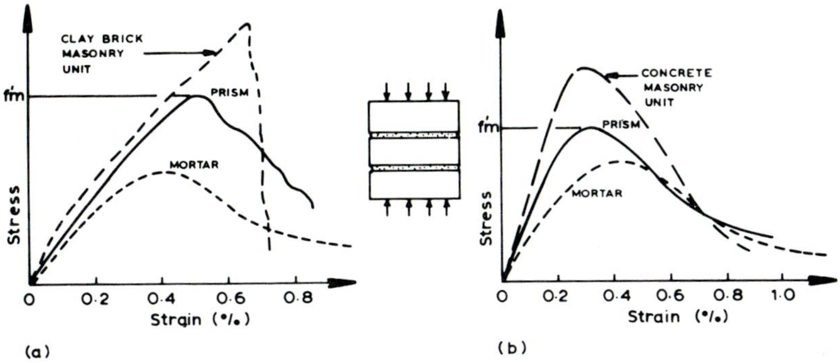

Fig. 6.3 Compression stress–strain curves for masonry prisms; (a) clay brick masonry, (b) concrete masonry

compression, and includes the stress–strain curves of the constituents. Brick units (Figure 6.3(a)) may have compression strengths exceeding 60 MPa, and generally exhibit an extremely brittle failure mode after reaching the maximum stress. Similar behaviour, though with lower crushing stresses and a somewhat more ductile failure, results from compression tests on concrete masonry units (Figure 6.3(b)). The mortar has a lower modulus of elasticity and ultimate compression strength than brick or concrete, but has a comparatively flat falling branch portion of the curve. Prism stress–strain curves tend to be intermediate between the constituent curves. Crushing stresses exceed that of mortar, the weakest link, and the falling-branch portion of the curve is flatter and extends further than for the brick or concrete masonry unit. Tests on grouted prisms indicate further modification of the stress–strain curve away from the brittle unit behaviour.

6.3.2 Failure mechanism under uniaxial compression

Failure of prisms of three or more courses under uniaxial compression is almost always initiated by vertical tension cracking of the relatively stronger brick or concrete masonry unit, rather than by apparent crushing of the weaker mortar or grout. This behaviour has been explained by Hilsdorf[6.4] as being a consequence of the lower modulus of elasticity and crushing strength of the mortar, and of the rapid increase in the mortar Poisson's ratio as the crushing stress is approached (see Figure 6.4(a)). If the mortar could deform freely, its lateral strain, as a result of the combination of the above effects, would exceed that of the brick or concrete unit. Friction, and strain compatibility constrain the lateral strains in the mortar and unit to be the same, inducing high lateral compression in the mortar, and lateral tension in the brick or concrete unit, particularly adjacent to the mortar bed. Since the mortar is effectively in triaxial compression, the prism crushing stress exceeds that of the unconfined mortar. As the crushing strength of the mortar is approached the lateral tension in the masonry units increases rapidly as a

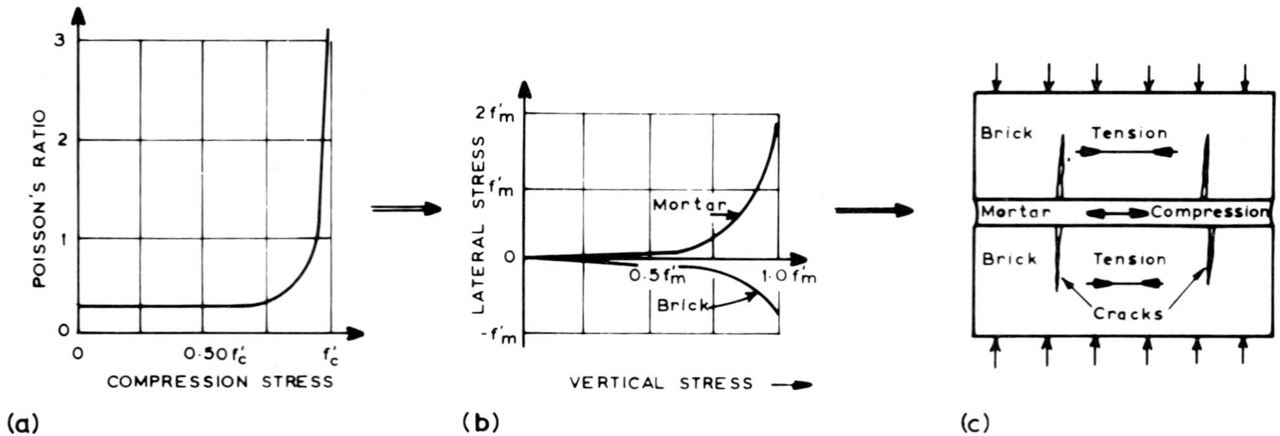

Fig. 6.4 Failure mechanism for masonry prisms. (a) Variation of mortar Poisson's ratio with compression stress. (b) Lateral stress versus vertical stress. (c) Tension cracking of masonry unit

consequence of the Poisson's ratio increase (Figure 6.4(b)), and eventually the biaxial tensile strength of the unit is exceeded resulting in vertical cracking, initiating close to the mortar bed (Figure 6.4(c)). Failure may occur immediately in an explosive manner due to the sudden reduction in lateral confinement of the mortar, or may occur gradually as more vertical cracks form.

The masonry crushing strength f'_m depends on several variables that are difficult to measure, including the biaxial tension strength of the brick, and the degree of nonuniformity of normal stress within the prism[6.4]. It is thus desirable to measure f'_m directly from prism tests. Foster and Bridgeman[6.5] have shown that prisms should be at least 3 units and preferably 5 units high if results independent of test machine platen restraint are to be obtained. Measurement of f'_m on site-formed prisms has a secondary but most important function as a quality control influence on the brick or block layer.

The mechanism of masonry compression failure profoundly influences the behaviour of masonry shear walls subjected to in-plane seismic forces. At ultimate flexural capacity of the wall, the formation of vertical cracks at the compression end results in a number of unstable vertical blocks subjected to high compression and shear and with greatly reduced capacity to transfer the ultimate shear force to the foundation. Explosive compression toe failures may result, and have been noted in tests[6.6] and earthquake damage[6.1] (see Figure 6.1(b), for example.)

Even if toe failure does not immediately follow vertical crack formation, lateral support of vertical steel near the wall end, or jamb is lost, and the compression steel may buckle, dislodging masonry at the toe. This behaviour is particularly characteristic of reinforced grouted brick masonry shear walls[6.6]. Where high ductility is required, toe collapse can be eliminated by inclusion of thin corrosion resistant confinement plates in the mortar beds within potential crushing areas. These plates are equivalent to confining stirrups in reinforced concrete columns or shear walls, and reduce the lateral expansion of the mortar. Tests performed on reinforced grouted brick masonry panels incorporating the plates, and subjected to high ductility racking displacements, showed substantial improvement in toe behaviour over unprotected panels[6.6]. Similar improvement has been noted by Mayes, *et al*[6.7] in the behaviour of reinforced hollow unit concrete masonry pier units with critical areas protected by confinement plates.

Figure 6.5 shows typical $\frac{1}{8}$ in (3 mm) stainless steel confinement plates inserted in mortar beds at wall ends. The design is such that minimum obstruction to grout core results, and outside dimensions are kept about $\frac{1}{2}$ in (13 mm) inside the wall to permit normal pointing of mortar joints. Circular holes are provided in plates for reinforced grouted masonry to improve bond between mortar and plates. Confinement plates, when incorporated in wall designs, need only extend from the extreme compression fibre to the ultimate neutral axis position, and to reinforce those mortar joints where crushing is likely. In the case of cantilever shear walls, this would mean providing plates at each end of the wall from the base to a height of about 5–10% of the total wall height.

Fig. 6.5 Mortar bed confining plates for masonry shear walls. (a) Reinforced grouted masonry construction (Courtesy of N. Z. Pacra). (b) Reinforced hollow unit masonry construction (Courtesy of N. Z. Pacra)

6.4 BEHAVIOUR AND DESIGN OF MASONRY SHEAR WALLS

6.4.1 Design philosophy

There is still a tendency, unfortunately incorporated in many codes, to design masonry structures for seismic resistance using working stress methods on the assumption that this is a 'safe' design technique, and that response under seismic attack should be elastic, to prevent damage in a material known to exhibit brittle failure. This is a misleading assumption, unless the structure is designed for the true level of force corresponding to elastic response, rather than codified levels, which assume ductile inelastic response. Typically, masonry structures are stiff, and have low natural periods. The elastic response of such structures to damaging earthquakes of the size of El Centro 1940 NS may be close to 1.0 g[6.8], even for viscous damping of 5%. Design to this level of lateral forces would in most cases be uneconomic.

Elastic design to a code coefficient of 0.10 g–0.20 g will imply an ultimate capacity of about 0.20 g–0.40 g, well below the peak elastic response of about 1.0 g. Thus, far from remaining elastic, substantial ductility will still be required from masonry structures designed by working stress methods.

A further disadvantage of elastic design is that it encourages the designer to concentrate the flexural reinforcement in shear walls close to the wall ends. As discussed later in this chapter, this is undesirable from the viewpoint of satisfactory behaviour under inelastic response.

As with other materials it is much more satisfactory to recognise that inelastic deformations will occur under severe seismic attack, and to design accordingly.

This section considers strength and deformation characteristics of simple load-bearing cantilever shear walls designed to provide the major seismic resistance of buildings. Figure 6.6 illustrates typical examples of structures consisting of a number of cantilever shear walls linked by flexible floor slabs. Analysis of such structures is comparatively simple, and follows the principles described in Section 5.6.2. It is initially assumed below that any openings within each shear wall are sufficiently small so that no significant modification of the basic cantilever action results. Effects of larger openings are discussed separately in Section 6.4.8.

6.4.2 Flexural stiffness and building period

The modulus of elasticity of structural masonry depends on the type and proportion of constituent materials. Values measured for masonry prisms and walls under uniform compression[6.3,6.9] tend to fall in the range 1.0–2.5 $\times 10^6$ psi (6.8–17.0 GPa). However, for seismic design purposes the most significant influence of the modulus of elasticity is its effect on the flexural stiffness, and thus on the natural period of vibration. For this purpose, the stiffness should be based on the cracked section. Tests on squat masonry shear walls[6.10] have indicated that use of the uncracked section and an

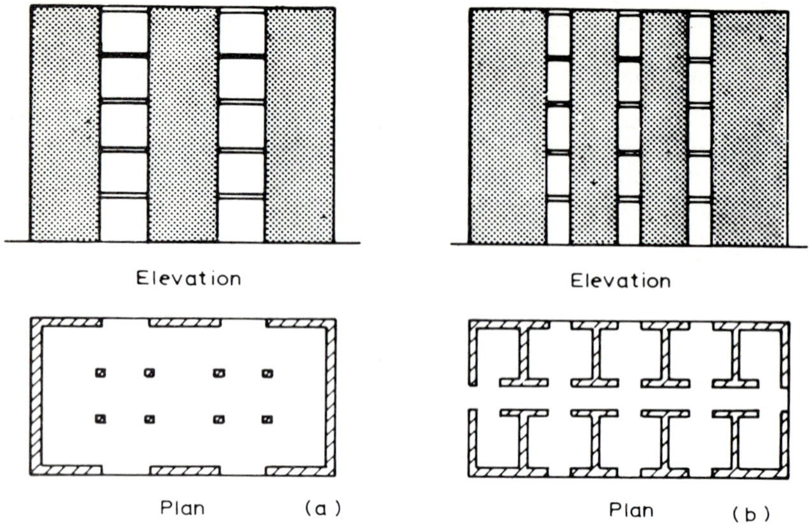

Fig. 6.6 Typical example of masonry cantilever shear walls. (a) Exterior masonry shear walls with interior columns. (b) Masonry exterior and interior shear walls

effective modulus of elasticity $E_e = 7.5 \times 10^5$ psi (5.2 GPa) gives a good representation of the cracked stiffness for both reinforced grouted, and hollow unit masonry, provided shear deformation is considered. For other forms of masonry construction, values for E_e should be based on local data. When calculating the natural period, soil compliance effects should be included, as for typically stiff, rather squat masonry structures the elastic lateral displacements resulting from soil compliance can be many times greater than flexural displacements.

6.4.3 Ultimate flexural capacity

The results of a wide range of test programmes show that the strength of reinforced masonry shear walls failing in flexure is conservatively predicted by reinforced concrete ultimate strength theory. For a rectangular wall length l_w containing uniformly distributed vertical reinforcement of total area A_s and subjected to axial force N_u, the ultimate moment capacity can be shown[6.11] to be approximated by

$$M_u = 0.5 A_s f_y l_w \left(1 + \frac{N_u}{A_s f_y}\right)\left(1 - \frac{c}{l_w}\right) \tag{6.1}$$

where c is the distance from the neutral axis at ultimate to the extreme compression fibre.

Masonry walls subjected to monotonic loading to failure generally exceed

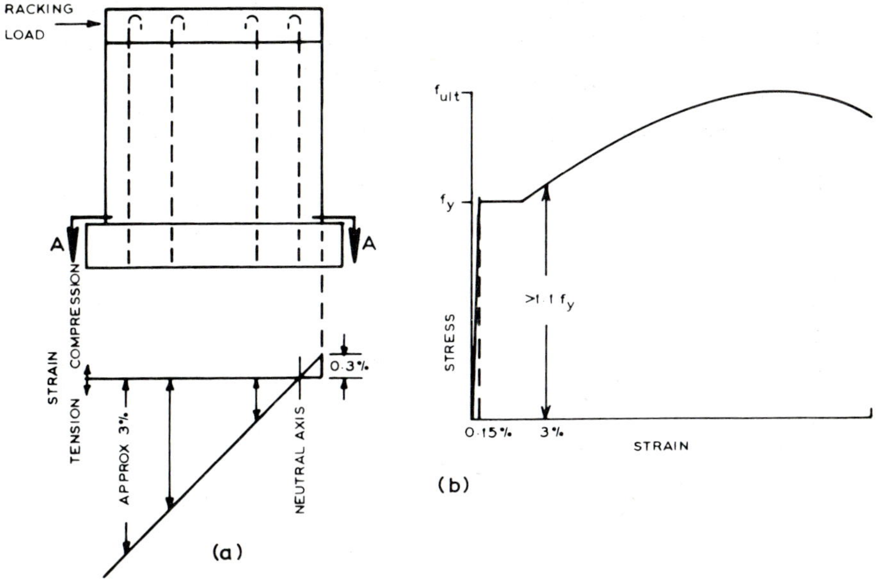

Fig. 6.7 Strain hardening enhancement of ultimate load; (a) strain distribution at AA, (b) steel stress–strain curve

the predicted ultimate moment by 10–30%. As shown in Figure 6.7, the low steel ratios common for masonry walls will typically result in steel strains in the vicinity of 3–5% at this stage. Since strain-hardening commences between 0.5 and 2% strain, depending on the steel properties, steel stresses of 10–30% above yield can be expected at strains of 3–5%, resulting in the observed overstrength. However, under cyclic load reversals structural degradation will reduce the ultimate capacity, and probable overstrength capacity should not be used when calculating flexural steel requirements. To account for unpredictable quality of workmanship, and dimensional and strength tolerance, a flexural capacity reduction factor of between $0.7 < \varphi_f < 0.85$ is appropriate.

6.4.4 Distribution of reinforcement for flexural strength

The ultimate flexural capacity of walls with low axial load and reinforcement ratio is effectively independent of the distribution of the vertical reinforcement, provided it is symmetrically placed about the wall centroid. A wall of effective length l with total vertical steel area A_s divided into two groups of $A_s/2$, one at each end of the wall, has an ultimate flexural capacity of

$$M_u \sim \frac{A_s}{2} f_y l$$

If the same content is evenly distributed along the wall, virtually all steel will yield in tension due to the small neutral axis depth, resulting in a tension force of $A_s f_y$ at a lever arm of $l/2$, and the same flexural capacity as above. There is thus no significant advantage in terms of ultimate capacity in concentrating the steel at the wall ends, or distributing it evenly along the wall. However, large amounts of steel within comparatively small grout areas at wall ends create bond and anchorage problems, and are difficult to restrain against buckling in compression. Uniform distribution of reinforcement reduces these problems, and has the added advantages of providing a clamping action across the wall-base construction joint, and providing greater energy absorption at lower levels of seismic attack.

6.4.5 Shear design

Design shear force Shear failures of masonry shear walls are sudden, and result in almost total loss in lateral load carrying capacity. Figure 6.8(a) shows a typical shear failure of a reinforced grouted brick masonry panel, unreinforced for shear, subjected to maximum shear stress, based on gross cross-sectional area, of 90 psi (0.62 MPa). After opening of the diagonal shear crack, the lateral load capacity reduced to 10% of the initial capacity. Because of the brittleness of masonry shear failure it is essential to design masonry shear walls to carry the true shear force, based on flexural overcapacity, to which they will be subjected. As shown in the following example, these may substantially exceed the code specified base shear.

Using the capacity approach outlined in Chapter 5, Section 5.5.1, the required design shear force V_D will be related to the code base shear V_B by

$$V_D = \frac{\varphi_o}{\varphi_f} V_B \tag{6.2}$$

For example, taking the capacity reduction factor for flexure, $\varphi_f = 0.7$, and the overcapacity factor $\varphi_o = 1.25$, gives $V_D = 1.79 V_B$ as the likely maximum shear force when the overstrength of the wall is reached in flexure. If a shear capacity reduction factor of $\varphi_s = 0.85$ is adopted, the area of shear steel required would be given by

$$A_V = \frac{2.10 \times V_B s}{f_y d} \tag{6.3}$$

assuming all shear to be carried by the steel, where s = spacing of shear steel, d = effective depth of section.

For masonry buildings higher than three storeys, higher mode effects may further increase the maximum shear force above the code level[6.12]. This is dealt with in more detail in Chapter 5, Section 5.6.2.

Fig. 6.8 ***Influence of horizontal shear steel on shear failure mode.*** *(a)* ***RGM wall not reinforced for shear.*** *Failure at 90 psi (0.62 MPa) (Courtesy of N. Z. Pacra). (b)* ***RHUM wall reinforced for shear.*** *No shear failure at 380 psi (2.62 MPa)*

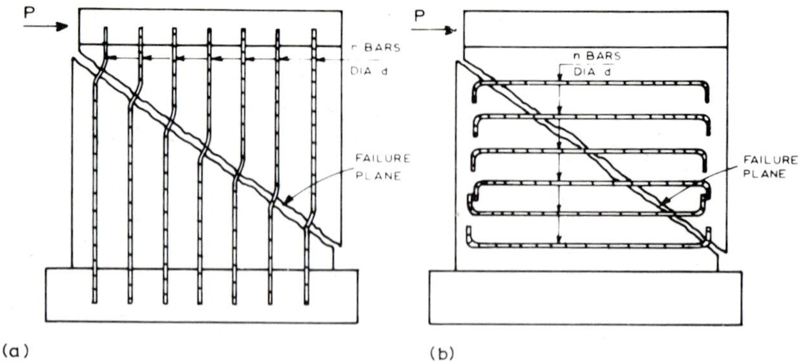

Fig. 6.9 Relative effectiveness of vertical and horizontal shear steel. (a) Steel carries shear by dowel action ($P_{max} = 0.24\ nd^2f_y$)*. (b) Steel carries load by tension* ($P_{max} = 0.785\ nd^2f_y$)

Masonry without shear reinforcement Although a substantial body of research has been carried out into the shear strength of masonry unreinforced for shear, the results are variable and shear capacity of the masonry cannot be relied on where flexural yielding can occur. Flexural cracking in masonry occurs primarily along mortar joints, in the form of smooth cracks. With tensile steel yield and reversed cyclic loading, these cracks may be open for the full length of the wall until sufficient moment is applied to yield the compression steel and close the crack at the compression end. It is to be expected that the masonry shear capacity of the compression zone will degrade more rapidly than with concrete because of the potential instability resulting from vertical splitting of the compression zone. Shrinkage cracks, poor bond between mortar and masonry unit at headers, and the high variability of workmanship common in masonry construction, all conspire to make the shear capacity of masonry unreinforced for shear difficult to predict by laboratory tests, and highly unreliable. Thus, unless the structure is designed to respond elastically to the design earthquakes, all seismic shear can be carried by properly detailed shear reinforcement.

Masonry with shear reinforcement It might be supposed that uniformly distributed vertical reinforcement crossing potential 45° failure planes would be equally as efficient in shear transfer as horizontal reinforcement, and thus flexural steel near the centre of the wall, which may be at comparatively low stress, could be utilised to carry part or all of the shear load, as shown in Figure 6.9(a). However, once cracking has initiated the shear force tends to displace the upper portion of the wall horizontally, past the lower portion, rather than open the crack perpendicular to the crack plane. Under these conditions horizontal steel, as in Figure 6.9(b), resists the shear force by direct tension but vertical steel resists the force by dowel action, which can only be mobilised if substantial horizontal displacements occur across the crack. Under optimum conditions, the force carried by dowel action will be about 30% of that carried by the same steel placed horizontally. Since dowel capacity reduces as crack width increases, horizontal steel should be used to carry the entire shear force.

Many design codes[6.13,6.14] specify low maximum allowable shear stresses because of the brittle nature of shear failure, and because early tests[6.15,6.16] apparently indicated that shear reinforcement did not significantly influence the shear capacity of shear walls. More recent tests[6.6,6.10] have shown that insufficient horizontal reinforcement was provided in the earlier tests, and that very high shear stresses can be sustained during inelastic cycling by both reinforced hollow unit masonry and reinforced grouted masonry walls, when properly detailed horizontal shear reinforcement designed to carry the full shear associated with flexural overstrength is provided. Figure 6.8(b) shows the condition of a reinforced hollow unit masonry wall after inelastic cycling at ultimate flexural capacity, corresponding to a maximum shear stress of 380 psi (2.62 MPa) based on the gross cross-sectional area. As the prism strength f'_m was 2650 psi (18.3 MPa), this corresponds to a shear stress of

$$v_u = 7.4\sqrt{f'_m} \text{ psi} \qquad (0.62\sqrt{f'_m} \text{ MPa})$$

After completion of testing the well distributed shear cracks closed up, indicating excellent control by the horizontal shear reinforcement.

In fact there seems little justification at present in a limiting ultimate shear stress on the basis of the prism strength. Severe structural degradation due to sliding shear along the base construction joint will generally occur at shear stresses well below those capable of being supported by masonry shear walls, as discussed further in Section 6.4.7.

For walls of aspect ratio (height/length) greater than unity, the total area of shear steel crossing each potential 45° crack should be

$$A_v = \frac{V}{\varphi_s f_y} \tag{6.4}$$

where V is the shear force associated with flexural overcapacity and φ_s is the shear capacity reduction factor. When the aspect ratio is less than unity, the approach outlined in Section 5.6.3, should be adopted.

6.4.6 Bond and anchorage

Because of construction methods it is difficult to avoid lapping all vertical bars in a wall immediately above the wall base, thus locating the lap splice within the potential plastic hinge region. Conditions are therefore not good for development of reinforcement and may be aggravated by the small size of the grout core, poor bond between grout and masonry unit, and by eccentricity of the lapping bars. Under these conditions, laps should be at least equivalent to ACI 318-77[6.29] Class D splices. The new ATC3 proposals for seismic design of masonry[6.17] require a minimum lap length of

$$l_d = 0.08 d_b^2 f_y / \sqrt{f_g} \text{ in} \tag{6.5}$$

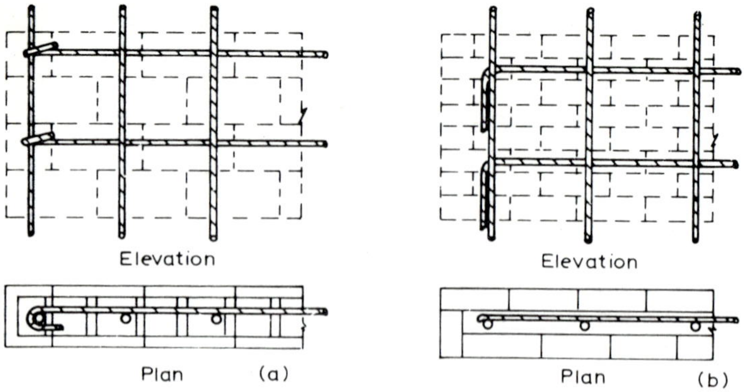

Fig. 6.10 Anchorage of horizontal shear steel. (a) Hook round jamb bar in RHUM. (b) Bend vertically in RGM

where f_g = grout crushing strength in psi and d_b = bar diameter in inches.

Since it is difficult to estimate the strength of the grout in situ, it is probably more realistic to relate lap length directly to bar diameter by the conservative requirement

$$l_d = 0.001 f_y d_b \text{ in} \qquad (0.145 f_y d_b \text{ mm}) \tag{6.6}$$

but not less than 20 in (500 mm). Tension tests on lap splices in 6 in (150 mm) reinforced hollow unit masonry using deformed Grade 40 bar sizes up to No. 8 (25 mm dia) showed laps designed to Equation 6.6 capable of transmitting reinforcement stresses consistently well into the strain-hardening range of the stress–strain curve. Shear panels designed to this criterion[6.10] and subjected to inelastic cycling at high ductilities have also performed well, with no indication of bond failure. However, because of the dimensions of the test panels, a substantial reduction in moment applied over the lap length resulted, and conditions were less severe than might occur in practice. Consequently it is conservatively suggested that bond conditions should be further eased by relating the maximum bar diameter to wall thickness as follows:

For reinforced hollow unit masonry: $$d_b \leq \frac{b_w}{10} \tag{6.7(a)}$$

For reinforced grouted masonry: $$d_b \leq \frac{b_g}{4} \tag{6.7(b)}$$

where b_w = overall wall width and b_g = width of the grout core. No more than one bar should be placed in any cell of hollow unit masonry.

Horizontal shear reinforcement should be hooked around the extreme vertical bars, as shown in Figure. 6.10(a), or if the grout-gap width in reinforced grouted masonry precludes this, a 90° bend and extension parallel to the flexural reinforcement capable of developing the full yield force should be used (Figure 6.10(b)). In no cases should horizontal shear reinforcement be lapped in potential plastic hinge zones. Plain reinforcing bar should not be used as flexural or shear reinforcement in shear walls.

Minimum reinforcement A minimum quantity of well distributed reinforcement needs to be placed in walls to ensure control of shrinkage and settlement cracks. This steel also provides 'basketing' of the masonry, reducing the tendency for masonry shedding under seismic loads. At least 0.15% of reinforcement, based on the gross cross-section area, should be placed both vertically and horizontally. Spacing between adjacent bars should not exceed 32 in (800 mm) horizontally nor 24 in (600 mm) vertically.

6.4.7 Ductility of masonry shear walls

Available ductility Tests on shear panels of various construction types[6.6,6.10,6.18] under gradually increasing displacement-controlled cyclic loading have indicated that substantial ductility is available from masonry shear walls over a wide range of maximum shear stresses. Figure 6.11 shows load deflection curves for a reinforced grouted brick masonry wall with moderate maximum shear force, and a reinforced hollow unit concrete masonry shear wall of higher shear capacity. Both walls had masonry confining plates in the bottom three mortar courses at each end of the wall. For these and similar walls designed in accordance with the provisions of earlier sections of this chapter, cycling at displacement ductility factors less than 4 resulted in reduction in peak loads from the initial maxima (which exceeded or were close to the theoretical yield loads predicted on the basis of measured material properties), to stable values at about 80–90% of yield, with a narrowing, or pinching of the hysteresis loops. Displacement ductility factor is defined here as the ratio of maximum displacement to displacement found from extrapolation of the post-cracking elastic load-deflection curve to the theoretical ultimate load. On increasing the amplitude of cyclic displacements the initial load maxima would be regained or exceeded, indicating that the earlier stiffness degradation had not been accompanied by moment degradation. The stiffness degradation apparent in Figure 6.11, and other tests, was invariable found to be the result of sliding of the entire wall along the base-course. The extent of base-course sliding increased with increasing ductility factor, and the hysteresis loops resemble the Clough degraded stiffness model[6.19].

On the basis of these tests, the following conclusions may be drawn for masonry cantilever shear walls of similar construction with uniformly distributed flexural reinforcement, where shear reinforcement is provided in accordance with Section 6.4.5.

(1) Significant energy absorption is available below theoretical yield displacement due to the width of the hysteresis loops.

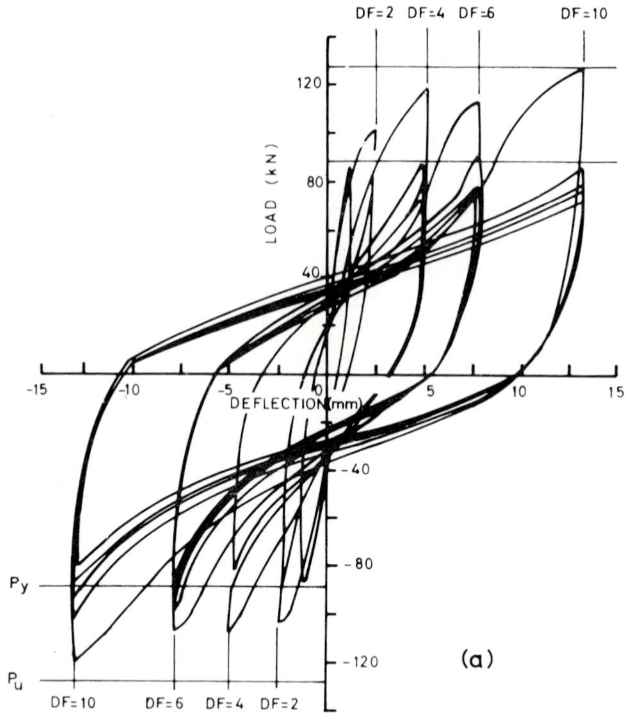

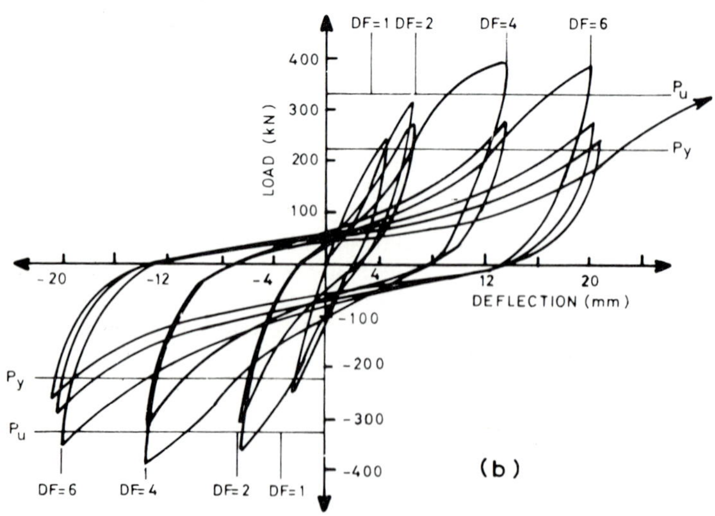

Fig. 6.11 Typical load displacement loops for masonry walls (P_u = *theoretical ultimate load*, P_y = *load at first yield of reinforcement. (a) RGM wall with confining plates*, $v_u = 56$ *psi* (0.39 *MPa*). *(b) RHUM wall with confining plates*, $v_u \sim 191$ *psi* (1.32 *MPa*)

(2) Satisfactory behaviour occurs for displacement ductility factors of 2 for walls with applied shear stress up to 350 psi (2.4 MPa) with only minor load and stiffness degradation, whether or not the compression zones are reinforced with confining plates. Behaviour is very similar to that expected from reinforced concrete shear walls.
(3) For walls whose compression zones are protected by confining plates, ductility factors exceeding 4 can be satisfactorily sustained if shear stresses do not exceed 200 psi (1.4 MPa).
(4) Reinforced grouted brick masonry walls without confining plates exhibit rapid structural degradation at ductility factors exceeding 2.
(5) Stiffness degradation resulting from base sliding occurs regardless of the level of base shear, but is more rapid for ultimate shear stresses exceeding 200 psi (1.4 MPa).

These conclusions can only be taken to apply to good quality materials with the following minimum crushing strengths: clay brick units 4000 psi (27.5 MPa); concrete block units 2500 psi (17.2 MPa); grout 2000 psi (13.8 MPa); mortar 1500 psi (10.3 MPa). Where lower strength materials are locally common, maximum shear stress levels should be based on equivalent local test data.

Required ductility Results of the above tests have indicated that well designed masonry shear walls will exhibit dependable, but limited, ductility. Design should reflect this limitation, and be based on higher base shear coefficients than appropriate for, say, a reinforced concrete or steel frame of the same fundamental period. Also the typically short fundamental period of a masonry structure may result in higher ductility demand than imposed on more flexible structures. For example, inelastic analyses by Thompson[6.20] indicate that a masonry building with a period of 0.43 s and 5% damping design in accordance with New Zealand code requirements[6.21] to a base shear coefficient of 0.288 g and exhibiting degrading stiffness hysteresis behaviour would be subjected to displacement ductility demands in the vicinity of 4.0 under the El Centro May 1940 NS record.

It is recommended that in the absence of specific code requirements for masonry shear walls, seismic base shear coefficients should be those required for reinforced concrete shear walls of the same stiffness, multiplied by a factor 1.2 for walls with maximum shear stress (associated with flexural overcapacity) up to 180 psi (1.25 MPa), increasing linearly to 2.4 for shear stresses of 300 psi (2.05 MPa). Reinforced grouted brick masonry shear walls without confining plates in potential crushing zones should adopt the upper limit of 2.4 for all levels of shear stress.

6.4.8 Shear walls with openings

Figure 6.12 shows two possible failure modes for walls whose openings are sufficient in size to destroy the basic cantilever action assumed in the previous section. In Figure 6.12(a), characterised by deep, stiff spandrels and

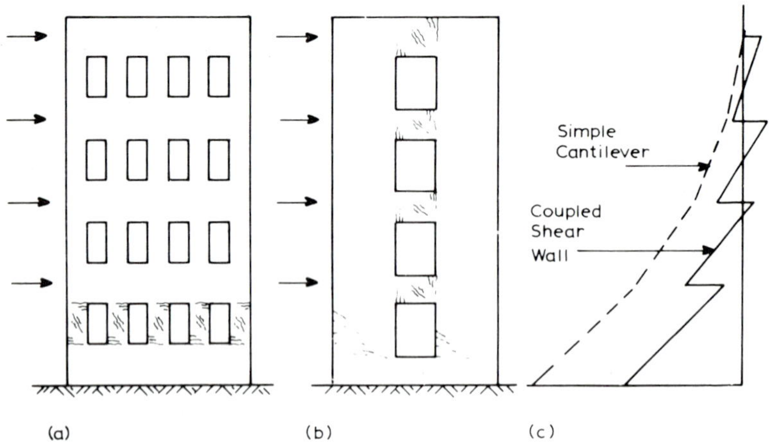

Fig. 6.12 Failure of shear walls with openings; (a) pier failure, (b) spandrel failure, (c) bending moments for (b)

comparatively slender piers, behaviour is similar to shear frames, with ductility resulting from flexural plastic hinging or shear failure of the piers. For this type of system it is virtually impossible to avoid concentration of plastic hinging in the piers of one storey, generally the lower, with consequential extremely high ductility demand at that level. As with reinforced concrete frames, use of columns or piers as energy dissipating elements is highly undesirable for structures more than two storeys in height.

Figure 6.12(b) shows a system where spandrels are weaker than the piers, resulting in coupled shear wall characteristics. Energy is dissipated by hinging or shear failure in the spandrels. Although this system is very suitable for reinforced concrete coupled shear walls if properly detailed diagonal reinforcement is incorporated in the spandrels (see Section 5.6.4), it is unsuitable for masonry. In masonry spandrels it is impractical to place and confine diagonal reinforcement, and difficult to adequately anchor it in the shear walls. Under seismic loading very high ductility is required of the spandrels, and it is probable that under inelastic cycling, the masonry spandrel capacity would rapidly degrade to a fraction of the initial value. The coupled shear walls would then behave effectively as individual cantilever shear walls linked by pin-ended members, with greatly increased base bending moments as illustrated in Figure 6.12(c). Alternatively, for a given base moment capacity, the ductility required of the base hinges would be many times greater than assumed on the basis of coupled shear wall action, and failure could result.

Under these circumstances it is best not to rely on the spandrels as part of the energy dissipating mechanism. Spandrels should be lightly reinforced, and the shear walls designed to carry the full seismic load as cantilevers. The systems shown in Figures 6.12(a) and (b) may be adopted as fully integrated structures only if designed to resist the true seismic response elastically. If this approach is adopted, it would be reasonable to assume a comparatively high damping ratio, say 10% of critical, to account for the limited ductility capacity.

For smaller openings in shear walls where the size of opening is such that basic cantilever action of the entire wall is not jeopardised, special reinforcement should be placed close to all edges of the opening, similar to that required for openings in reinforced concrete shear walls, to transmit the shear forces.

6.5 MASONRY-INFILLED FRAMES

6.5.1 Influence of infill on seismic response

It is a common misconception that masonry infill in structural steel or reinforced concrete frames can only increase the overall lateral load capacity, and therefore must always be beneficial to seismic performance. In fact, there are numerous examples of earthquake damage that can be traced to structural modification of the basic frame by so-called non-structural masonry partition and infill panels. Glogau[6.1] notes that even if relatively weak, as in the Macuto Sheraton Hotel, masonry panels can drastically alter the intended structural response, attracting forces to parts of the structure which have not been designed to resist them.

Incorporation of masonry infill results in a decrease in fundamental period, and an increase in seismic shears, frequently resulting in shear failure of the frame columns. Brittle failure of unreinforced infill has resulted in shedding of masonry into streets below, or into stairwells, with great hazard to life.

When masonry infill is to be used there are two alternatives open to the designer. He may effectively isolate the panel from frame deformations, or ensure intimate and continuous contact between the panel and frame, and design both for the seismic forces to which they will be subjected.

6.5.2 Isolation of non-structural infill

To ensure normal frame action, infill panels may be structurally separated around the sides and top to allow free deformation of the frame relative to the panel. Flexible couplings between panel and frame must be adopted in this case to provide adequate lateral support at the panel top. Figure 6.13 shows possible connection details. Calculations should be carried out to ensure that the separation provided between frame and panel is sufficient to allow the required interstorey frame drift to develop. If the details of Figure 13(a) and 13(b) are adopted and the partition is above or close to the frame, the stiffening effect of the partition on the beam may significantly alter the frame properties, and must be considered in analysis.

Newmark and Rosenblueth[6.22] report advantages of a connection detail using a peripheral metal band consisting of small metal bents at close

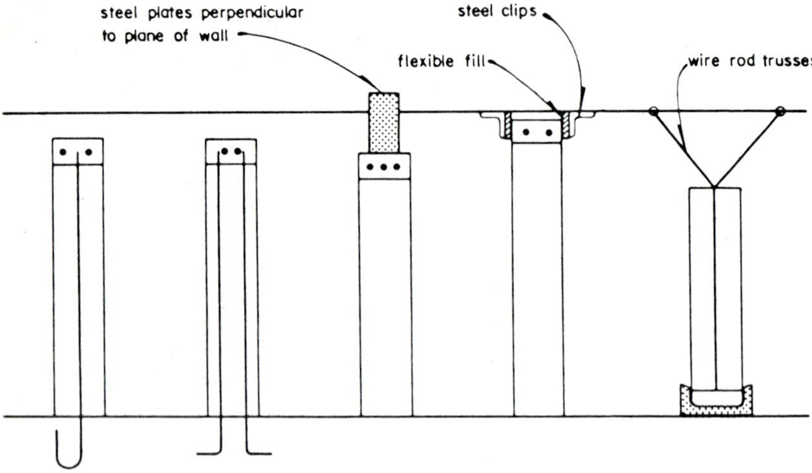

Fig. 6.13 Common connection detail for non-structural infill (after Newmark and Rosenblueth[6.22])

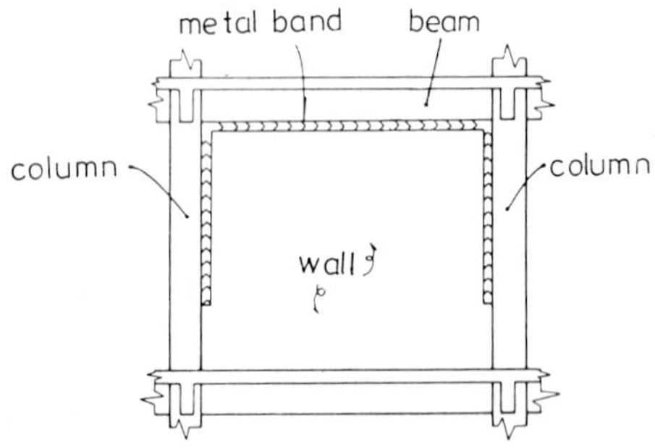

Fig. 6.14 Energy dissipating connection between panel and frame (after Newmark and Rosenblueth[6.22])

spacing between frame and panel, as shown in Figure 6.14. The bents can be designed to yield at a load low enough to protect the partitions from serious damage under relative frame-panel movement, and in so doing, to provide a significant amount of energy dissipation. (See also Section 3.2.3.)

Because of the danger of total collapse of unreinforced panels under a major earthquake, it is recommended that even isolated infill panels contain nominal reinforcement.

6.5.3 Structural infill in contact with frame

Stiffness The behaviour of structural infill where continuous contact is provided on all sides between frame and panel, has been the subject of

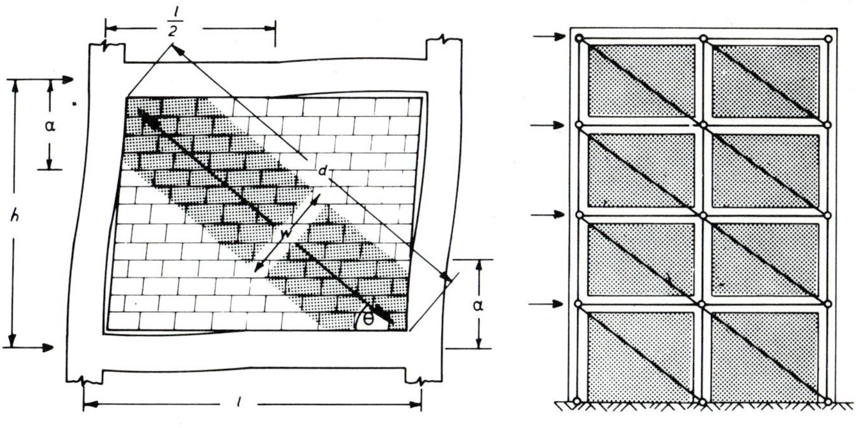

Fig. 6.15 Equivalent diagonal bracing action of infill. (a) Deformation under shear load. (b) Equivalent braced frame for 2-bay, 4-storey infilled wall

considerable recent research activity[6.18,6.23,6.28]. Although beam theory can predict the elastic response of infilled frames with adequate accuracy, behaviour becomes more complex as deformations increase, and separation between frame and panel occurs due to differences between flexural deformations of the frame and shear deformations of the panel (see Figure 6.15(a)). This separation may occur at 50–70% of the ultimate lateral shear capacity of the infill for reinforced concrete frames, and at very much lower loads for steel frames. After separation occurs, the panel acts as a diagonal strut with an effective width, w, less than that of the full panel, as shown in Figure 6.15(a).

Natural period calculations should be based on the structural stiffness after bond separation. This may be found by assuming the frame member to be pin-jointed and braced by the equivalent diagonal masonry struts. Figure 6.15(b) shows the equivalent system for a 2-bay, 4-storey frame. Stafford-Smith and Carter[6.23] have developed analytical expressions based on a beam on elastic foundation analogy, modified by experimental results, showing that the effective width w of the diagonal strut depends on the relative stiffnesses of frame and panel, the stress–strain curves of the materials, and the load level. However, since a high value of w will result in a stiffer structure, and therefore potentially higher seismic response, it is reasonable to take a conservatively high value of

$$w = 0.25d \tag{6.8}$$

which agrees reasonably well with Smith and Carters' charts, assuming typical masonry infill properties and a load level of 50% of the ultimate capacity of the infilled frame.

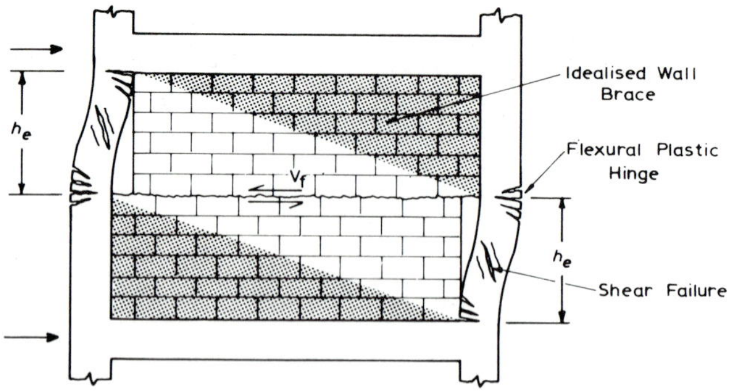

Fig. 6.16 Knee braced frame concept for sliding shear failure

Strength There are several different possible failure modes for masonry infilled frames, including

(1) tension failure of the tension column resulting from applied moments,
(2) sliding shear failure of the masonry along horizontal mortar courses (generally at or close to midheight of the panel),
(3) diagonal tensile cracking of the panel. This does not generally constitute a failure condition, as higher lateral loads can be supported until:
(4) compression failure of the diagonal strut,
(5) flexural or shear failure of the columns.

In many cases the failure may be a sequential combination of some of the above failure modes. For example, flexural or shear failure of the columns will generally follow a sliding shear failure, or diagonal compression failure, of the masonry. For a particular infilled frame, the strength associated with the various possible failure modes should be calculated, and the lowest value used as the basis for design.

Tension failure mode For infilled frames of high aspect ratio, the critical failure mode may be flexural, involving tensile yield of the tension column steel and of any vertical steel in the tension zone of the infill panel. Under these conditions the frame is acting as a cantilever shear wall, and a reasonably ductile failure mode can be expected. Design can then be in accordance with the recommendations for masonry shear walls in Section 6.4. Special confining steel will be required for compression columns in reinforced concrete frames (see Section 5.5.3).

Sliding shear failure If sliding shear failure of the masonry infill occurs, the equivalent structural mechanism changes from the diagonally braced pin-jointed frame of Figure 6.15(a) to a knee-braced frame shown in Figure 6.16[6.27]. The support provided by the masonry panel forces column hinges to form at approximately midheight and top or bottom of the columns, or may result in column shear failure. The lateral load capacity is initially governed by the combined actions of frictional resistance to sliding shear in the panel, and shear in the columns. Thus the failure load v_u for the panel in Figure 6.16 may be estimated as

$$V_u = \frac{2}{h_e}(M_{u_t} + M_{u_c}) + V_f \tag{6.9}$$

where M_{u_t} and M_{u_c} are the ultimate moment capacities of the tension and compression columns respectively, including the effects of axial forces resulting from gravity loads and overturning moments.

Shear friction force V_f degrades rapidly with cycling[6.27] and should conservatively be ignored in calculating the shear capacity of this failure mode. The effective column height h_e between column hinges is approximately half the storey height h, both for exterior columns and for columns between adjacent infill panels, as in the latter case hinges tend to form close to the quarter points. Thus for a knee-braced frame of n bays wide, and $n+1$ columns, the ultimate shear capacity is

$$V_u = \frac{4}{h}\sum_{i=1}^{n+1} M_{u_i} \tag{6.10}$$

where M_{u_i} is the ultimate moment capacity of the ith column, including axial force effects. Column shear steel should be based on a capacity approach using overstrength column moments to avoid column shear failure.

The equivalent diagonal strut compression force to initiate horizontal shear sliding depends on the infill masonry type, the shear bond strength f_{bs} between the masonry and mortar, and the ratio between the maximum shear and the corresponding vertical stress in the panel[6.23]. From these data, a conservative estimate of the diagonal strut force (Figure 6.15(a)) when sliding initiates for masonry infill frames of typical dimensions and strength may be taken as

$$R_s = \left(0.9 + 0.3\frac{l}{h}\right) f_{bs} h t \tag{6.11}$$

where l/h is the ratio of panel length to height, and t is the panel thickness. Equation 6.11 should be used to define the force required to initiate this failure mode, for comparison with the flexural failure moment and the diagonal crushing force, as below, to establish the probable failure mode. However, for ductile response, the value predicted by Equation 6.10 should be used as the design capacity.

Compression failure of diagonal strut For typical masonry infill panels, Stafford-Smith and Carter[6.23] show that diagonal tensile splitting will precede diagonal crushing. However, the final panel failure force will be dictated by the compression strength, which may thus be used as the ultimate capacity. Trigo[6.28] and Leuchars and Scrivener[6.26] found that the following modified form of the diagonal compression failure force predicted by Stafford-Smith and Carter gave a conservative agreement with test results:

$$R_c = \frac{2}{3}\alpha t f'_m \sec \theta \tag{6.12}$$

where f'_m = masonry prism strength and α defines the contact length between panel and column (see Figure 6.15(a)) given by

$$\alpha = \frac{\pi}{2}\left(\frac{4E_f I_f h_m}{E_m t \sin 2\theta}\right)^{1/4} \tag{6.13}$$

where E_f and I_f are the modulus of elasticity and moment of inertia of the frame columns, E_m and h_m are the modulus of elasticity and height of the infill, and θ is the angle between the diagonal strut and the horizontal. On inelastic cycling the capacity of the diagonal strut will degrade, and behaviour will approximate the knee-braced frame of Figure 6.16.

Ductility The flexural failure mode involving tensile column yield possesses adequate ductility potential but is an uncommon failure mode. Klingner and Bertero[6.25] with work on carefully designed infilled frames with closely spaced vertical and horizontal reinforcement in the infill have shown that high ductilities can be obtained from panel shear mode failure, with good energy absorption characteristics. However, the knee-braced frame concept involves energy dissipation by column hinging in the bottom storey. To provide the required total structure ductility, the level of ductility required of the hinging 'soft storey' will be very large for frames more than two storeys high. It is thus recommended that infilled frames with either sliding shear failure or diagonal compression failure should be designed to respond elastically under seismic attack. A limited structural ductility would be appropriate for frames of moderate height provided well detailed and distributed horizontal and vertical reinforcement is placed in the infill panels, spliced to dowel starters in the frame columns and beams.

Meli[6.18] reports results of tests of a common form of masonry construction consisting of walls of solid clay brick masonry units strengthened by peripheral masonry columns, or pilasters, and bond beams. This form of construction falls in the range between in-filled panel and reinforced masonry shear wall. These tests[6.18] indicate satisfactory behaviour with limited, but significant ductility, provided a flexural failure mode is assured, though performance under combined racking and face loading during an earthquake may be suspect. The reinforced tie columns help to confine the compression corner of the wall. delaying the onset of compression crushing and degradation. It is considered reasonable to design such walls for a ductility factor of 2 [6.18]. It is important, however, that solid masonry units be used for the infill material with this form of construction, or where concrete or steel peripheral frames are provided, as crushing and spalling of shells of light hollow clay brick masonry units results in rapid and severe structural degradation under seismic attack.

6.6 MASONRY COLUMNS

It is inadvisable to use masonry columns as part of a ductile seismic resistant frame because of difficulty in providing adequate ductility, and in designing beam-column joints. Exceptions may be made if the frame is designed to remain fully elastic under seismic attack, or where the strength and ductility of the columns are based on the grout section alone. In the latter case, the brickwork or blockwork is considered only to function as permanent formwork, and is ignored in strength calculations. Care must be taken to ensure that masonry units are well tied into the column core to avoid shedding of the masonry.

REFERENCES

6.1 Glogau, O. A., 'Masonry Performance in Earthquakes', *Bull. N. Z. natn. Soc. Earthq. Engng*, Vol. 7, No. 4, pp. 149–166 (1974)

6.2 Jennings, P. C., Ed., 'Engineering Features of the San Fernando Earthquake of Feb. 9, 1971', Publication EERL 71-02, California Institute of Technology (1971)

6.3 Scrivener, J. C. and Williams, D., 'Compressive Behaviour of Masonry Prisms', *Proc. Third Australasian Conference on the Mechanics of Structures and Materials*, Vol. II, Session B4, Univ. of Auckland (1971)

6.4 Hilsdorf, H. K., 'Investigation into the Failure Mechanism of Brick Masonry Loaded in Axial Compression', *Proc. International Conference on Masonry Structural Systems*, Univ. Texas, 1967, published as *Designing, Engineering and Constructing with Masonry Products*, Johnson, F. B., Ed., Gulf Pub. Co. Houston (1969)

6.5 Foster, P. K. and Bridgeman, D. O., 'Prism Tests for the Design and Control of Brick Masonry', New Zealand Pottery and Ceramics Research Association, Technical Report No. 22 (1973)

6.6 Priestley, M. J. N. and Bridgeman, D. O., 'Seismic Resistance of Brick Masonry Walls', *Bull. N. Z. natn. Soc. Earthq. Engng*, Vol. 7, No. 4, pp. 167–187 (1974)

6.7 Mayes, R. L., Omote, Y. and Clough, R. W., 'Cyclic Shear Tests on Masonry Piers', *Proc. Sixth World Conference on Earthquake Engineering*, New Delhi (1977)

6.8 Skinner, R. I., 'Earthquake-generated Forces and Movements in Tall Buildings', New Zealand Department of Scientific and Industrial Research, *Bulletin No.* 166 (1964)

6.9 Sahlin, S., *Structural Masonry*, Prentice-Hall, New Jersey (1971)

6.10 Priestley, M. J. N., 'Cyclic Testing of Heavily Reinforced, Concrete Masonry Shear Walls', University of Canterbury Research Report 76-12 (1976)

6.11 Cardenas, A. E. *et al.*, 'Design Provisions for Shear Walls', *J. Am. Concr. Inst.*, Vol. 70, No. 3, pp. 221–230 (1973)

6.12 Blakeley, R. W. G., Cooney, R. C. and Megget, L. M., 'Seismic Shear Loading at Flexural Capacity in Cantilever Wall Structures', *Bull. N. Z. natn. Soc. Earthq. Engng*, Vol. 8, No. 4, pp. 278–290 (1975)

6.13 'DZ4210 – Masonry Design and Construction', Draft Code, Standards Association of New Zealand (1974)

6.14 ACI Committee 531, 'Concrete Masonry Structures – Design and Construction', *J. Am. Concr. Inst.*, Vol. 67, No. 6, pp. 442–460 (1970)

6.15 Scrivener, J. C., 'Static Racking Tests on Concrete Masonry Walls', Proc. International Conference on Masonry Structural Systems, Univ. Texas, pp. 185–191 (1967)

6.16 Schneider, R. R., 'Lateral Load Tests on Reinforced Grouted Masonry Shear Walls', Univ. Southern California Eng. Centre Report 70-101 (1959)

6.17 'Working Draft of Recommended Comprehensive Seismic Design Provisions for Buildings', Applied Technology Council, San Francisco (1976)

6.18 Meli, R., 'Behaviour of Masonry Walls Under Lateral Loads', *Proc. Fifth World Conference on Earthquake Engineering*, Rome, pp. 853–862 (1974)

6.19 Clough, R. W., 'Effect of Stiffness Degradation on Earthquake Ductility Requirements', Univ. California Structural Engineering Lab. Report No. 66-16 (1966)

6.20 Thompson, K. J., 'Ductility of Concrete Frames Under Seismic Loading', Ph.D. Thesis, Univ. Canterbury Civil Engineering Department Research Report No. 75-14 (1975)

6.21 *General Structural Design and Design Loading for Buildings*, NZS4203:1976, Standards Association of New Zealand (1976)

6.22 Newmark, N. M. and Rosenblueth, E., *Fundamentals of Earthquake Engineering*, Prentice-Hall Inc., Englewood Cliffs, N.J. (1971)

6.23 Stafford-Smith, B. and Carter, C., 'A Method of Analysis for Infilled Frames', *Proc. Inst. Civ. Engrs*, Vol. 44, pp. 31–48 (1969)

6.24 Esteva, L., 'Behaviour Under Alternating Loads of Masonry Diaphragms Framed by Reinforced Concrete Members', *Proc. International Symposium on the Effects of Repeated Loading on Materials and Structural Systems*, RILEM, Vol. 5, Mexico (1966)

6.25 Klingner, R. E. and Bertero, V. V., 'Infilled Frames in Aseismic Construction', *Proc. Sixth World Conference on Earthquake Engineering*, New Delhi (1977)

6.26 Leuchars, J. M. and Scrivener, J. C., 'Masonry Infill Panels Subjected to Cyclic In-Plane Loading', *Bull. N. Z. natn. Soc. Earthq. Engng*, Vol. 9, No. 2, pp. 122–131 (1976)

6.27 Fiorato, A. E., Sozen, M. A. and Gamble, W. L., 'An Investigation of the Interaction of Reinforced Concrete Frames with Masonry Filler Walls', Structural Research Service No. 370, Univ. Illinois (1970)

6.28 Trigo, J. d'A., 'Estructuras de Paineis sob a Acçaõ de Solicitaçoes Horizontais', Laboratório Nacional de Engenharia Civil, Lisbon (1968)

6.29 ACI Committee 318, *Building Code Requirements for Reinforced Concrete* (ACI 318-77), American Concrete Institute, Detroit (1977)

7 Foundations

Robert V. Whitman and Jacobo Bielak

7.1 GENERAL CONSIDERATIONS

This chapter deals with two interrelated problems: the effect of a structure's foundation upon motions and forces induced in the structure during an earthquake, and design of foundations to resist the base shear and overturning moment from the structure. The first problem is referred to as *soil-structure interaction.* Foundation design involves choice of type of foundation and proportioning of its elements, and may include ensuring that the soil beneath the structure does not liquefy or compact excessively.

Interaction is closely related to design: the design of a foundation influences the magnitude of forces induced in the structure, and, in turn, these forces influence the design. This principle, which applies to the whole of the structure, is especially important in the case of the foundation because structure and foundation designs are often carried out by separate engineering organizations.

Definition of soil-structure interaction If the motion at any point on the soil-structure interface differs from the motion that would occur at this point in the free field if the structure were not present, there is soil-structure interaction (Figure 7.1). This is a very general definition: if the interface moves or distorts differently than the corresponding surface in the free field, there is interaction. Average horizontal and vertical translation, rocking about a horizontal axis, and twisting about a vertical axis are all included in the definition.

In many cases modification to free-field motions will be insignificant compared to uncertainties in the characteristics of these motions, or the motions of the structure as a result of foundation rocking and twisting will be small compared to the distortions within the structure itself. These cases can be treated as if there were no interaction; that is, the structure is imagined to rest on a rigid base which moves as the ground motion.

Mats and interconnected footings (and interconnected pile caps) have considerable in-plane rigidity with respect to horizontal translation. This may not be true for vertical translation, rocking, and twisting, although additional stiffening is provided by the structural system supported by the elements. The side walls of embedded structures may also have limited rigidity. Lack of rigidity at the interface greatly complicates the analysis of

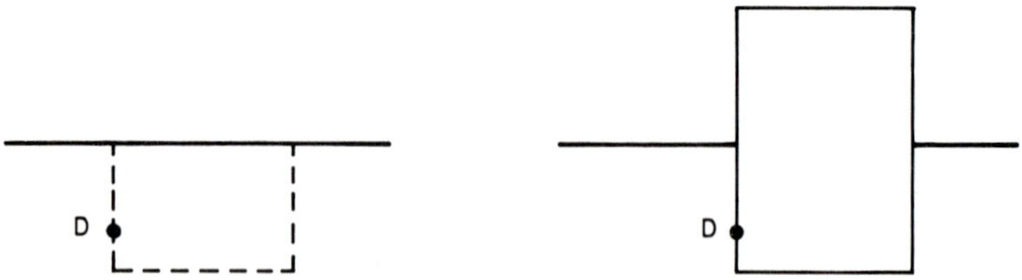

Fig. 7.1 Definition of soil-structure interaction. If motions at points D differ, there is soil-structure interaction

interaction. In practice, however, most structures for which interaction is an important consideration have foundations which behave in an essentially rigid manner.

Aims of the chapter Many techniques for analyzing soil-structure interaction have been developed. This chapter strives to tell when and in what way interaction may be important, and to introduce some methods of analysis that may be used when interaction must be considered.

It is useful to consider two types of interaction —kinematic and inertial — as discussed in the next section.

7.2 FORMULATION OF THE PROBLEM OF DYNAMIC SOIL-STRUCTURE INTERACTION

Let us adopt a particular, common way of specifying the problem of analyzing the seismic response of a building: given an input ground motion at some elevation in the free field, find the structural response. Ground motion specification would typically take the form of giving the ground motion at a point plus stating that this motion is caused by some form of wave. In many cases it would be assumed that this motion is caused by vertically propagating plane SH or P waves.

To carry out the analysis in a single step, we may take our system to consist of the structure plus the rest of the earth, the latter idealized as a homogeneous or stratified halfspace. It is generally necessary to consider nonlinear behaviour of soil. This is almost always done by using a quasi-linear iterative analysis[7.1]. A linear solution technique is used and repeated iteratively, adjusting soil properties for the current iteration consistently with the strains computed in the previous iteration, until attaining a satisfactory degree of convergence (see Chapter 2).

A typical set of results appears in Figure 7.2. The input ground motion was specified at ground surface away from the structure. The figure shows, in the form of response spectra, the computed motions at the structure's base and top and at a point in the soil, well away from the structure, at the elevation of the structure's base.

Understanding of such results is aided by noting that each linear analysis

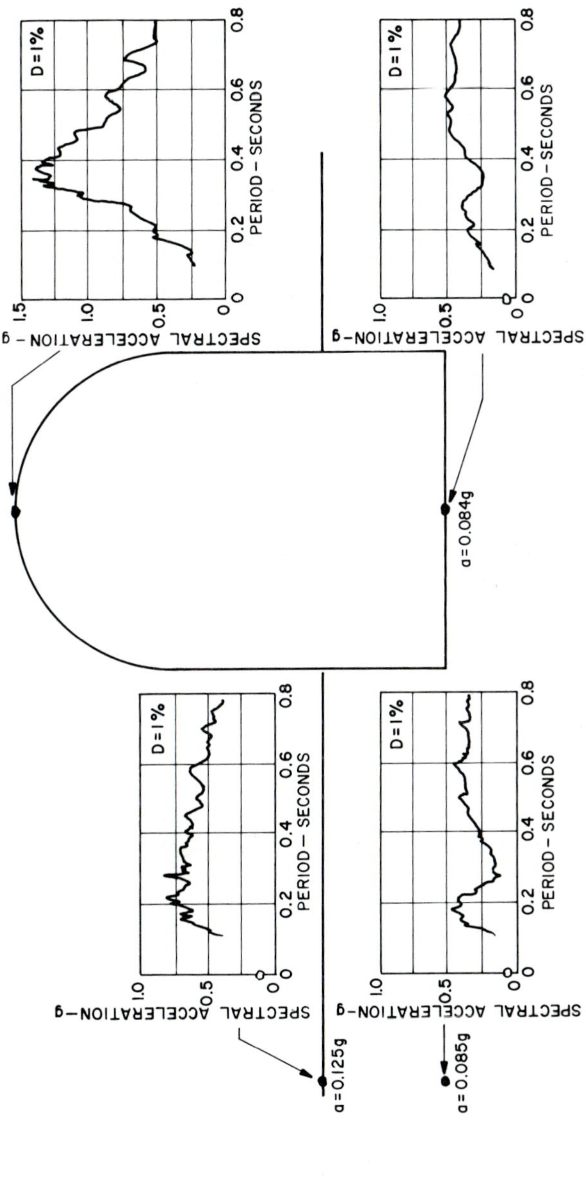

Fig. 7.2 Typical results of soil–structure interaction analysis (after Idriss[7.25]*)*

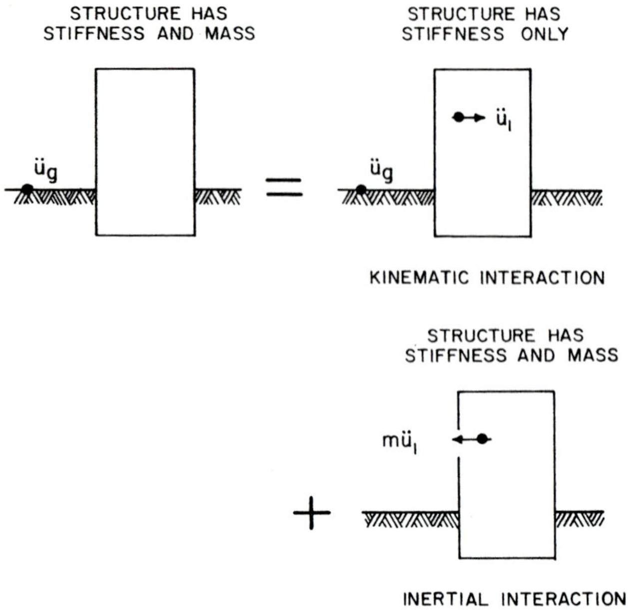

Fig. 7.3 Two-step solution for soil–structure interaction problem

may be broken into the two steps depicted in Figure 7.3, according to the superposition principle for dynamic soil-structure interaction[7.2,7.3].

(1) The structure's mass is ignored but its stiffness is included. The resulting motion of the structure at each mass point is computed.
(2) At each mass point, the acceleration computed in Step 1 is multiplied by the mass to obtain a dynamic force, and the structural response to this set of forces is computed.

Total motions and stresses are found by adding those computed in the two steps. The first step introduces variation of free-field ground motions with elevation (and with horizontal distance, if there is horizontal propagation of the input), plus interaction with these motions of a deformable, but weightless, structure; this is called *kinematic interaction.* The structure's mass is introduced in the second step, resulting in *inertial interaction.*

Validity of this superposition principle can be proved by partitioning the matrix equation for a finite element representation of the soil-structure system. The basic equations of motion are

$$\mathbf{M}\ddot{\mathbf{u}} + \mathbf{C}\dot{\mathbf{y}} + \mathbf{K}\mathbf{y} = \mathbf{0} \tag{7.1}$$

in which **M, C,** and **K** are the system mass, damping, and stiffness matrices; and **u** and **y** are the absolute and relative displacemments with respect to some ground reference system. The solution of this equation is equivalent to

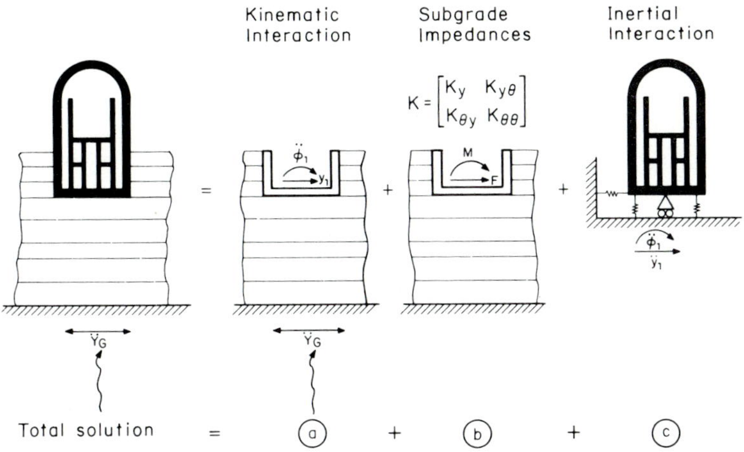

Fig. 7.4 Three-step solution with rigid foundation base (after Kausel and Roesset[7.4])

the solution of the matrix equations

$$\mathbf{M}_{so}\ddot{\mathbf{u}}_1 + \mathbf{C}\dot{\mathbf{y}}_1 + \mathbf{K}\mathbf{y}_1 = 0 \tag{7.2}$$

and

$$\mathbf{M}\ddot{\mathbf{y}}_2 + \mathbf{C}\dot{\mathbf{y}}_2 + \mathbf{K}\mathbf{y}_2 = -\mathbf{M}_{st}\ddot{\mathbf{u}}_1 \tag{7.3}$$

in which $\mathbf{u}_1 = \mathbf{y}_1 + \mathbf{u}_g$; $\mathbf{u} = \mathbf{u}_1 + \mathbf{y}_2$; $\mathbf{y} = \mathbf{y}_1 + \mathbf{y}_2$; and $\mathbf{M} = \mathbf{M}_{so} + \mathbf{M}_{st}$. The value of $\mathbf{M}_{so}$ excludes the structure's mass, while $\mathbf{M}_{st}$ excludes the soil mass. Equivalence of Equations 7.2 and 7.3 with Equation 7.1 follows from simple addition.

If the foundation is rigid near the soil-structure interface, then the structure itself is not required in Step 1 — only the rigid massless interface. On the other hand, since fictitious forces are applied on the structure alone in Equation 7.3, it is possible to eliminate the degree of freedom in the soil and express the subgrade effect in terms of impedance functions (dynamic stiffness functions) for a solution in the frequency domain. In addition, the fictitious inertia forces in the structure can be expressed in terms of the rigid translations and rotations defined by the kinematic interaction. It follows that $\mathbf{u}_1$ in Equation 7.3 can be regarded as a vector of displacements relative to a fictitious support, while the rigid body translations and rotations of Step 1 are the support motion. For a rigid foundation, it is therefore valid to break the solution into steps (a), (b) and (c), as shown in Figure 7.4. Steps (b) and (c) in Figure 7.4 amount to the inertial interaction formulation described by Equation 7.3.

Alternative formulation for deformable foundations The three-step procedure developed for analysis of soil-structure interaction with rigid

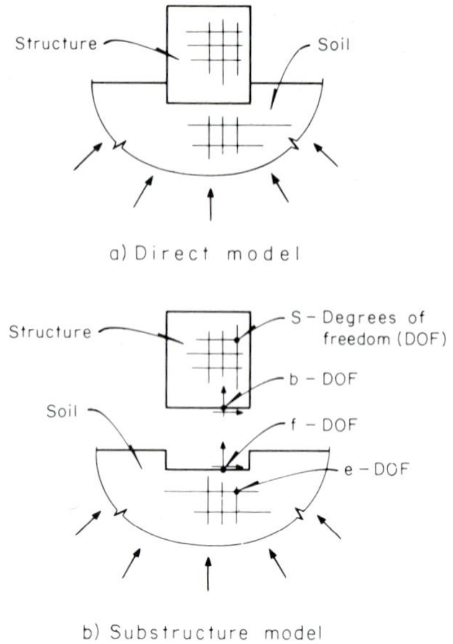

Fig. 7.5 Substructure method applied to soil–structure interaction problem (after Gutierrez[7.6])

foundations (Figure 7.4) has been extended[7.5,7.6] to deformable foundations by the substructure method. In it, structure and soil are regarded as substructures of the complete system, with $\mathbf{R}_b(t)$ and $\mathbf{R}_f(t)$ representing the interaction forces at the interface. With reference to Figure 7.5, the frequency domain equations for the structure can be written as

$$\left(-\omega^2\begin{bmatrix}\mathbf{M}_s & \mathbf{0}\\ \mathbf{0} & \mathbf{M}_b\end{bmatrix}+i\omega\begin{bmatrix}\mathbf{C}_{ss} & \mathbf{C}_{sb}\\ \mathbf{C}_{bs} & \mathbf{C}_{bb}\end{bmatrix}+\begin{bmatrix}\mathbf{K}_{ss} & \mathbf{K}_{sb}\\ \mathbf{K}_{bs} & \mathbf{K}_{bb}\end{bmatrix}\right)\begin{Bmatrix}\bar{\mathbf{u}}_s^t(\omega)\\ \bar{\mathbf{u}}_b^t(\omega)\end{Bmatrix}=\begin{Bmatrix}\mathbf{0}\\ \bar{\mathbf{R}}_b(\omega)\end{Bmatrix} \tag{7.4}$$

and, for the foundation, as

$$\left(-\omega^2\begin{bmatrix}\mathbf{M}_f & \mathbf{0}\\ \mathbf{0} & \mathbf{M}_e\end{bmatrix}+i\omega\begin{bmatrix}\mathbf{C}_{ff} & \mathbf{C}_{fe}\\ \mathbf{C}_{ef} & \mathbf{C}_{ee}\end{bmatrix}+\begin{bmatrix}\mathbf{K}_{ff} & \mathbf{K}_{fe}\\ \mathbf{K}_{ef} & \mathbf{K}_{ee}\end{bmatrix}\right)\begin{Bmatrix}\bar{\mathbf{u}}_f^t(\omega)\\ \bar{\mathbf{u}}_e^t(\omega)\end{Bmatrix}=\begin{Bmatrix}\bar{\mathbf{R}}_f(\omega)\\ \mathbf{0}\end{Bmatrix} \tag{7.5}$$

where ω = frequency variable, an overbar denotes Fourier transform, and superscript t indicates total nodal point displacements measured with respect to a fixed reference.

The prescribed excitation is taken as the free-field motion $\mathbf{u}_f^o(t)$ that would have occurred at the excavated soil-structure interface had there been no building. Kinematic interaction is thus automatically incorporated into the formulation, since the effective input motion includes the perturbation of the original excitation due to excavation for the foundation. Clearly, $\mathbf{u}_f^o$ depends only on the geometry of the soil-structure interface, the soil properties, and the original unperturbed excitation, and must satisfy the homogeneous equations of motion

$$\left(-\omega^2\begin{bmatrix}\mathbf{M}_f & \mathbf{0}\\ \mathbf{0} & \mathbf{M}_e\end{bmatrix}+i\omega\begin{bmatrix}\mathbf{C}_{ff} & \mathbf{C}_{fe}\\ \mathbf{C}_{ef} & \mathbf{C}_{ee}\end{bmatrix}+\begin{bmatrix}\mathbf{K}_{ff} & \mathbf{K}_{fe}\\ \mathbf{K}_{ef} & \mathbf{K}_{ee}\end{bmatrix}\right)\begin{Bmatrix}\bar{\mathbf{u}}_f^0(\omega)\\ \bar{\mathbf{u}}_e^0(\omega)\end{Bmatrix}=\begin{Bmatrix}\mathbf{0}\\ \mathbf{0}\end{Bmatrix} \quad (7.6)$$

where the displacements $\mathbf{u}_e^o$ are the total free-field displacements within the soil.

With the aid of Equations 7.5 and 7.6, and the compatibility conditions

$$\bar{\mathbf{u}}_b^t(\omega)=\bar{\mathbf{u}}_f^t(\omega) \quad \text{and} \quad \bar{\mathbf{R}}_b(\omega)=-\bar{\mathbf{R}}_f(\omega) \quad (7.7)$$

imposed to ensure continuity of displacements and tractions across the soil-structure interface, it can be shown that Equation 7.4 leads to

$$\left(-\omega^2\begin{bmatrix}\mathbf{M}_s & \mathbf{0}\\ \mathbf{0} & \mathbf{M}_b\end{bmatrix}+i\omega\begin{bmatrix}\mathbf{C}_{ss} & \mathbf{C}_{sb}\\ \mathbf{C}_{bs} & \mathbf{C}_{bb}\end{bmatrix}+\begin{bmatrix}\mathbf{K}_{ss} & \mathbf{K}_{sb}\\ \mathbf{K}_{bs} & \mathbf{K}_{bb}+\mathbf{S}_f(\omega)\end{bmatrix}\right)\begin{Bmatrix}\bar{\mathbf{u}}_s^t(\omega)\\ \bar{\mathbf{u}}_b^t(\omega)\end{Bmatrix}=\begin{Bmatrix}\mathbf{0}\\ \mathbf{S}_f(\omega)\,\bar{\mathbf{u}}_f^0\end{Bmatrix} \quad (7.8)$$

where $\mathbf{S}_f(\omega)$ = complex-valued impedance matrix relating the forces and displacements at the structure-soil interface[7.6]. Elements of this matrix depend only on soil properties and the interface geometry and — as for rigid foundations — can be interpreted as frequency-dependent stiffness and damping coefficients of a set of coupled springs and dampers.

Equation 7.8 governs the motion of the structure due to an arbitrarily prescribed free-field excitation at the structure-soil interface. This equation shows that analysis for any linear soil-structure interaction system may be carried out by a three-step procedure similar to the one described for rigid foundations.

Solution of Equation 7.8 yields the combined effects of the two types of interaction — kinematic and inertial. (Pure inertial interaction is governed also by Equation 7.8, with the effective excitation $\bar{\mathbf{u}}_f^o$ replaced by the corresponding unperturbed free-field displacements.) Having determined the structural response, $\bar{\mathbf{u}}_s^t(\omega)$ and $\bar{\mathbf{u}}_b^t(\omega)$ from Equation 7.8, displacements $\bar{\mathbf{u}}_e^t(\omega)$ in the soil's interior are obtained directly from Equations 7.4, 7.5 and 7.7. Displacements in the time domain are then found by inverting the corresponding Fourier transforms.

7.3 KINEMATIC INTERACTION

The problem of kinematic interaction is one of diffraction — determining the effect a region with mechanical properties differing from those of unperturbed soil (i.e. the excavation and massless foundation) has on the excited motion — in which the incident and reflected waves make up the unperturbed soil motion and the diffracted wave is the motion that must be added to the former to get the effective input. Only when the foundation is supported on the ground surface and the incoming excitation is represented by plane waves travelling vertically upward will there be no wave diffraction. Any other assumption concerning the input motion and geometry of the foundation will result in an effective excitation that will not coincide with the unperturbed motion; hence, kinematic interaction. Changes in the unperturbed motion will occur as a result of inclined incidence of the incoming waves and foundation embedment. Effects of these factors, which include rocking and twisting of the foundation and suppression of high-frequency components of the ground motion, will be automatically incorporated into the analysis — whether the corresponding wave diffraction problem is solved as the first step of the superposition principle, or from the explicit formulation as required by the substructure method — once the unperturbed motion has been specified.

For engineering purposes, motions are often obtained by introducing several simplifying assumptions. To better understand the implications and degree of approximation of various assumptions in the analysis, it is convenient to study the individual factors separately. This is helpful also in developing alternative methods for evaluating kinematic interaction which, though approximate, may be used in practice without resorting to a full wave-diffraction analysis.

Effects of foundation geometry Observations of the behavior of buildings during earthquakes indicate that the maximum horizontal amplitude of a foundation is always smaller than that of the adjacent ground, and that the difference becomes larger as the wavelength becomes shorter. As a consequence of nonvertical incidence of seismic waves, motions at different points of the ground surface are not in phase during an earthquake, as illustrated in Figure 7.6. The accelerograms depicted were recorded simultaneously at several locations on the ground of the Hachinohe Technical College (Figure 7.7). Because the building has no basement, the ground-floor motion is that of the foundation. Movements at two points, A and B, away from the building are almost in phase for long wavelengths (long periods) but out of phase for shorter ones. The motions of shorter periods on the ground scarcely appear in the motion on the building floor (location C). Since buildings, in general, have enough strength and rigidity to resist stresses caused by ground-motion phase differences, the foundation slab will average out the ground-motion vectors.

The average amplitude of the building foundation may be obtained by integrating the displacements beneath the foundation slab over the length, with the result

$$\eta = \frac{\xi}{\pi} \sin \frac{\pi}{\xi} \tag{7.9}$$

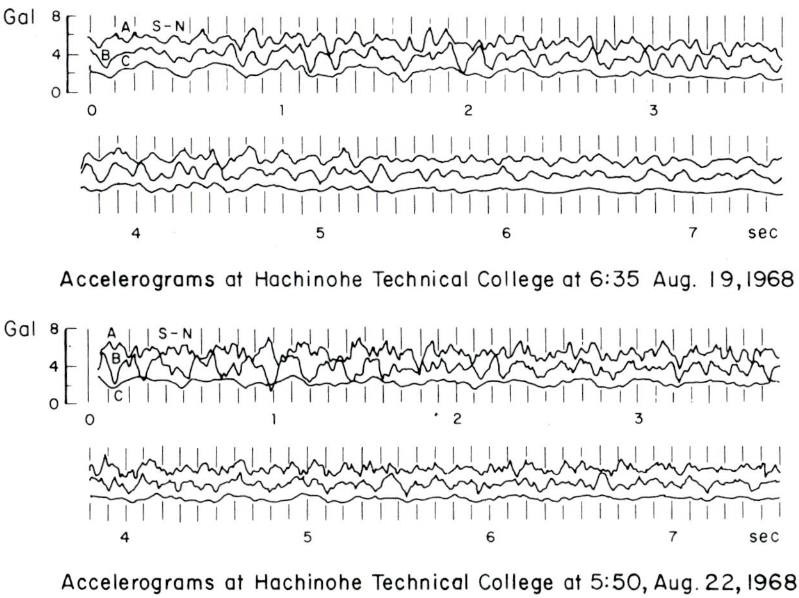

Fig. 7.6 Accelerograms of three ground locations in Hachinohe Technical College (after Yamahara[7.7])

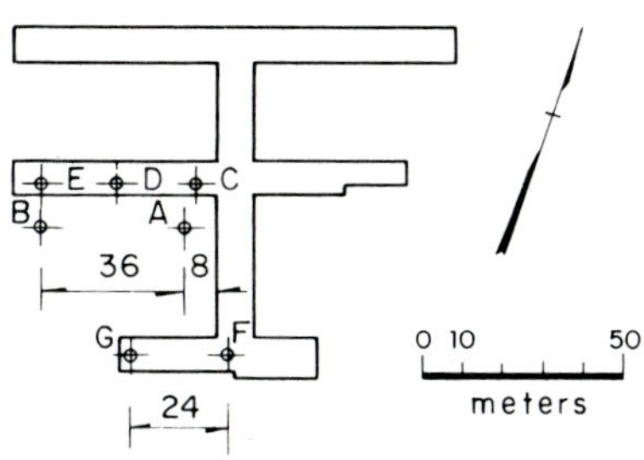

Fig. 7.7 Locations of measured points in Hachinohe Technical College (after Yamahara[7.7])

in which η denotes the ratio of mean to particle amplitude, and ξ is the ratio of wavelength, $L = VT$, to the length of the building, l. This equation is valid for rectangular foundations and for harmonic waves of period T traveling parallel to one side of the foundation with apparent surface velocity V[7.7]. A graphic representation of Equation 7.9 (Figure 7.8) shows that the reduction factor η depends strongly on the relative wavelength developed on the ground surface.

For actual earthquakes recorded in the free field, averaging of the digitized version of the ground motion $f(t)$,

$$f(t) = f(k\Delta t) = f_k \tag{7.10}$$

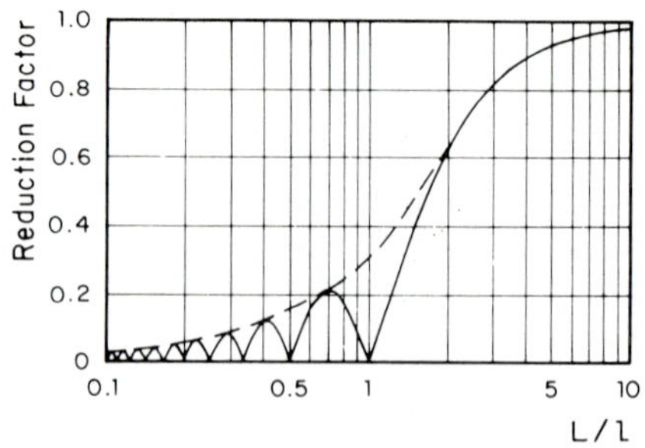

Fig. 7.8 Ratio of mean to particle amplitude due to presence of rigid base (after Yamahara[7.7])

in which Δt is the digital time increment has been suggested[7.7] according to the rule

$$g_k = \tfrac{1}{n}\left[f_{k-(n-1)/2} + \dots + f_{k-1} + f_k + f_{k+1} \dots f_{k+(n-1)/2}\right] \tag{7.11}$$

This equation, in which g_k is the average of f_k over the transit time $\tau = n\Delta t$ that it takes for a wave to traverse the building length, provides a filtering effect for discretized functions similar to that of Equation 7.9 for harmonic functions. The value of n is determined from

$$n = \frac{\tau}{\Delta t} = \frac{l}{\Delta t V} \tag{7.12}$$

To illustrate the effect of smoothing out the acceleration input on the response of structures, acceleration spectra corresponding to a damping ratio $\zeta = 0.05$ have been presented[7.7] — both for the free-field record of the Tokachi-oki earthquake and for the modified records obtained by varying the transit factor n (Figure 7.9). Similar results have been reported by other authors[7.8–7.10].

Torsion A massless foundation supported on the ground surface will experience, in general, both torsion and rocking as a consequence of the nonuniform distribution of free-field displacements beneath the foundation. Torsion is potentially the more serious problem.

Torsional excitation generated by obliquely-incident seismic waves has been evaluated by approximating the twist of the foundation by the rotation of the free-field motion about a vertical axis[7.11]. This leads to a foundation twist proportional to the velocity of the free-field motion. Input rotation at the level of the foundation has also been approximated by the average second polar moment of the free-field tangential displacement under the foundation[7.12]. In addition, the problem of torsional steady-state response of a circular massless rigid disc on an elastic halfspace excited by an obliquely-

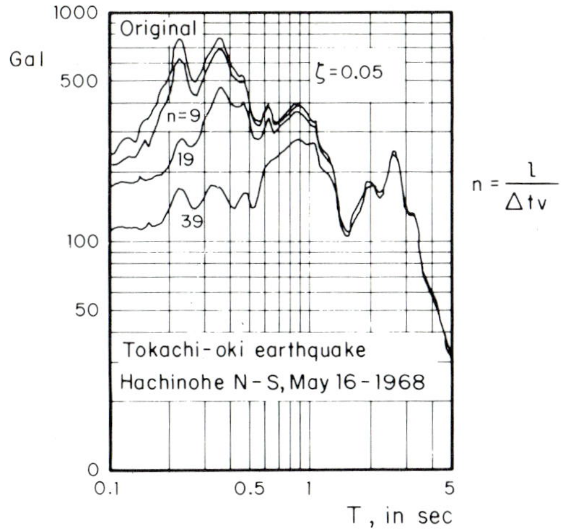

Fig. 7.9 Response spectra for original and modified accelerograms (after Yamahara[7.7]*)*

incident plane SH harmonic wave has been formulated exactly and solved numerically.[7.13] Results are shown in Figure 7.10. The real and imaginary parts of the normalized induced twist $a\alpha^*/u_{go}$ (u_{go} = amplitude of unperturbed surface motion) are shown as functions of the dimensionless frequency $a_o = \omega a/V_s$ for several values of the angle of incidence. Angles $\theta = 0°$ and 90° correspond, respectively, to horizontal and vertical incidence. Input twist vanishes identically for $\theta = 90°$ and increases with decreasing θ for values of $a_o \leqq 3$. The maximum input twist occurs when $a_o \simeq 2\cos^{-1}\theta$, i.e. when the apparent wavelength on the soil surface, $\lambda = 2\pi V_s/(\omega\cos\theta)$, is approximately three times the foundation radius. For these particular frequencies, the amplitude of the tangential component of displacement on the disc perimeter, $a\alpha^*$, attains 60–70% of the free-field amplitude.

Figure 7.10 applies only to plane SH oblique incident waves. To evaluate the twist at the base of foundations of arbitrary shape under actual earthquake excitations, the averaging rule suggested by Iguchi[7.12] may be used as an approximation. This rule has been shown to give satisfactory results for the simplified problem considered by Luco[7.13].

Induced excitation can produce significant torsional earthquake effects even in symmetrical building-foundation systems; it should be taken into account in design. However, since actual ground motions do not consist entirely of coherent traveling waves, direct use of theoretical solutions tends to overestimate the effect.

Effects of foundation embedment Earthquake accelerograms recorded at a number of locations show that the motion applied at a given site is generally greatest on the ground surface and decreases with depth. This suggests that the effective excitation at the soil-structure interface resulting from kinematic

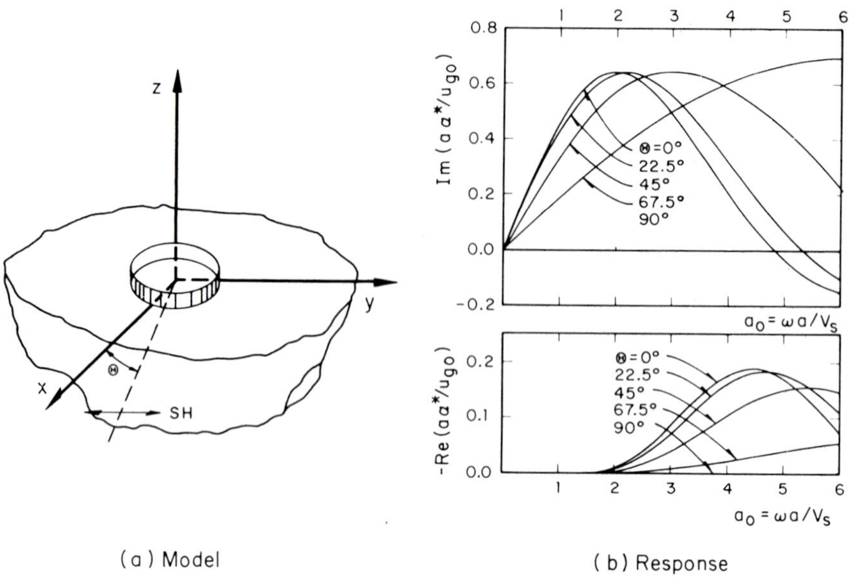

Fig. 7.10 Torsional response of rigid disc to oblique incident waves (after Luco[7.13])

interaction will be a decreasing function of foundation depth.

To investigate embedment effects, the motion of a rigid foundation of elliptical cross-section embedded in an elastic halfspace has been considered subject to plane SH harmonic waves with incident angle θ [7.14]. In Figure 7.11 is shown the base displacement normalized amplitude $|u/u_{go}|$ as function of the dimensionless frequency of excitation a_o for several values of the embedment ratio b/a and of the angle of incidence. Embedment had the general effect of reducing the response at the rigid soil-foundation interface with respect to the free-field surface motion. The extent of this reduction varies with the frequency and incidence angle of the incoming waves. To describe a_o in terms of actual physical quantities, a value of 1.5, for instance, might correspond to a frequency excitation of 5 Hz acting on a base of halfwidth $a = 20$ m founded on a soil with a shear-wave velocity $V_s = 420$ m/s. Effects of the angle of incidence are most pronounced for shallow foundations and become less significant as the base approaches a circular cross-section.

Embedment also causes rocking of the foundation. Approximate formulas have been presented[7.15] on the basis of one-dimensional wave propagation theory for calculating decrease of input motion and amount of rocking. These formulas are given as frequency-dependent correction factors for the Fourier transform of the free-field surface accelration. This yields satisfactory results for a wide range of embedment ratios.

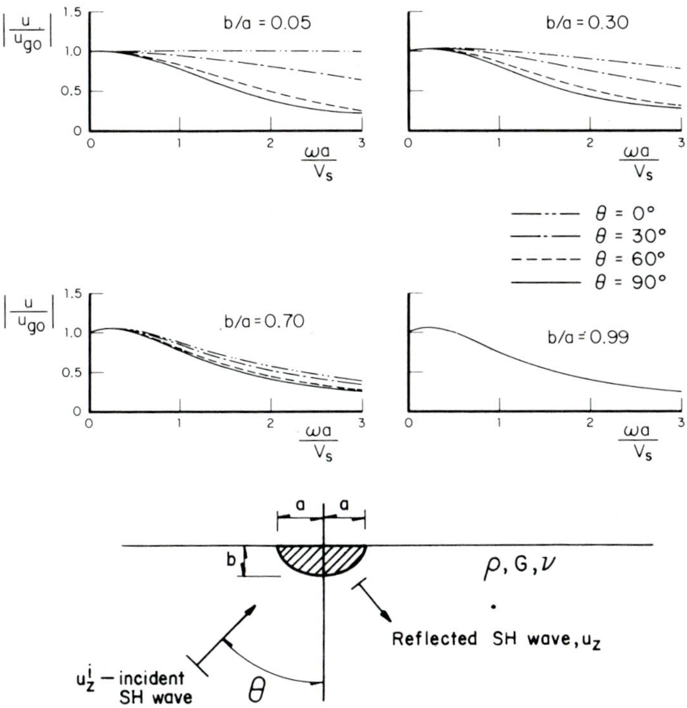

Fig. 7.11 Response of embedded foundation to oblique incident waves (after Wong[7.14]*)*

7.4 INERTIAL INTERACTION

Having determined the kinematic interaction between the foundation and the soil medium, the problem of inertial interaction reduces to finding the motion of the system caused by application of prescribed fictitious inertia forces on the structure alone (Equation 7.3). This, in general, involves the simultaneous computation of the coupled response of structure and soil in one step, unless use is made of the substructure method (Equation 7.8). If the foundation is rigid, the procedure is greatly simplified and the resulting motion of the structure and the ground can be obtained separately. Frequency-dependent subgrade impedances for the relevant degrees of freedom are evaluated first. This is followed by computation by direct extensions of standard methods of structural dynamics (see Chapter 2)[7.16–7.21] of the response of the actual structure, supported on springs and dampers and subjected to the motion calculated under dynamic excitation[7.4]. This exciting motion generally will consist of both translations and rotations of the base (Figure 7.4(c)).

Though actual building foundations are not rigid, results of a study[7.22] on the forced vibration of a nine-story concrete building with a rather common

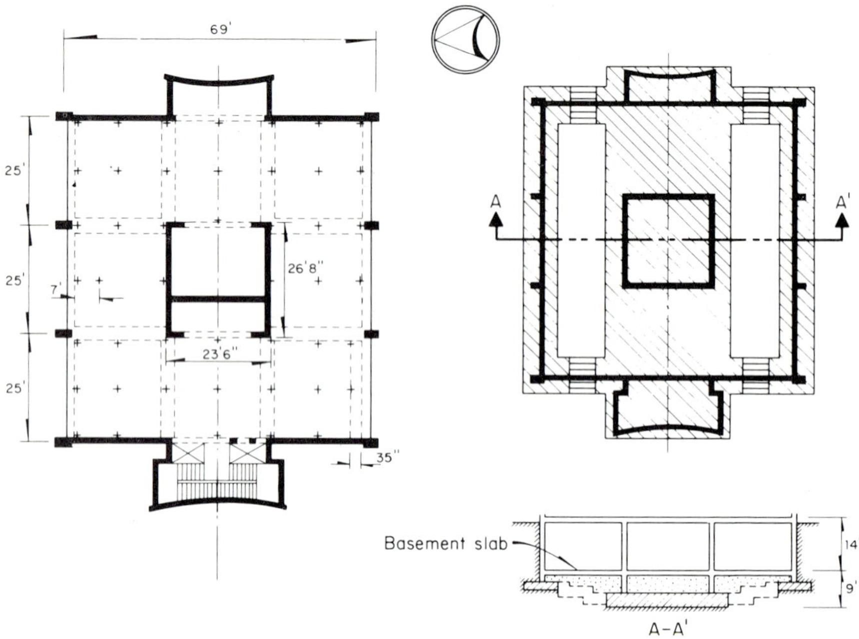

Fig. 7.12 Typical floor and foundation plan and cross section of Millikan Library (after Wong, Luco and Trifunac[7.22])

configuration of superstructure and flexible foundation (Figure 7.12) suggest that the rigid-foundation assumption is appropriate in evaluating the overall dynamic response of many buildings.

The vertical, N–S, and E–W components of motion at fifty locations in the basement were recorded during shaking of the building. Records showed that the basement slab did not behave as a rigid plate (Figure 7.13). Theoretical studies were made assuming both a rigid foundation and a more realistic model incorporating the flexibility of the foundation. It was found that the base shear and moment acting at the contact and the average motion of the basement slab were almost the same for the flexible and rigid models. Thus, once base shear and base overturning moment have been determined from the simplified rigid-mat assumption, strains and stresses in the foundation and adjoining soil may be calculated by using a more realistic model of the foundation.

To aid in understanding the way in which base shear and overturning moments of structures founded on deformable soil differ from the corresponding values for rigid ground, consider the idealized single-story interaction system shown in Figure 7.14 which is a linear structure with a rigid circular base mass of radius a resting on the surface of a homogeneous, elastic halfspace. The free-field motion of the ground surface is taken as a horizontally-directed simple harmonic motion with a period T_o and

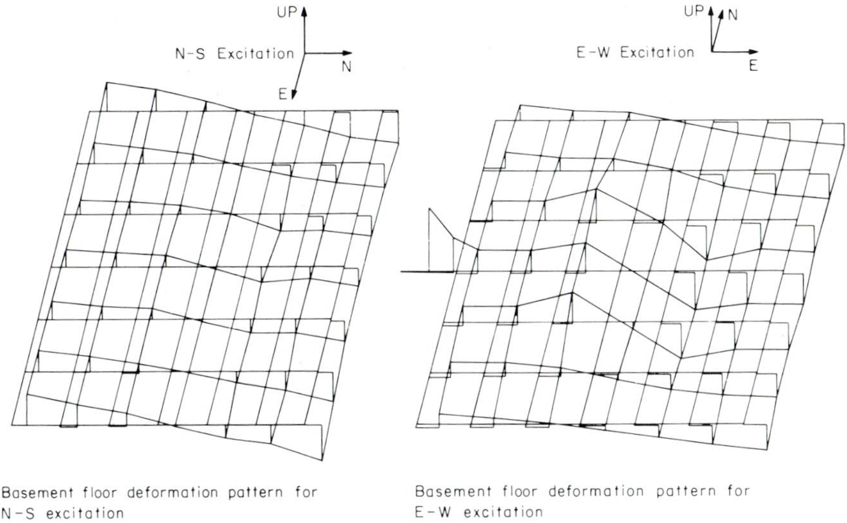

Fig. 7.13 Millikan Library basement floor deformation (after Wong, Luco and Trifunac[7.22])

acceleration amplitude $\ddot{v}_g$. With these specifications, the problem is one of pure inertial interaction. The system, which has a single degree of freedom when supported on rigid soil and three degrees of freedom when flexibly supported, may be viewed either as a direct model of a one-story building frame or as a one-mode approximation of a multi-story structure. In the latter case, m_1, k_1, c_1, and h_1 must be replaced with the corresponding modal parameters[7.16].

Solid lines in Figure 7.15 represent response spectra for the steady-state amplitude Q of the base shear in the columns[7.23]. The relative flexibility parameter for the soil and the structure, φ_o, is defined as

$$\varphi_o = \frac{h_1}{V_s T} \tag{7.13}$$

in which T is the natural period of the fixed-base structure. $\varphi_o = 0$ corresponds to a rigidly supported structure.

Comparison of results in this figure reveals that effects of inertial interaction are most strikingly reflected in a shift to the right of the peak of the response spectrum and in a change of the magnitude of the peak. These changes can be conveniently expressed by an increase in the natural period of the system over its fixed-base value and by a change in its damping ratio.

Shown in dotted lines are response spectra for single-degree oscillators whose natural period and damping have been adjusted so that the absolute maximum (resonant) value of the base shear and the associated period are in each case identical to those of the actual interacting systems. With these

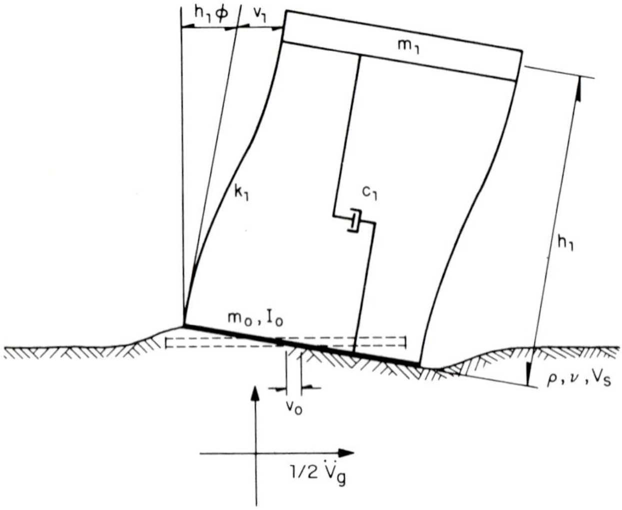

Fig. 7.14 Model of building-foundation system

properties of the replacement oscillators, response spectra for the actual and the replacement systems are in excellent agreement over wide ranges of the excitation period on both sides of the resonant period. This suggests that no significant differences are to be expected in the response to earthquake motion of the system analyzed, as calculated with the replacement oscillator method or by a more rigorous approach, such as that provided by the Fourier operational method[7.16,7.19].

The natural period $\tilde{T}$ of the replacement oscillator is given approximately by[7.23]

$$\tilde{T} = T\left[1 + \frac{k_1}{k_y} + \frac{k_1 h_1^2}{k_\theta}\right]^{1/2} \tag{7.14}$$

in which k_y is the dynamic translational stiffness of the foundation in harmonic motion, and k_θ is the associated rocking stiffness.

The corresponding damping ratio $\tilde{\zeta}$ may be estimated from[7.23]

$$\tilde{\zeta} = \left(\frac{T}{\tilde{T}}\right)^3 \zeta + \zeta_o \tag{7.15}$$

where ζ is the critical damping factor of the superstructure if rigidly supported and ζ_o (Figure 7.16) measures the energy dissipated by the soil in the form of wave radiation and material damping. Even with a purely elastic soil, the value of $\tilde{\zeta}$ is greater than ζ, except for a small range of parameters for

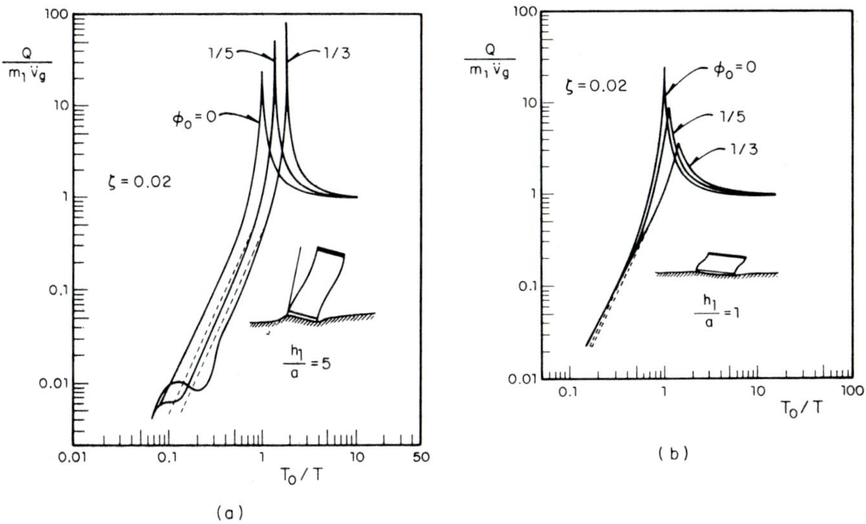

Fig. 7.15 *Steady-state harmonic response of single story building-foundation system*[7.23]

which the net effect is a decrease. When realistic values of internal damping in soil are added, $\tilde{\zeta}$ always exceeds ζ. This increase in damping, along with the lengthening of the natural period of vibration, indicates that interaction tends to *reduce* the maximum response whenever the ordinates in the design spectrum do not increase with period in the region of interest. Net *increase* in the response is likely to occur, however, if the design acceleration spectra are increasing functions of period in the period range of interest, and in areas where acceleration spectra systematically have one or more pronounced peaks.

Impedance functions Dynamic force–displacement relationships have been evaluated for a variety of rigid bases of different geometric configurations and for various idealizations of the soil medium. A sampling of typical studies is presented in the following paragraphs. More comprehensive bibliographies will be found, for instance, in References 7.3, 7.24 and 7.25.

The simplest case of foundation vibration corresponds to the different modes of excitation — vertical, horizontal, torsional, and rocking — of a rigid disk placed on the surface of an elastic halfspace[7.26,7.27]. This model of the soil medium considers the energy dissipated by wave radiation but does not account for losses due to internal friction. Material damping has been incorporated into the elastic soil model in the form of viscous damping and, more in accordance with results from dynamic tests, by hysteretic frequency-independent damping[7.28]. Both numerical and simple closed-form approximations are available for these problems. Numerical results are available as well for the case of a disk supported on a layer underlain by a viscoelastic halfspace[7.29].

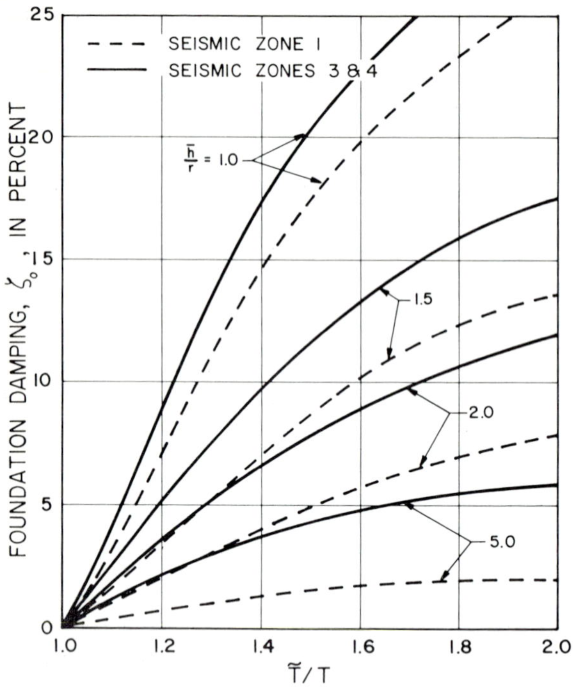

Fig. 7.16 Factor ζ_0 in Equation 7.15 for effect of foundation interaction upon apparent damping of structure[7.23]

The finite element method has been used to calculate the impedance functions of embedded foundations with circular cross-section supported on a viscoelastic layer fixed at its base[7.30]. With this model a parametric study was conducted to arrive at simplified formulas for practical application[7.31]. Contrary to the case of surface foundations, for which the coupling stiffness terms $k_{y\theta}$ and $k_{\theta y}$ and corresponding damping coefficients $c_{y\theta}$ and $c_{\theta y}$ can be neglected without significantly affecting the system response, their effects for embedded foundations are significant.

An approximate numerical procedure has been developed for the vibration of a rigid foundation of arbitrary shape on an elastic halfspace[7.32,7.33], and this procedure has been used to evaluate the vertical, rocking, and horizontal impedance functions for rectangular foundations, and the vertical impedance for a square base with an internal hole[7.33].

Experiments conducted with vertical vibration of surface and embedded model footings of various shapes indicate that the harmonic force-displacement relationship is a function of the cross-sectional area of the foundation and is practically independent of footing shape[7.34]. This property is useful in applications for it allows results that were developed originally for circular bases to be used in foundations of different cross-sections by the expedient of an equivalent radius[7.23,7.35,7.36]. A similar approach can be used for the remaining vibration modes of the foundation, with moments of inertia replacing cross-sectional area for motions involving rotations.

7.5 PILE FOUNDATIONS

Because of the complexity of the pile-soil interaction problem, present knowledge and understanding of the dynamic behavior of pile foundations is less complete than for mat foundations. General and simple design procedures for practicing engineers have only begun to be developed.

Dynamic response of an embedded pile to a prescribed excitation can be predicted if the soil reactions to the motion of the body are known. In the past, soil reactions have been defined by a subgrade modulus or Winkler continuous spring constant, a static stiffness density coefficient distributed over the length of the pile. This formulation does not take into consideration the frequency variability of the soil reactions which results from soil inertia nor the out-of-phase (damping) reactions caused by elastic wave radiation and internal friction. To incorporate these dynamic characteristics into the problem formulation, it is convenient to use a generalized, complex-valued version of the subgrade modulus.

Consider, for instance, the case in which a point-bearing pile embedded in a uniform soil layer of thickness H is excited by steady-state horizontal translation and rotation of its head in the vertical plane. The soil layer is modeled by a set of independent infinitesimally thin horizonal viscoelastic layers with material damping of the frequency-independent hysteretic type. When a pile element dz undergoes a horizontal displacement $u(z)\exp(i\omega t)$ at height z, it meets a soil reaction

$$Q = G(S_{u1} + \mathrm{i}S_{u2})u(z)\exp(i\omega t)\mathrm{d}z \tag{7.16}$$

in which G = shear modulus of soil and the term $G(S_{u1} + \mathrm{i}S_{u2})$ defines the complex-valued subgrade modulus. Separately, S_{u1} and S_{u2} (times G) represent the effective stiffness and damping density coefficients, and are functions of the dimensionless frequency of excitation $a_o = \omega r_0/V_s$, material damping, and Poisson's ratio[7.37,7.38]. G, S_{u1}, and S_{u2} do not vary with depth for the case of a uniform soil, but may depend on z for more general models[7.39 – 7.43].

Continuous lines in Figure 7.17 depict functions S_{u1} and S_{u2} evaluated for a circular pile of radius r_o embedded in an elastic soil ($D = 0$), whereas dotted lines are used to represent the same functions for a soil model with material damping[7.43]. It is seen that internal friction in the soil produces an increase in the damping coefficient S_{u2} beyond the fundamental natural frequency of the soil layer, and also a reduction in S_{u1}. This reduction may be viewed either as a decrease in the soil stiffness or as an effective mass of soil moving along with the pile. Curves for S_{u1} show, in addition, dips around particular values of a_o which correspond to the fundamental frequencies of the layers in question.

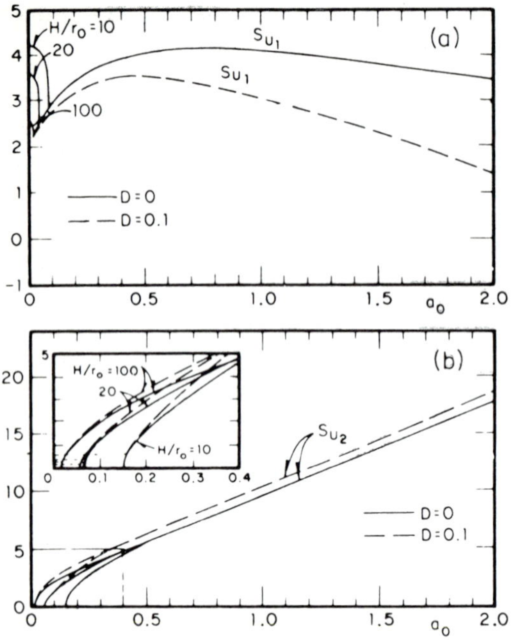

Fig. 7.17 Complex-valued subgrade modulus for viscoelastic soil (after Bielak and Palencia[7.43])

Presence of these dips shows that the effective stiffness of a pile foundation during an earthquake can be less than the corresponding static value whenever the natural frequency of the superstructure is close to that of the soil layer.

With soil reactions defined by Equation 7.16 for horizontal translation, and a similar equation for vertical motion of the pile, the problem of determining the dynamic stiffness functions at the pile head becomes equivalent to that of analyzing a beam-column on a Winkler foundation with a complex-valued subgrade modulus.

To obtain the overall impedance functions of the complete pile foundation, interaction between the individual piles must be considered. The total foundation stiffness is less than the sum of stiffnesses of the individual piles. Coupling effects between piles are usually incorporated into the analysis by introducing a group reduction factor[7.44] or an average displacement within the pile group[7.40].

The finite element method also has been used for evaluating the dynamic stiffnesses of a pile embedded in an elastic layer under a horizontal force or a moment applied at the top. From this model approximate formulas have been suggested[7.42]. An alternative formulation[7.45] uses the concept of an equivalent cantilever, in which the pile and soil reactions are replaced by a cantilever having an equivalent axial length for resisting axial loads and an equivalent bending length for resisting lateral loads and moments. It is necessary to select a nominal value of damping.

Effective excitation Again, it is useful to separate kinematic and inertial interactions. Kinematic interaction now involves determining the motions atop the piles if no structure is present. Results of computations carried out with continuous[7.46] and finite element models[7.42] indicate that relative motions between piles and the soil become important only in the case of stubby piles. This is in agreement with observations[7.47,7.48] of the field behavior of piles in recent major earthquakes which show that, in general, piles follow the soil motion. Hence, in prescribing the earthquake excitation at the base of a structure at the surface, it is valid to use the free-field motion as the effective excitation. Thus, the problem of structure-pile–soil interaction essentially reduces to one of inertial interaction. Effects of kinematic interaction must be included if there is significant embedment of the structure.

7.6 COMPARISON OF TECHNIQUES FOR ANALYSIS

By separating the treatment of kinematic and inertial interaction in Sections 7.3 and 7.4, much has been learned concerning the nature of the interaction problem. It is, of course, possible to incorporate both types of interaction into a direct analysis. Such direct analyses are almost invariably accomplished using finite element methods. Computer programs are available for both plane and axi-symmetrical cases[7.4,7.49,7.50], and a pseudo three-dimensional program is also available[7.51]. The more recent programs use transmitting boundaries at the sides of the finite element mesh. The various programs have used both time domain and frequency domain solution techniques; the latter is preferred for a linear analysis because it permits better modelling of damping by use of damping coefficients which vary inversely with frequency.

Often a first step in a direct analysis is to deconvolute the prescribed ground motion so as to find the corresponding motion at the elevation used for the bottom of the finite element mesh (see Figure 7.18), using techniques described in Chapter 1. Then this computed base motion is used as input to the finite element mesh. This procedure can lead to the appearance of very strong high frequencies in the base motion, which sometimes must be removed arbitrarily. An alternative approach, if one is working in the frequency domain, is to compute transfer functions relating the motion at the point of the free field at which the design motion is prescribed to motions at points of interest on the structure. This approach avoids having to determine a base motion. If there is no obvious elevation, such as hard rock, at which to set the bottom of the finite element mesh, it can then simply be set at an elevation such that further lowering of the bottom of the mesh has negligible effect upon the transfer functions.

It is, of course, possible to break a direct analysis down into separate steps

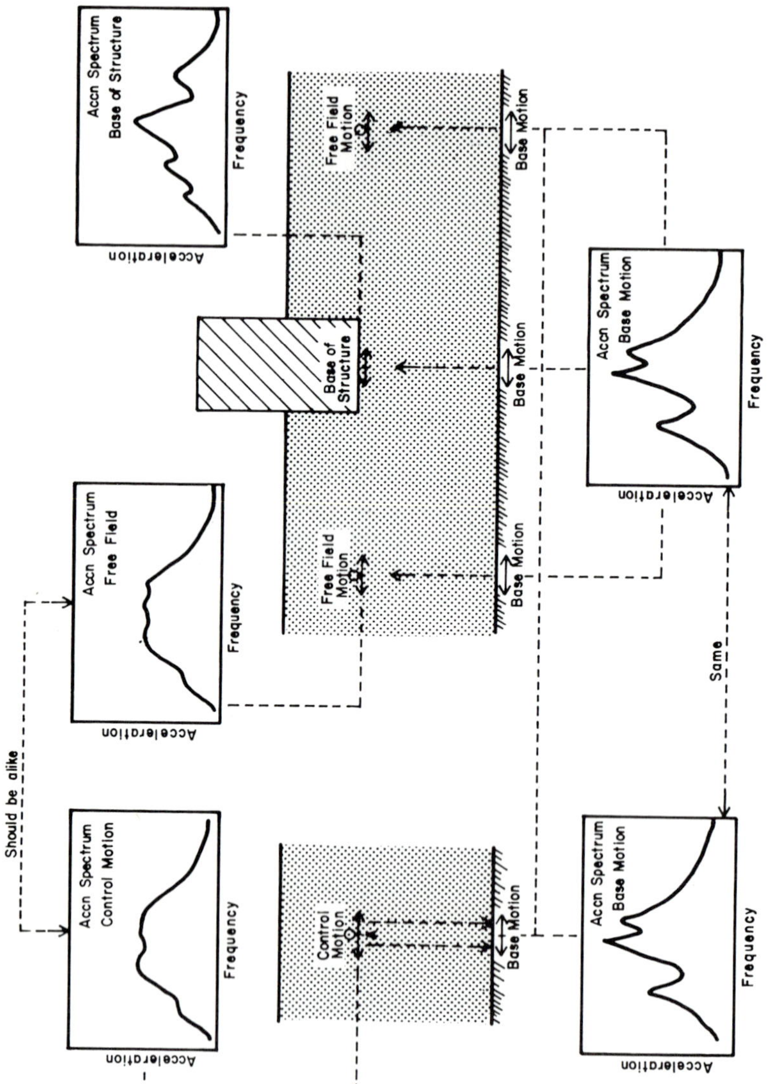

Fig. 7.18 Schematic representation of soil-structure interaction analysis (after Idriss[7.25]*)*

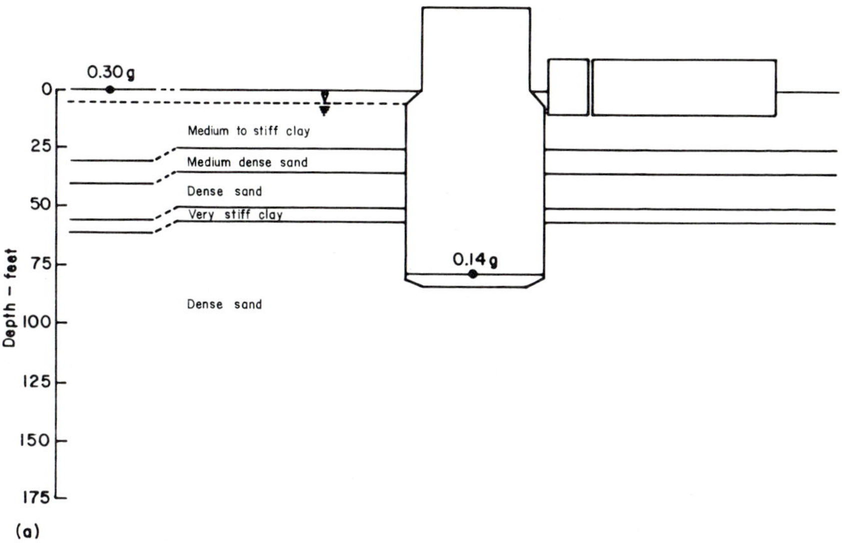

Fig. 7.19 Observed and computed motions at Humboldt Bay Power Plant (*after Valera* et al.[7.53])

which first consider kinematic interaction and then inertial interaction. Such a multi-step analysis may be advantageous if a number of different internal configurations of the structure are to be investigated; then only the inertial interaction part of the analysis need be repeated. A multi-step approach also permits use of the approximate simplifications discussed in the preceding sections.

Comparisons among methods During the past half-decade, there has been argument over the relative merits of using a complete interaction analysis, in the sense that both kinematic and inertial forms of interaction are incorporated in the analysis *vs* using simply an inertial interaction analysis for embedded structures.[7.52] There have been technical errors or shortcomings to the specific techniques advanced by both schools of thought. There is also a basic philosophical difference as to how the ground motion input to any analysis should be prescribed.

With a complete interaction analysis, one must be quite specific as to just how and where the input motion applies. As mentioned in Section 7.2, it is most common to specify that the input motion applies at the surface of the free field, and results from vertically-propagation SH waves. (Other assumptions can be made, such as saying that the specified motion results from horizontally-propagated surface waves. For most structures, the computed motions in the structure are not greatly affected by the assumed nature of the motions.) If the motion at the surface really is known, then a complete interaction analysis clearly is the best method for predicting the response of an embedded structure; if only an inertial interaction analysis is made, then

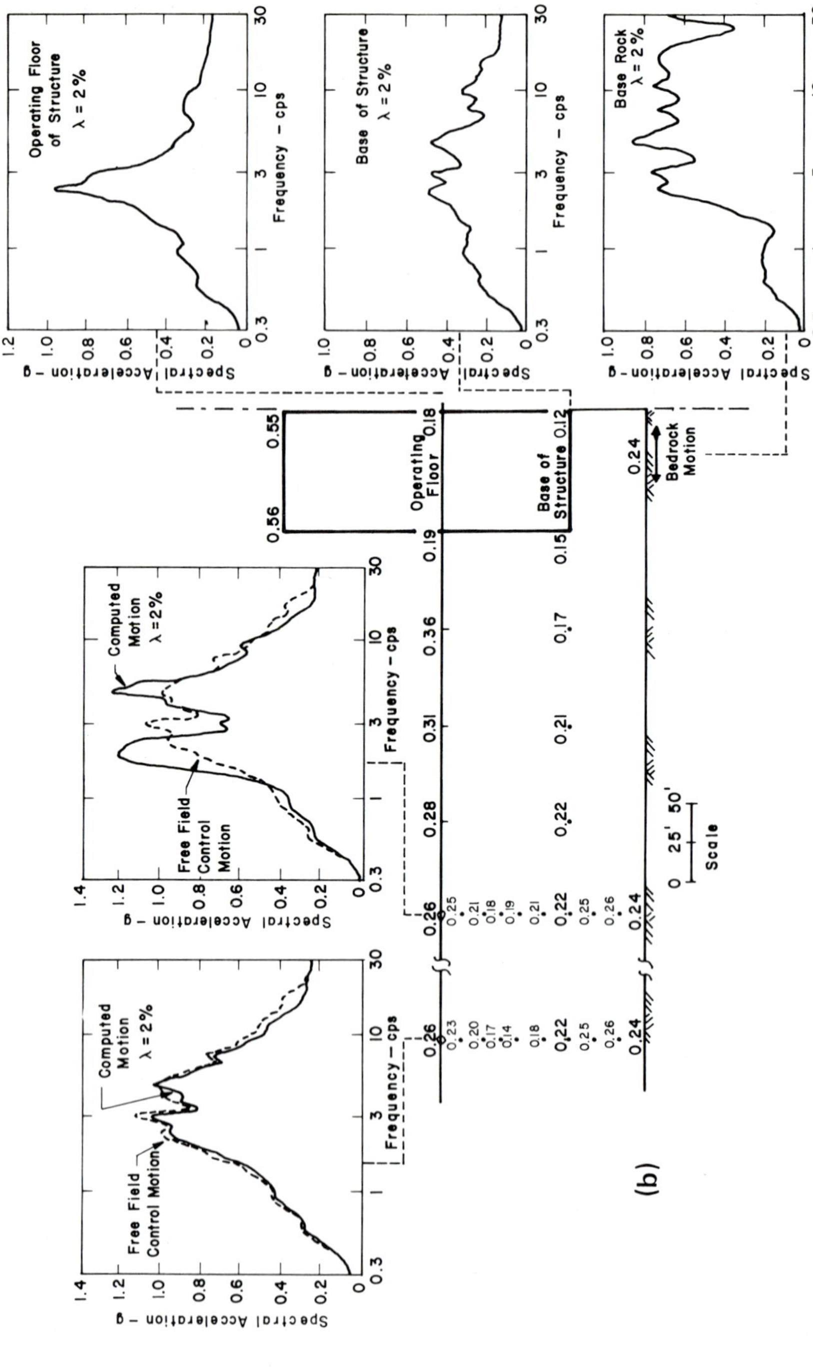

Fig. 7.19 (contd.)

TABLE 7.1 Comparison of complete and inertial interaction analysis for embedded structures

	If site amplification effects have been considered fully in choice of input motion	*If nominal site-independent input motion is used*
Complete interaction analysis	Theoretically correct	Potentially unconservative
Inertial interaction analysis	Conservative	Factors being ignored tend to compensate

an important aspect of the problem — kinematic interaction — is ignored and the result generally will be a conservative prediction of the response of the structure. This point is illustrated by the motions measured at the Humboldt Bay nuclear plant[7.53], as shown in Figure 7.19(a). The considerable decrease in peak acceleration from the ground surface to the base of the structure is predicted by a complete interaction analysis (see Figure 7.19(b)), but would not be predicted by only an inertial interaction analysis.

However, the design motions assigned for buildings or plants do not necessarily have frequency characteristics that would be computed considering site effects. A kinematic interaction analysis for an embedded structure in effect removes part of the site amplification, and unless the amplification was included when determining the surface motion, the overall result may be unconservative. Under such circumstances, it may well be better to use the assigned design motion as input to an inertial interaction analysis, assuming in effect that it is the motion in the general vicinity of the structure.

These conclusions are summarized in Table 7.1. Considering the cost of site amplification and complete interaction analyses, and the need for specialized computer programs, it seems appropriate to use only inertial interaction analyses in all but the most important projects. For some projects, kinematic interaction may be accounted for approximately using methods mentioned in Section 7.3.

Linear versus non-linear analysis In current practice, most analyses employ quasi-linear iterative techniques (see Chapter 1). Such solutions appear to be adequate as long as there is no concern about permanent displacement. If large permanent displacements or bearing capacity failures are of concern, true non-linear analyses are essential. Such analyses have been made[7.54–7.56] and techniques are still under development, but inevitably will remain complex and costly.

7.7 SITE STABILITY

At sites with poor soils, strong ground-shaking may cause permanent horizontal displacements due to lateral spreading or subsidence of the ground. Either type of movement can be damaging to structures and horizontal displacements, if large enough, can lead to collapse or total

Fig. 7.20 Failures caused by liquefaction during Niigata earthquake of 1964 (Courtesy of H. Kishida). (a) Bearing capacity failure of apartment building; (b) excessive movement of bridge piers; (c) empty tank which "floated" to surface; and (d) failure of quay wall, causing collapse of wharf building

destruction. There is also the possibility of an earthquake-induced slide from an adjacent hillside onto the site of a proposed building. This can happen in areas of low relief where the soils are poor. However, the problem is most common in steep mountainous country where rockfalls are a feature of any significant earthquake. Where conditions are especially unfavorable, such sliding or spreading begins to become a problem when peak accelerations exceed about 0.15 g during at least several cycles of motion. With much larger peak accelerations, the problem may become quite widespread.

In general, the dangers are greatest when the earth material tends to lose strength when shaken. The most troublesome situations involve loose- to medium-dense saturated sand which may liquefy and loosely jointed rock which may shake apart and lose its coherence. With such situations, flow slides and avalanches may develop where the terrain is steep, or lateral spreading of several meters or more where the terrain is gentle. On the other hand, clays and clayey granular soils which tend to deform plastically will generally experience only modest downslope movements unless the static safety factor before the earthquake is essentially unity.

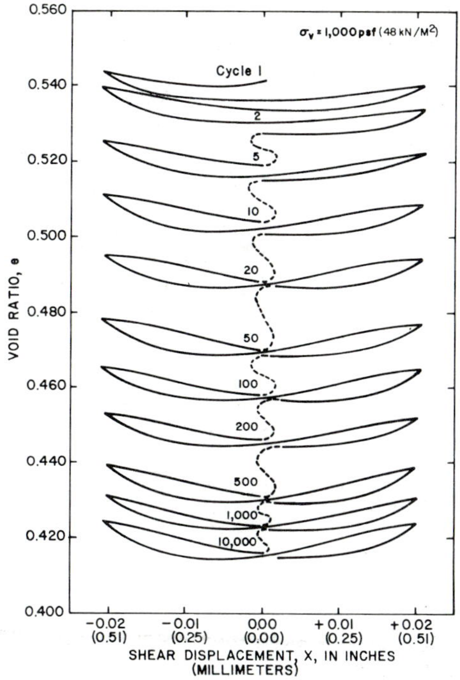

Fig. 7.21 Densification of sand by back-and-forth straining (after Youd[7.60])

Services of a specialist in soil mechanics or rock mechanics are required to assess the danger of most site stability problems. This section will attempt to provide some general guidance concerning the most widespread problem: subsidence and/or lateral spreading of loose, saturated granular soils. Figure 7.20 illustrates several failures resulting from this phenomenon; other examples are discussed in References 7.57–7.58. Seed[7.59] has provided a detailed state-of-the-art assessment.

Settlement of dry or partially-saturated granular soils Back-and-forth shearing is an effective method for densifying granular soils, perhaps the most effective of all methods. Sands having a relative density of 100%, as determined by standard tests, can often be densified further by such shearing. Figure 7.21 gives a typical result from a laboratory test; note that the amount of densification per cycle decreases as the test progresses and, eventually, an equilibrium density is reached. While this particular test involved a dry sand, the same process occurs in a partially-saturated granular soil and even in a fully-saturated soil, provided that there is time for the excess pore water to get out of the sand.

The amount of densification that occurs as the result of a fixed number of cycles increases with increasing shear strain and with decreasing density

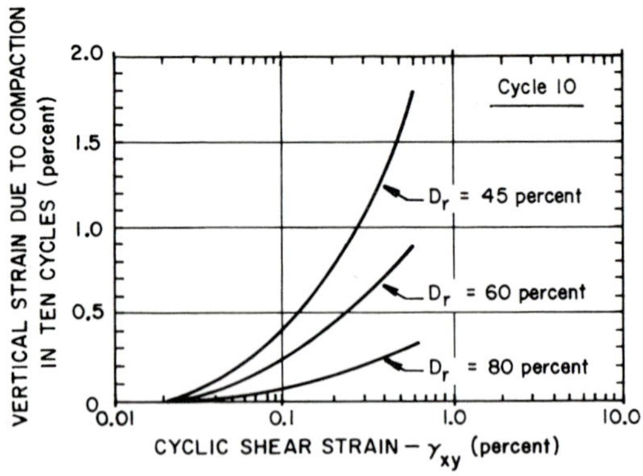

Fig. 7.22 Densification as function of relative density and shear strain for fixed number of cycles (after Silver and Seed[7.61])

prior to shearing. Figure 7.22 shows some typical results. Strains less than about 0.01% cause very little or no densification. On the other hand, strains of the order of 0.5% — which can develop during moderate-to-severe earthquakes — can cause vertical strains of more than 0.2%, even in dense soils. For a soil with a thickness of 20 m, a vertical strain of 0.2% corresponds to a settlement of 40 mm. Simultaneous straining in two horizontal directions increases the densification and, hence, settlement. Further details concerning the effects of the several variables may be found in References 7.60–7.62.

The best documented examples of earthquake-induced settlement occurred at the Jensen Filtration Plant during the San Fernando earthquake of 1971 [7.62]. Here a well-compacted fill, 17 m thick experienced settlement of 90 to 100 mm. Subsidence of fills for the approaches to bridges is a common occurrence during major earthquakes, and at least part of this subsidence must be the result of densification.

A method for calculating the densification of a soil deposit during an earthquake is available[7.63]. For such calculations, shearing strain is replaced by a related parameter $\tau_d/\bar{\sigma}_v$ in which τ_d is the cycle shear stress and $\bar{\sigma}_v$ is the vertical effective stress.

Increase in pore pressure in saturated sand If a sand is saturated and water is unable to escape from the pores, then densification cannot occur (neglecting the very small compressibility of the water). To offset the tendency for densification by a cycle of straining, there must be a decrease in the effective stress within the soil. For typical conditions, in order to counteract a densification of 0.2%, the effective stress must decrease by about 100 kN/m^2 (15 psi). Since the total stress is remaining constant, this means that the pore pressure must rise by this same amount[7.64].

When cyclic shear strain is applied to a saturated sand, the pore water pressure will vary as shown in Figure 7.23(a). During each cycle, the pore pressure fluctuates corresponding to the volume change tendencies shown in

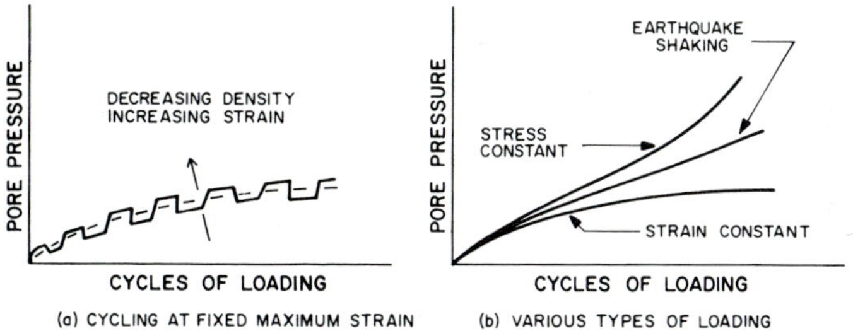

Fig. 7.23 Increase in pore pressure during cyclic loading

Figure 7.21. However, there is a steady rise in the average pore pressure, and a residual pore pressure remains when the shear straining stops. Situations that produce more rapid densification in dry sands — low relative density and large shear strains — cause rapid build-up of pore pressures in saturated sands.

When a cyclic shear stress is applied, a somewhat more complex situation develops. As the effective stress decreases, the shear modulus decreases so that the shear strain increases which, in turn, means that the rate of pore pressure increase tends to accelerate — leading to the type of curve suggested in Figure 7.23(b). This pattern of increase has frequently been observed in laboratory tests[7.65] During an earthquake ground motion, an even more complicated situation exists. Now, as the pore pressure increases and the shear modulus decreases, there is a decrease in the shear stress caused by the earthquake[7.66]. The result is a curve of pore pressure increase intermediate between that for fixed shear strain and fixed shear stress.

There are a number of factors which affect the rate of pore pressure increase in a given soil at a given relative density[7.59]. Testing apparatus and testing procedures, and also methods of sample preparation, are of major importance. Also important are such factors as the ratio of initial horizontal to vertical effective stress, the length of time that the consolidation stress has been applied, the past history of shaking, whether or not a condition of zero shear strain occurs during the cycling, and multi-directional effects. (The frequency and detailed time history of the loading function are of less importance.) These considerations, plus the great difficulty in obtaining truly undisturbed samples of saturated soils, mean that it is quite difficult to make confident predictions of pore pressure increases in actual field situations.

As yet there have been few, if any, direct observations of increased pore pressures during actual earthquakes, inasmuch as there are very few installations of piezometers to measure such increases. Increases were measured as a result of an underground nuclear explosion[7.67]. Temporary increases in flow from shallow wells have often been recorded and, no doubt, result from the processes just described.

Liquefaction If the shearing action is strong enough, the pore pressure may increase to the point where the effective stress becomes zero, at least during a

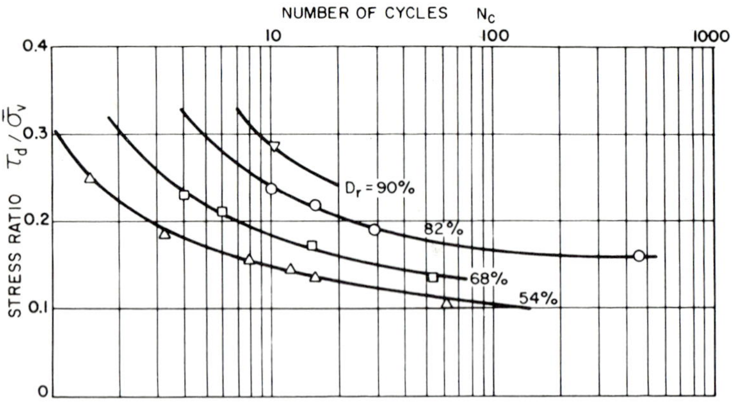

Fig. 7.24 $\tau_d/\bar{\sigma}_v$ *versus number of cycles to cause initial liquefaction* (*after DeAlba* et al.[7.66])

part of each loading cycle. Such an occurrence is termed *initial liquefaction*. Whether or not initial liquefaction occurs depends both upon the magnitude of the cyclic shear stress and upon the number of cycles of this stress, as indicated by the typical data shown in Figure 7.24. Thus, a ground motion with a high peak acceleration but a very short duration may not cause initial liquefaction, whereas such liquefaction may occur during ground motion with only moderate peak accelerations but a very long duration.

With a loose sand, initial liquefaction leads immediately to a dramatic loss of strength. Up until the point of initial liquefaction, the strains caused by the cyclic loading are small, but within just one or two additional cycles, huge strains suddenly appear, indicating that the soil has taken on liquid-like characteristics. This is *liquefaction* in every sense of the word.

With a dense sand, however, the situation is more complex. Initial liquefaction causes a change in the nature of hysteresis loop, as shown in Figure 7.25; now small strains can occur without increase in the shear stress, but with larger strains the sand begins to dilate and pore pressure drop and shear resistance builds up. With constant applied cyclic stress, the maximum strain increases slowly with additional cycles beyond initial liquefaction; with earthquake-shaking, a limiting strain may be reached[7.66]. Thus, the soil cannot flow as a liquid, although potentially important strains may occur. Names such as *cyclic mobility* and *liquefaction with limited strain potential* have been suggested to describe this situation. It is difficult to make accurate assessments of the post-initial-liquefaction condition by means of laboratory tests because non-uniform strain conditions tend to develop.

All of the many factors which affect the rate of pore pressure increase, as mentioned in the previous subsection, also affect the number of cycles required to cause initial liquefaction or any level of strain after initial liquefaction.

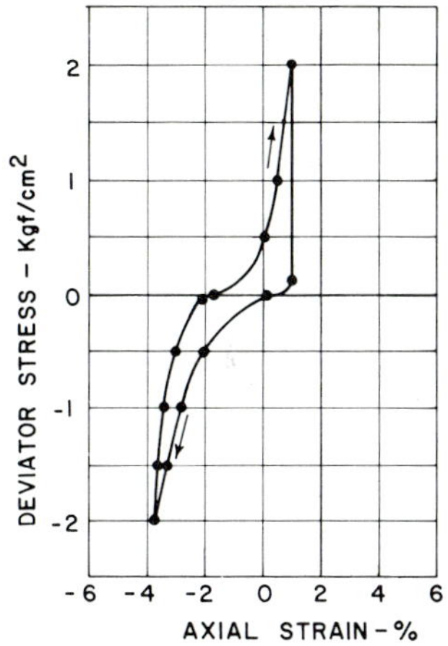

Fig. 7.25 Hysteresis loop for dense sand following initial liquefaction

Theoretical methods for evaluating liquefaction potential A substantial body of theory has developed for calculating whether or not a condition of liquefaction or cyclic mobility will develop in a given soil during a given earthquake.

The older methods presume that an element of soil in the field will be undrained and unaffected by neighboring elements. These procedures involve computation of the shear stresses that will develop in the ground during the specified earthquake and comparison with the shear stress required to cause liquefaction or cyclic mobility in the soil during the same number of cycles of shaking[7.68–7.69]. Earthquake-induced shear stresses may be determined either by analysis of waves passing vertically through the soil (see Chapter 1) or by simplified methods derived from such analyses. Resistance to liquefaction or cyclic mobility may be determined either directly by laboratory tests upon soil samples or indirectly through use of penetration resistance. In such analyses, it is necessary to convert the expected pattern of shear stress versus time into an equivalent number of uniform cycles[7.70].

More recently, analyses have been developed which take into account dissipation and spreading of pore pressures from liquefied zones[7.65,7.71]. Movement of pore water from a liquefied zone will strengthen that zone, but may lead to increased pore pressures and, hence, less strength in adjacent zones.

While there are many uncertainties in the evaluation of input parameters for these analyses, they are especially useful in helping to understand the complicated interplay among the many factors which influence the buildup of pore pressures and the resulting decreases in strength.

Empirical methods for evaluating liquefaction potential Because of the considerable difficulties involved in the direct evaluation of liquefaction

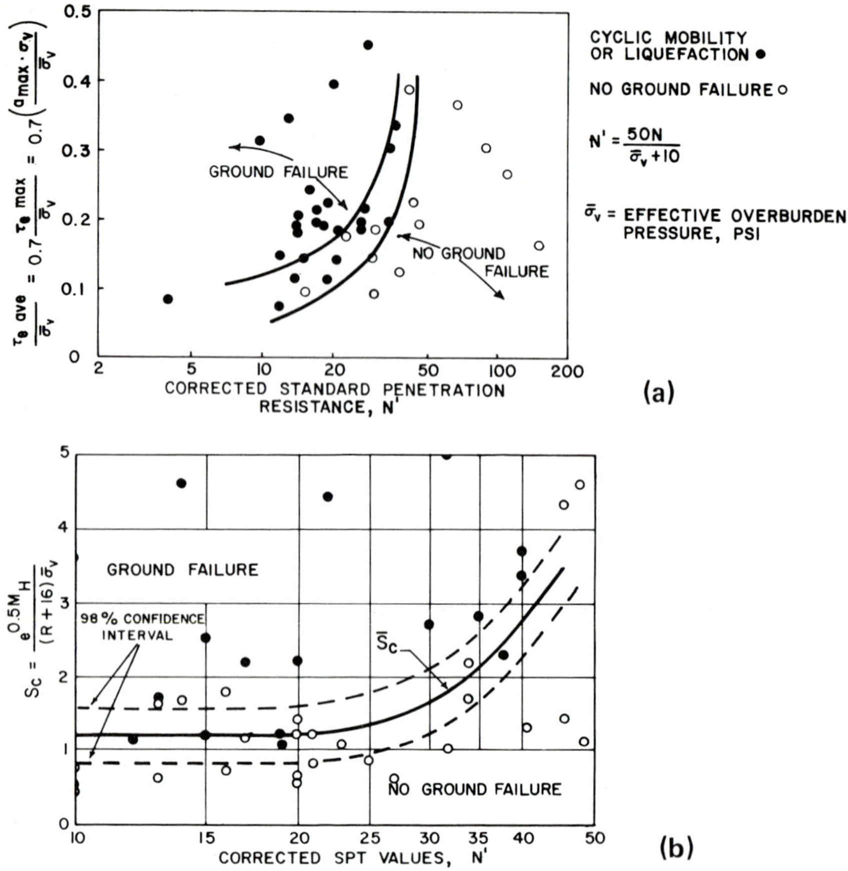

Fig. 7.26 Correlations between resistance to liquefaction and corrected standard penetration resistance. (a) After Castro[7.73] *and (b) Yegian and Whitman*[7.75]

potential using laboratory tests, it has been of considerable value to develop correlations between liquefaction potential and penetration resistance. The standard penetration test does not necessarily provide a good measurement of relative density; on the other hand, penetration resistance is determined by many of the same factors that determine liquefaction potential. Several correlations have been suggested;[7.72–7.75] Figure 7.26 shows two of them. The horizontal scale is the standard blow count N corrected for overburden stress $\bar{\sigma}_v$ (in psi), according to

$$N' = \frac{50N}{\bar{\sigma}_v + 10}$$

In Figure 7.26(a) the cyclic shear stress to cause liquefaction is given as a ratio to the vertical effective stress $\bar{\sigma}_v$; hence, the effect of the depth to the water table is taken into account. The influence of duration may be introduced by raising $\tau_d/\bar{\sigma}_v$ for a very short duration and decreasing it for a very long duration[7.72]. An alternative correlation is given in Figure 7.26(b) where now liquefaction potential is represented by a function involving magnitude and epicentral distance. Since duration is related to magnitude, the effect of duration is implicitly accounted for in this correlation.

Plots such as these must, of course, be used with judgement. They are especially useful for screening those cases where liquefaction either clearly is or clearly is not a problem, and form a basis for provisions in building codes.

Design actions Where liquefaction is found to be a problem and yet a site must still be used, the most common step is to densify the soil —either by in-place densification or by replacement and recompaction. Ohsaki[7.76] has provided a vivid account of the effectiveness of this approach: a mill founded upon densified soil came through a major earthquake well while there were large settlements in adjacent undensified areas. In cases where possible subsidence are not large enough to be of concern but initial liquefaction must be prevented, it may be feasible to use drains which will speed the dissipation of excess pore pressures[7.77]

In marginal cases, a minimum step is to adopt a type of foundation and structure that will minimize damage resulting from differential foundation movements.

7.8. DESIGN CONSIDERATIONS

There are two important general principles to keep in mind when providing foundations for earthquake-resistant structures. The first is to use as simple a foundation scheme as possible. For example, mixtures of piles and footings should be avoided. The second principle is to tie together foundations elements, such as pile caps, where necessary to ensure that relative horizontal displacements between bottoms of columns are held to an absolute minimum.

Allowable bearing stresses for direct foundations When proportioning footings for a building, the engineer usually is concerned about settlement rather than a bearing capacity failure. With partially-saturated soils, and in cases where shear deformations of the soil contribute significantly to the settlement, movement of a footing under an earthquake load is similar to that under the same load applied statically. However, for saturated soils where settlements result primarily from consolidation, movement may be much less for an earthquake loading than for the same loading if static.

With regard to earthquake loadings, it is the settlements caused by overturning moments that are of concern. In general, rotations at the foundation level act to decrease the forces experienced by a structure (although displacements in the upper stories of the structure may increase).

Moreover, soil usually is a ductile material. It must also be kept in mind that the design earthquake is an extreme event, during which a structure may well experience minor non-structural damage, and residual settlements larger than normal permissible settlements should be tolerable. For all these reasons, overstressing at the foundation level during the design earthquake is not necessarily to be feared.

However, most building codes take little recognition of these considerations. Recently-recommended design provisions[7.25] permit a foundation to be proportioned for 90% of the overturning moment required in the lowest part of the structure. It would seem that a larger reduction might be permitted in many situations. This is a very important problem, especially with regard to foundations for shear wall buildings, and further research is urgently needed.

*Deep foundations** Where surface soils are soft enough to require the use of piles or caissons, it is especially necessary to provide a foundation which acts as a unit and does not permit one column or wall to move appreciably with respect to another. Therefore, pile caps or caissons are tied together with nominal ties designed (in both tension and compression) for some percentage — usually 10% in moderately-seismic areas — of the heavier of the column loads at the end of each tie. Alternate methods of tying foundations are permitted, such as using a properly reinforced slab that can take both tension and compression. Lateral soil pressure on pile caps is *not* a recommended method because if the soil is soft enough to require piles, little reliance can be placed upon passive pressure to restrain relative displacement under dynamic conditions.

Ground-shaking is accompanied by relative horizontal displacements between different elevations in a soil deposit and, as a result, piles are distorted. Care must be taken to ensure that piles are not thereby so damaged that they cannot safely carry the dead load of the structure. Hence, unreinforced concrete piles are undesirable for earthquake-resistant design. Nominal longitudinal reinforcement — of the order of $\frac{1}{4}$ to $\frac{1}{2}$% with a minimum of four #5 bars — is specified to reduce this hazard, together with ties. Experience has shown that concrete piles tend to hinge or shatter immediately below the pile cap, so tie spacing should be reduced in this area to better contain the concrete. At the present time, there is little or no information concerning the ductility of precast-prestressed piles; pending the availability of additional data, care should be taken in the use of such piles for earthquake-resistant construction.

An obvious point is that friction piles should be avoided in materials that might liquefy during an earthquake. Also, the performance of batter pile systems has generally been poor during earthquakes, and the use of batter piles should be avoided where lateral soil movements might occur.

---

* Much of the following material is taken from Reference 7.23.

REFERENCES

7.1 Seed, H. B. and Idriss, I. M., 'Influence of Soil Conditions During Earthquakes', *J. Soil Mech. Fnds. Div. Am. Soc. civ. Engrs*, Vol. 95, No. SM1, 99–137 (1969)

7.2 Whitman, R. V., Discussion presented at *First ASCE Speciality Conference on Structural Design in Nuclear Plant Facilities*, Chicago (1973)

7.3 Seed, H. B., Whitman, R. V. and Lysmer, J., 'Soil Structure Interaction Effects in the Design of Nuclear Power Plants', *Proc. Symposium on Structural and Geotechnical Mechanics*, Univ. Illinois, Urbana (1975)

7.4 Kausel, E. and Roesset, J. M., 'Soil-Structure Interaction for Nuclear Containment Structures', *Proc. ASCE Power Division Specialty Conference*, Boulder, Colorado (1974)

7.5 Vaish, A. K. and Chopra, A. K., 'Earthquake Finite Element Analysis of Structure-Foundation Systems', *J. Engng Mech. Div. Am. Soc. civ. Engrs*, Vol. 100, No. EM6, 1101–1116 (1974)

7.6 Gutierrez, J. A., 'A Substructure Method for Earthquake Analysis of Structure–Soil Interaction', Report EERC 76–9, Earthquake Engineering Research Centre, Univ. California, Berkeley, California (1976)

7.7 Yamahara, H., 'Ground Motions During Earthquakes and the Input Loss of Earthquake Power to an Excitation of Buildings', *Soils and Foundations*, Japanese Society of Soil Mechanics and Foundation Engineering, Vol. 10, No. 2, 145–161 (1970)

7.8 Rosenblueth, E., 'The Six Components of Earthquakes', *Proc. Australian and New Zealand Regional Conference on Planning and Designing of Tall Buildings*, Sydney (1973)

7.9 Scanlan, R. H., 'Seismic Wave Effects on Soil–Structure Interaction', *Earthq. Engng. struct. Dynamics*. Vol. 4, 379–388 (1976)

7.10 Newmark, N. M., Hall, W. J. and Morgan, J. R., 'Comparison of Building Response and Free Field Motion in Earthquakes', *Proc. Sixth World Conference on Earthquake Engineering*, New Delhi (1977)

7.11 Newmark, N. M., 'Torsion in Symmetrical Buildings', *Proceedings, Fourth World Conference on Earthquake Engineering*, Santiago, Chile, Vol. 2, 19–32 (1969)

7.12 Iguchi, M., 'Seismic Response With Consideration of Both Phase Differences of Ground Motion and Soil-Structure Interaction', *Proc. Japanese Earthquake Engineering Symposium*, Tokyo, Japan (1973)

7.13 Luco, J. E., 'Torsional Response of Structures to Obliquely Incident Seismic SH Waves', *Earthq. Engng struct. Dynamics*, Vol. 4, 207–219 (1976)

7.14 Wong, H. L. and Trifunac, M. D., 'Interaction of a Shear Wall with the Soil for Incident Plane SH Waves: Elliptical Rigid Foundation', *Bull. seism. Soc. Am.*, Vol. 64, No. 6, 1825–1842 (1974)

7.15 Kausel, E., Whitman, R. V., Morray, J. P. and Elsabee, F., 'The Spring Method for Embedded Foundations', *Nucl. Engng Des.*, Vol. 48, 377–392 (1978)

7.16 Jennings, P. C. and Bielak, J., 'Dynamics of Building-Soil Interaction', *Bull. seism. Soc. Am.*, Vol. 63, No. 1, pp. 9–48 (1973)

7.17 Roesset, J. M., Whitman, R. V. and Dobry, R., 'Modal Analysis for Structures with Foundation Interaction', *J. struct. Div. Am. Soc. civ. Engrs*, Vol. 99, No. ST3, 339–416 (1973)

7.18 Bielak, J., 'Modal Analysis for Building–Soil Interaction', *J. Engng mech. Div. Am. Soc. civ. Engrs*, Vol. 102, No. EM5, 771–786 (1976)

7.19 Liu, S. C. and Fagel, L. W., 'Earthquake Interaction by Fast Fourier Transform', *J. Engng mech. Div. Am. Soc. civ. Engrs*, Vol. 97, No. EM4, 1223–1237 (1971)

7.20 Chopra, A. K. and Gutierrez, J. A., 'Earthquake Analysis of Multi-story Buildings Including Foundation Interaction', *Earthq. Engng. struct. Dynamics*, Vol. 3, 65–77 (1974)

7.21 Parmelee, R. A., Perelman, D. S. and Lee, S. L., 'Seismic Response of Multistory Structures on Flexible Foundations', *Bull. seism. Soc. Am.*, Vol. 59, 1061–1070 (1969)

7.22 Wong, H. L., Luco, J. E. and Trifunac, M. D., 'Contact Stresses and Ground Motion Generated by Soil-Structure Interaction', *Earthq. Engng. struct. Dynamics*, Vol. 5, 67–79 (1977)

7.23 Applied Technology Council, *Tentative Guidelines for Earthquake Resistant Provisions*, Palo Alto, California (1977)

7.24 Scavuzzo, R. J. and Raftopoulos, D. D., 'A Review of Soil-Structure Interaction Effects in the Seismic Analysis of Nuclear Power Plants', *Nucl. Engng. Des.*, Vol. 28, 400–413 (1974)

7.25 Idriss, I. M., (Chairman), 'Analysis for Soil-Structure Interaction Effects for Nuclear Power Plants', Reported by the Ad Hoc Group on Soil-Structure Interaction, Nuclear Structures and Materials Committee of the Structural Division, *Am. Soc. civ. Engrs* (1976)

7.26 Veletsos, A. S. and Wei, Y. T., 'Lateral and Rocking Vibration of Footings', *J. Soil Mech. Fdns Div. Am. Soc. civ. Engrs*, Vol. 97, No. SM9, 1227–1248 (1971)

7.27 Luco, J. E. and Westmann, R. A., 'Dynamic Response of Circular Footings', *J. Engng. Mech. Div. Am. Soc. civ. Engrs*, Vol. 97, No. EM5, 1381–1395 (1971)

7.28 Veletsos, A. S. and Verbic, B., 'Vibration of Viscoelastic Foundations', *Earthq. Engng. struct. Dynamics*, Vol. 2, 87–102 (1973)

7.29 Luco, J. E., 'Vibrations of a Rigid Disc on a Layered Viscoelastic Medium', *Nucl. Engng. Des.*, Vol. 36, No. 3, 325–340 (1976)

7.30 Kausel, E. and Roesset, J. M., 'Dynamic Stiffness of Circular Foundations', *J. Engng. Mech. Div. Am. Soc. civ. Engrs*, Vol. 101, No. EM6, 771–785 (1975)

7.31 Elsabee, F., Kausel, E. and Roesset, J. M., 'Dynamic Stiffness of Embedded Foundations', *Proc. ASCE Second Annual Engineering Mechanics Division Specialty Conference*, North Carolina, pp. 40–43 (1977)

7.32 Elorduy, J., Nieto, J. A. and Szekely, E. M., 'Dynamic Response of Bases of Arbitrary Shape Subjected to Periodic Vertical Loading', *Proc. International Symposium on Wave Propagation and Dynamic Properties of Earth Materials*, Univ. New Mexico, pp. 105–121 (1967)

7.33 Wong, H. L. and Luco, J. E., 'Dynamic Response of Rigid Foundations of Arbitrary Shape', *Earthq. Engng. struct. Dynamics*, Vol. 4, 579–587 (1976)

7.34 Gupta, B. N., 'Effect of Foundation Embedment on the Dynamic Behavior of the Foundations-Soil System', *Geotechnique*, Vol. 22, No. 1, 129–137 (1972)

7.35 Whitman, R. V. and Richart, F. E., Jr., 'Design Procedures for Dynamically Loaded Foundations', *J. Soil Mech. Fdns Div. Am. Soc. civ. Engrs*, Vol. 93, No. SM6, 169–193 (1967)

7.36 Bielak, J. and Palencia, V. J., 'Approximate Dynamic Analysis of Noncircular Foundations', Internal Report, Instituto de Ingeneria, Universidad Nacional Autonoma de Mexico (1977)

7.37 Novak, M., 'Dynamic Stiffness and Damping of Piles', *Can. Geotech. J.*, Vol. 11, 574–598 (1974)

7.38 Novak, M., Nogami, T. and Aboul-Ella, F., 'Dynamic Soil Reactions for Plane Strain Case', Report BLWT-1-77, Faculty of Engineering Science, Univ. Western Ontario, Canada (1977)

7.39 Tajimi, H., 'Dynamic Analysis of a Structure Embedded in an Elastic Stratum', *Proc. Fourth World Conference on Earthquake Engineering*, Santiago, Chile, Vol. 3, 53–69 (1969)

7.40 Penzien, J., 'Soil-Pile Foundation Interaction', in *Earthquake Engineering* (Edited by R. L. Wiegel), Prentice Hall, Englewood Cliffs, N.J., 349–381 (1970)

7.41 Nogami, T. and Novak, M., 'Soil-Pile Interaction in Vertical Vibration', *Earthq. Engng. struct. Dynamics*, Vol. 4, 277–293 (1976)

7.42 Blaney, G. W., Kausel, E. and Roesset, J. M., 'Dynamic Stiffness of Piles', *Proc. Second International Conference on Numerical Methods in Geomechanics*, Blacksburg, Virginia, Vol. 2, 1001–1012 (1976)

7.43 Bielak, J. and Palencia, V. J., 'Dynamic Behavior of Structures with Pile-Supported Foundations', *Proc. Sixth World Conference on Earthquake Engineering*, New Delhi (1977)

7.44 Poulos, H., 'Lateral Load-Deflection Prediction for Pile Groups', *J. Geotech. Engng. Div. Am. Soc. civ. Engrs*, Vol. 101, No. GT1, 19–34 (1975)

7.45 Singh, J. P., Donovan, N. C. and Jobsis, A. C., 'Design of Machine Foundations on Piles', *J. Geotech, Engng. Div. Am. Soc. civ. Engrs*, Vol. 103, No. GT8, 683–877 (1977)

7.46 Flores Berrones, J. R., 'Dynamic Response of End-Bearing Piles Under Earthquake Excitation' (in Spanish), *Ph.D. Thesis*, Universidad Nacional Autónoma de México, México, D.F. (1975)

7.47 Margason, E. and Holloway, D. M., 'Pile Bending During Earthquakes', *Proc. Sixth World Conference on Earthquake Engineering*, New Delhi (1977)

7.48 Sugimura, Y., 'Earthquake Observation and Dynamic Analysis of a Building Supported on Long Piles', *Proc. Sixth World Conference on Earthquake Engineering*, New Delhi (1977)

7.49 Lysmer, J., Udaka, T., Seed, H. B. and Hwang, R., 'LUSH – A Computer Program for Complex Response Analysis of Soil-Structure Systems', *Report EERC* 74-4, Earthquake Engineering Research Center, Univ. California, Berkeley, California (1974)

7.50 Berger, E., Lysmer, J. and Seed, H. B., 'ALUSH – A Computer Program for Seismic Response Analysis of Axi-Symmetric Soil-Structure Systems', *Report EERC* 75-31, Earthquake Engineering Research Center, Univ. California, Berkeley, California (1975)

7.51 Lysmer, J., Udaks, T., Tsai, C.-F. and Seed, H. B., 'FLUSH – A Computer Program for Complex Response Analysis of Soil-Structure Systems', *Report EERC* 75-30, Earthquake Engineering Research Center, Univ. California, Berkeley, California (1975)

7.52 Seed, H. B., Lysmer, J. and Hwang, R., 'Soil-Structure Interaction Analyses for Seismic Response', *J. Geotech, Engng Div. Am. Soc. civ. Engrs*, Vol. 101, No. GT5, 439–457 (1975) (See also discussions of this paper in subsequent journals)

7.53 Valera, J. E., Seed, H. B., Tsai, C. F. and Lysmer, J., 'Seismic Soil-Structure Interaction Effects at Humboldt Bay Power Plant', *J. Geotech. Engng. Div. Am. Soc. civ. Engrs*, Vol. 103, No. GT10, 1143–1161 (1977)

7.54 Scaletti, H., *Non-Linear Effects in Soil-Structure Interaction*, Ph.D. Thesis, Dept. of Civil Engineering, Massachusetts Institute of Technology, Cambridge, Mass. (1977)

7.55 Prevost, J. H. and Hughes, T. J. R., 'Mathematical Modeling of Cyclic Soil Behavior', *ASCE Earthquake Engineering and Soil Dynamics Conf.*, Pasadena, Calif., June 1978

7.56 Isenberg, J., 'Interaction between Soil and Nuclear Reactor Foundations During Earthquakes', *USAEC Report*, ORO-3822-5, Agbabian-Jacobsen Associates, Los Angeles, California (1970)

7.57 Seed, H. B., 'Landslides during Earthquakes due to Soil Liquefaction', *J. Soil Mech. Fdns. Div. Am. Soc. civ. Engrs*, Vol. 94, No. SM5, 1053–1122 (1968)

7.58 Youd, T. L. and Hoose, S. N., 'Liquefaction during 1906 San Francisco Earthquake', *J. Geotech. Engng. Div. Am. Soc. civ. Engrs*, Vol. 102, No. GT5, 425–439 (1976)

7.59 Seed, H. B., 'State-of-the-Art Report in Liquefaction Problems in Geotechnical Engineering', ASCE Annual Convention, Philadelphia, Pa. (1976)

7.60 Youd, T. L., 'Compaction of Sands by Repeated Straining', *J. Soil Mech. Fdns. Div. Am. Soc. civ. Engrs*, Vol. 98, No. SM7, 709–725 (1972)

7.61 Silver, M. L. and Seed, H. B., 'Volume Changes in Sand during Cyclic Loading', *J. Soil Mech. Fdns. Div. Am. Soc. civ. Engrs*, Vol. 97, No. SM9, 1171–1182 (1971)

7.62 Pyke, R., Seed, H. B. and Chan, C. K., 'Settlement of Sands Under Multi-Directional Shaking', *J. Geotech. Engng. Div. Am. Soc. civ. Engrs*, Vol. 101, No. GT4, 379–398 (1975)

7.63 Seed, H. B. and Silver, M. L., 'Settlement of Dry Sands during Earthquakes', *J. Soil Mech. Fdns. Div. Am. Soc. civ. Engrs*, Vol 98, No. SM4, 381–397 (1972)

7.64 Martin, G. R., Finn, W. D. L. and Seed, H. B., 'Fundamentals of Liquefaction under Cyclic Loading', *J. Geotech. Engng. Div. Am. Soc. civ. Engrs*, Vol. 101, No. GT5, 423–438 (1975)

7.65 Seed, H. B., Martin, P. P. and Lysmer, J., 'Pore-Water Pressure Changes during Soil Liquefaction', *J. Geotech. Engng. Div. Am. Soc. civ. Engrs*, Vol. 102, No. GT4, 323–346 (1976)

7.66 DeAlba, P., Seed, H. B. and Chan, C. K., 'Sand Liquefaction in Large-Scale Simple Shear Tests', *J. Geotech. Engng. Div. Am. Soc. civ. Engrs*, Vol. 102, No. GT9, 909–927 (1976)

7.67 Banister, J. R., Pyke, R., Ellett, D. M. and Winter, L., 'In-Situ Pore-Pressure Measurements at Rio Blanco', *J. Geotech. Engng. Div. Am. Soc. civ. Engrs*, Vol. 101, No. GT10, 1073–1091 (1976)

7.68 Seed, H. B. and Idriss, I. M., 'Analysis of Soil Liquefaction: Niigata Earthquake', *J. Soil Mech. Fdns. Div. Am. Soc. civ. Engrs*, Vol. 93, No. SM3, 83–108 (1967)

7.69 Seed, H. B. and Idriss, I. M., 'Simplified Procedure for Evaluating Soil Liquefaction Potential', *J. Soil Mech. Fdns. Div. Am. Soc. civ. Engrs*, Vol. 97, No. SM9, 1249–1273 (1971)

7.70 Annaki, M. and Lee, K. L., 'Equivalent Uniform Cycle Concept for Soil Dynamics', *J. Geotech. Engng. Div. Am. Soc. civ. Engrs*, Vol. 103, No. GT6, 549–564 (1977)

7.71 Finn, W. D. L., Byrne, P. M. and Martin, G. R., 'Seismic Response and Liquefaction of Sands', *J. Geotech. Engng. Div. Am. Soc. civ. Engrs*, Vol. 102, No. GT8, 841–856 (1976)

7.72 Whitman, R. V., 'Resistance of Soil to Liquefaction and Settlement', *Soils and Foundations*, Tokyo, Japan, Vol. 11, No. 4, 59–67 (1971)

7.73 Castro, G., 'Liquefaction and Cyclic Mobility of Saturated Sands', *J. Geotech. Engng. Div. Am. Soc. civ. Engrs*, Vol. 101, No. GT6, 551–569 (1975)

7.74 Seed, H. B., Mori, K. and Chan, C. K., 'Influence of Seismic History on Liquefaction of Sands', *J. Geotech. Engng. Div. Am. Soc. civ. Engrs*, Vol. 103, No. GT4, 257–270 (1977)

7.75 Yegian, M. K. and Whitman, R. V., 'Risk Analysis for Ground Failure by Liquefaction', *J. Geotech. Engng. Div. Am. Soc. civ. Engrs*, Vol. 104, No. GT7, 921–938 (1978)

7.76 Ohsaki, Y., 'Effects of Sand Compaction on Liquefaction during the Tokachioki Earthquake', *Soils and Foundations*, Tokyo, Japan, Vol. 10, No. 4, pp. 61–73 (1970)

7.77 Seed, H. B. and Booker, J. R., 'Stabilization of Potentially-Liquefiable Sand Deposits Using Gravel Drain Systems', *Report EERC* 76-10, Earthquake Engineering Research Center, Univ. California, Berkeley, California (1976)

8 Nonstructural elements

Alfred Goldberg and Eduardo A. Rukos

8.1 TYPES OF NONSTRUCTURAL ELEMENTS

For many years, seismic design criteria provided for only limited consideration of the nonstructural elements of a building[8.1,8.2]. Such elements as parapets, nonbearing walls and partitions, ornamentation, and suspended ceilings have been the usual items subject to design consideration. The extensive damage of nonstructural elements in the 1971 San Fernando[8.3] and 1972 Managua[8.4,8.5] earthquakes, together with their effects on life safety and property damage, emphasized the need for more detailed and extensive attention to a wider range of nonstructural elements in buildings.

8.1.1 Architectural systems, components, or elements

Traditional elements previously considered subject to seismic design are among those classed as 'architectural' nonstructural systems, components, or elements. This class of nonstructural elements is much broader than just walls and ceilings. It encompasses roofing units such as tiles, concrete panels, and other individually attached and relatively heavy roof elements; all exterior and interior nonstructural walls and wall elements; stairway and elevator enclosures; architectural equipment mounted on walls or ceilings or floor supported. This latter grouping includes racks, shelving, laboratory equipment, storage units, and many similar components.

8.1.2 Mechanical and electrical systems, components, or elements

With few exceptions[8.6–8.8] there has been little or no attention given to the design of mechanical or electrical systems, components, or elements in the traditional structure. Nuclear power facilities and defense installations have been designed for these elements in order to assure the safety and continuity of their structures. The increasing reliance of modern buildings upon mechanical and electrical systems requires that these systems be properly integrated into seismic design considerations.

Those elements of a building which are classed as 'mechanical nonstructural' include:

(1) Equipment such as boilers and furnaces, which use combustible or high temperature energy sources

(2) Chimneys, flues, smokestacks
(3) Material conveyor systems
(4) Duct and piping systems
(5) Fire suppression systems
(6) Machinery, including manufacturing and process
(7) Reciprocating or rotating equipment
(8) Tanks, heat exchangers and pressure vessels

Those elements of a building which are classed as 'electrical non-structural' include:

(1) Electrical bus ducts and primary cable systems
(2) Emergency electrical systems
(3) Communication systems including public broadcasting
(d) Electrical motor centers, control devices, switchgear, transformers, and unit substations
(5) Electrical panelboards
(6) Fire and smoke detection systems
(7) Lighting fixtures

8.2 DISTINCTION OF ELEMENT IMPORTANCE

Nonstructural elements of a building constitute those components that directly satisfy human needs. Loss or failure of these elements can directly affect the life safety of the structure's occupants, or those of the public that are immediately outside the building. Furthermore, loss of these elements may endanger the general populace by causing an essential emergency facility to be unuseable or hazardous. It is therefore recognized that there are three distinct concerns that the designer must consider whenever nonstructural design criteria are to be developed:

(1) Life safety
(2) Property damage
(3) Continuity of operation

8.2.1 Life safety

The primary consideration of all building regulations, seismic design standards and comparable laws, ordinances, or other enactments, is to provide for the public welfare and safety.

Structural design criteria for seismic effects attempt to prevent the structure's collapse. This in turn provides protection to the public's health and welfare. However, within and without the structure, falling ceilings, walls, or glazing; loss of the means to fight fire, evacuate structures, contain fuels; all represent life safety hazards to the public. In some facilities such as hospitals, occupants are directly subject to high life hazards by failure of minor elements of the internal nonstructural systems. In dense urban areas

Fig. 8.1 Failure of exterior wall panel

Fig. 8.2 Pedestrian endangerment by failure of exterior nonstructural elements

with considerable pedestrian traffic, failure of an exterior wall panel can endanger many lives (Figures 8.1 and 8.2).

8.2.2 Property damage

It is traditional in regulations to place a low emphasis on property damage as a concern of the public. While this is a proper approach for the government, it is not proper for the designer to ignore such problems since his client may encounter extraordinary costs from such seismic induced damage. Further, widespread property damage, particularly the costly nonstructural damage such as was experienced in the Managua and other recent earthquakes, may create a serious economic disaster for the community. We have seen increasingly the satisfactory structural performance of a building accompanied by poor nonstructural performance such that the cost of repairs approaches the building's original cost[8.9,8.10]. Though governmental regulations have only made limited mandatory rules for nonstructural seismic performance, it can be expected that increasing attention will be directed toward expanding these mandatory provisions. Public safety and welfare during and after an earthquake cannot easily be separated from effects of nonstructural property damage and its impact on the public.

8.2.3 Continuity of operations

The main concept that has emerged in recent years, primarily but not exclusively related to nonstructural element performance, is that of continuity of a facility's operation. By this is meant the requirement that a particular use or occupancy of a building, or a particular system within a structure should be designed so as to be capable of experiencing a seismic excursion and still maintain performance throughout and after the episode.

For example a hospital or an emergency disaster center should be capable of operating in a normal manner during and especially immediately after an earthquake. This capability not only would save lives of those in the buildings at the time of the earthquake, but more importantly be available to save lives in the community after the earthquake by being capable of carrying out the emergency functions for which it was designed. Similarly, there are critical systems within most buildings that are needed during and after an earthquake so as to help provide for public safety. These include emergency lighting, fire sprinkler systems, exit sign illumination, and exit stairways.

The Uniform Building Code[8.2] provides for this consideration by use of an "importance" factor exceeding 1.0 for those uses classed as essential for the post-disaster period. Another proposal[8.11] refers to "Seismic Hazard Exposure" groups with special structural design criteria mandated for the higher ranked groups, and treats the nonstructural elements based upon an "occupancy–element–performance" concept which will be presented in detail later.

In the 1971 San Fernando earthquake, two hospitals were rendered unusable and their ambulances crushed under the collapsed garage roof. In Managua the central fire station doors became inoperable, trapping the vehicles. These are but two examples of the failure to design for continuity of operations. The concept is now becoming part of most regulations and an integral part of disaster planning. Designers should be aware of their responsibility to assure that the essential services needed in a building or by a community, in the event of disaster, will be operable and available.

As will be seen, design for continuity of operations requires close interaction among the design team, since the structure, architectural elements, and electrical and mechanical systems are inter-related when operability and functionability are required.

8.3 RIGID NONSTRUCTURAL ELEMENTS

When the nonstructural element is very stiff, the accelerations coming through the supports will produce in the element a simple type of dynamic behaviour. When the supporting devices are themselves very rigid, the element will reproduce at all points the incoming accelerations, and the attachment magnification factor equals unity. If supports have certain flexibility the secondary dynamic system, represented by the nonstructural element including restraints, behaves basically as a rigid body, but certain amplification of the motion can take place.

8.3.1 Flexible support condition

In order to gauge the importance of amplification on the overall response of this type of subsystem, we must compare its periods of vibration with the dominant periods of the supporting structure motion. The nonstructural element can have, in this case, up to six degrees of freedom, which for practical purposes may be considered as uncoupled. If the mass and support stiffness of the subsystem can be taken as concentrated the degrees of freedom are reduced to three. When the mass is clearly distributed, as in a wall or stair, the degrees of freedom may be selected to uncouple the response. Practical considerations will often reduce the number of significant degrees of freedom to one or two only, because of symmetry or dominant importance of behavior. For example a properly isolated nonstructural wall may need only be reviewed for out-of-plane forces, and a boiler or pump will probably require a check of responses to one horizontal and one rotational support component only. For a rigid nonstructural element with flexible support conditions, the response may thus be approximated in many practical cases as that of a single-degree-of-freedom system having the mass of the element and the stiffness of the support, both properties related to the direction under consideration. This element will be excited by the motion of the supporting structure.

The response will then be mainly a function of the dominant frequencies contained in the motion and the frequency of the model. The supporting structure motion has a frequency content resulting from the incoming earthquake frequencies and those of the structure itself.

At medium or high elevations, the structural frequencies and in particular the fundamental one, will usually dominate over those of the earthquake motion. The attachment magnification factor may then be computed as explained in Section 8.5.4. At low elevations, structural frequencies will usually not dominate. Use of Equation 8.8 in this case is acceptable, unless a better criterion can be adopted.

8.3.2 Upper bounds on magnification factors

A one-degree-of-freedom system subjected to sinusoidal base excitation gives the magnification factor[8.12]

$$M_e = \left[\frac{1+(2\xi r)^2}{(1-r^2)+(2\xi r)^2} \right]^{1/2} \tag{8.1}$$

where $r = T_a/T$, ξ is the fraction of critical damping for the supporting mechanism, T_a is the period of the element and T the fundamental period of the supporting structure. High values of M_e occur when $r = 1$, that is, when the fundamental period of vibration of the structure coincides with that of the nonstructural element and the support damping ξ is small. Support mechanism for certain types of equipment usually provide large damping ratios. For instance, when $r = 1$, a 2% damping ratio gives $M_e = 25$, a 5% damping ratio gives $M_e = 10$. To have $M_e = 2$, a 29% damping ratio is required.

Results obtained from Equation 8.1 are always conservative if r has been properly computed, for the resonant condition assumed implies harmonic excitation of infinite duration and linear behavior of the support mechanism. A second upper limit on the magnification factor is obtained using the following expression[8.13]

$$M_e = \frac{1}{\xi_e + \xi_s + (m/M_n)^{1/2}} \tag{8.2}$$

where ξ_e = damping ratio for the nonstructural element, ξ_s = damping ratio of the structure in its nth natural mode of vibration, m = mass of element, and M_n = generalized mass of structure in nth natural mode; M_n is computed with the modal displacement vector normalized to a unit participation factor. The mode n of the structure to assume in Equation 8.2 is the one having a period closest to T_a.

Since it is assumed that $r = 1$ in deriving Equation 8.2, the actual M_e can be much smaller than the bound thus obtained.

For illustration suppose $\xi_e = 0.02$, $\xi_s = 0.05$, $r = 1$, and $m/M_n = 1/400$. Then, according to Equation 8.1, $M_e = 25$ while according to Equation 8.2, $M_e = 8.33$. On the other hand, if $\xi_e = 0.05$, $\xi_s = 0.02$, $r = 1$, and $m/M_n = 1/10000$, Equation 8.1 gives $M_e = 10$ while according to Equation 8.2, $M_e = 12.5$. In either case the lower upper bound would govern, but, as will be seen in the next section, this bound may still be too conservative.

8.3.3 Floor response spectra

More information can be extracted from the floor response spectra than from the foregoing upper bounds. These spectra are obtained using the time record of acceleration at the point of interest in the supporting structure.

When the floor acceleration spectrum is divided by the maximum floor acceleration a magnification factor spectrum is obtained. Its maximum ordinate will be a bound for the magnification factor of a single-degree model for rigid nonstructural elements with flexible supports.

The maximum floor acceleration is obtained as the response for a rigid system, i.e. as the ordinate for $T=0$ in the floor response spectrum. A set of floor response spectra for an actual multistory structure with $\xi_s=0.02$ is depicted in Figure 8.3. These floor spectra were computed for the reactor building of a nuclear plant.

The floor response spectra have been smoothed out to $\pm 15\%$ of the structural periods of vibration to account for modeling errors[8.14]. Upper bound magnification factors for the different levels are calculated as shown in Table 8.1.

The theoretical factors obtained from Equation 8.1 are 25 for $\xi_e=0.02$ and 13 for $\xi_e=0.04$. Those for equation 8.2 with $m/M_n=1/10000$ are 20 and 14.3. These high values for M_e are not attained, in fact, as resonance is not really encountered. The tabulated values are, however, large enough to merit some additional comments.

The magnification factors when multiplied by the maximum floor acceleration will provide the maximum accelerations experienced by the linear single-degree model of the nonstructural element. The equivalent force obtained, however, will often be an impractical value to use for design. There are two reasons for this. For one thing, floor spectra assume that $(m/M_n)^{1/2}$ can be neglected relative to $\xi_e+\xi_s$; as may be constructed from Equation 8.2, when this is not the case, responses of linear single-degree systems will fall below the values computed from the spectra.

Much more importantly, though, account has not been given for ductility. In accordance with the principles set forth in Chapter 3, spectral accelerations for stress computations must be reduced to account for excursion in the nonlinear range of material behavior during large amplitude seismic events. Ductility factors vary over a certain range for different structures[8.2,8.11,8.15]. An equivalent criterion developed for nonstructural element attachments will result in a considerable reduction of the magnification factors to levels where the use of the proposed values for Equation 8.8 in the following can be considered a reasonable approach.

Reduced magnification factors are computed from elastic floor magnification spectra (Figure 8.3) by taking

$$M_{e,\,elastoplastic}=\max(1,\ M_{e,\,linear}/\mu_s\ \mu_e)$$

where M_e = the peak linear magnification factor, μ_s = structure ductility factor and μ_e = element ductility factor. For instance if $\mu_s=2$ and $\mu_e=4$ for level 7 of the above mentioned structure and $\xi_e=0.02$ then $M_e=1.38$ is the elastoplastic magnification factor.

8.4 FLEXIBLE NONSTRUCTURAL ELEMENTS

When the nonstructural element is flexible, its behavior during an earthquake can no longer be properly defined by a model having a single degree of

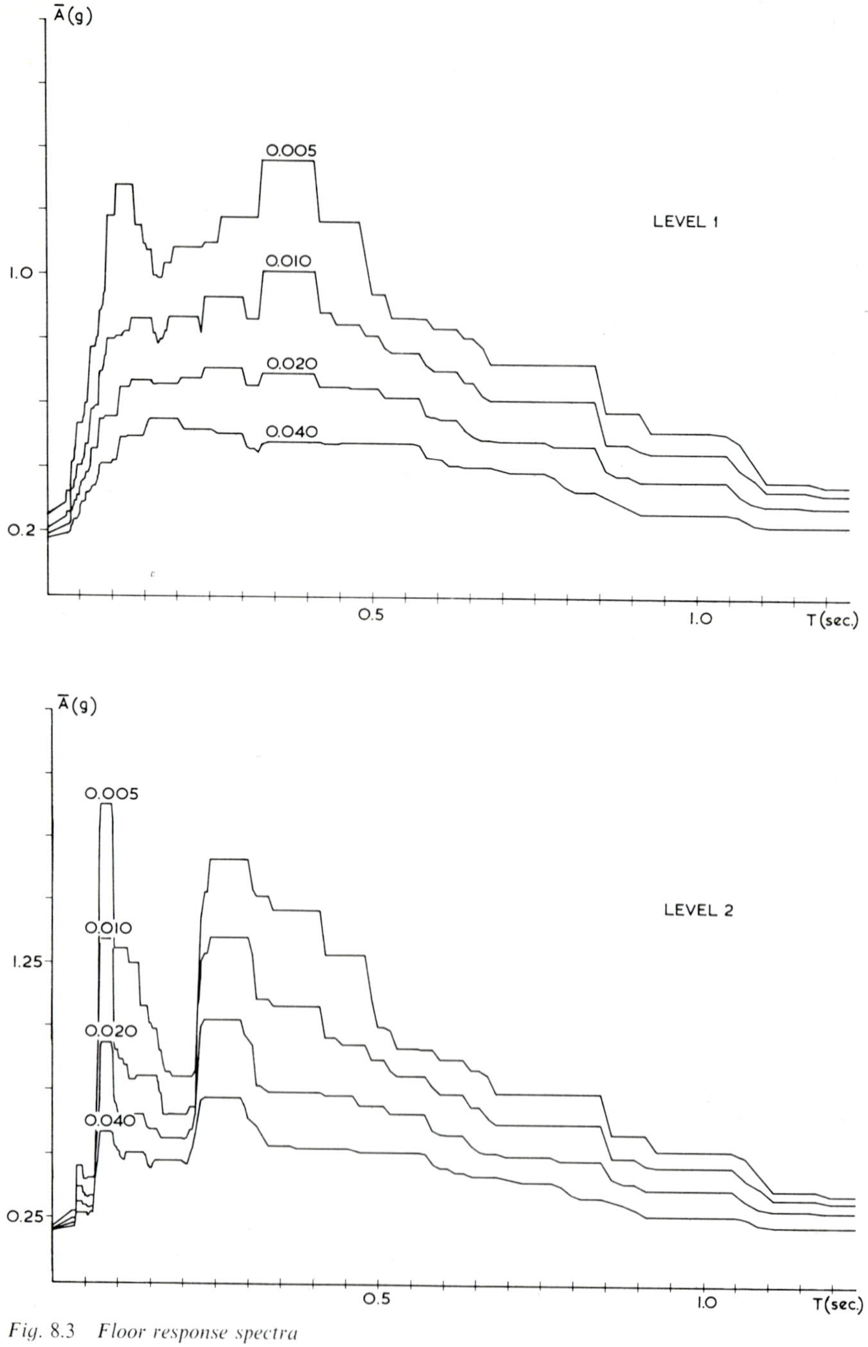

Fig. 8.3 Floor response spectra

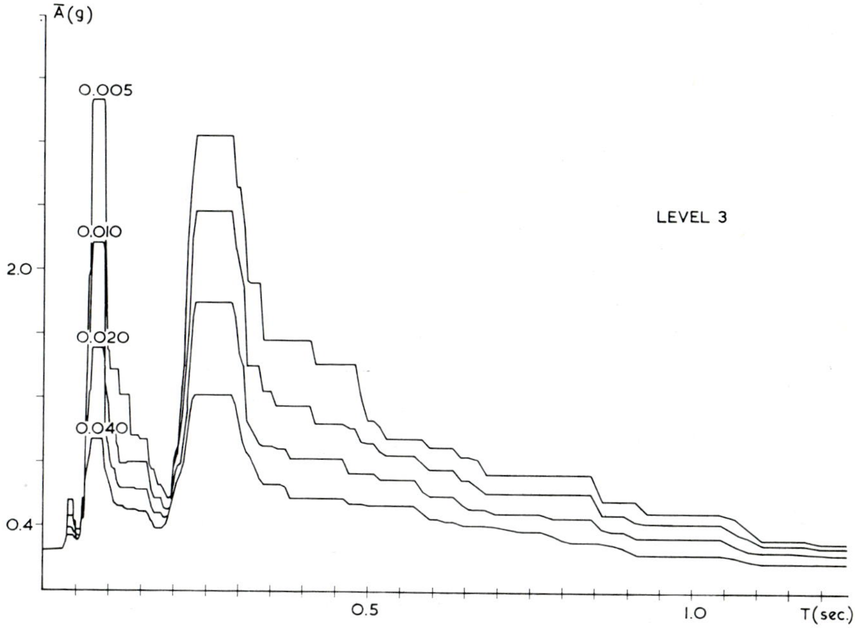
Ā(g)
0.005
0.010
0.020
0.040
2.0
0.4
LEVEL 3
0.5
1.0
T(sec.)

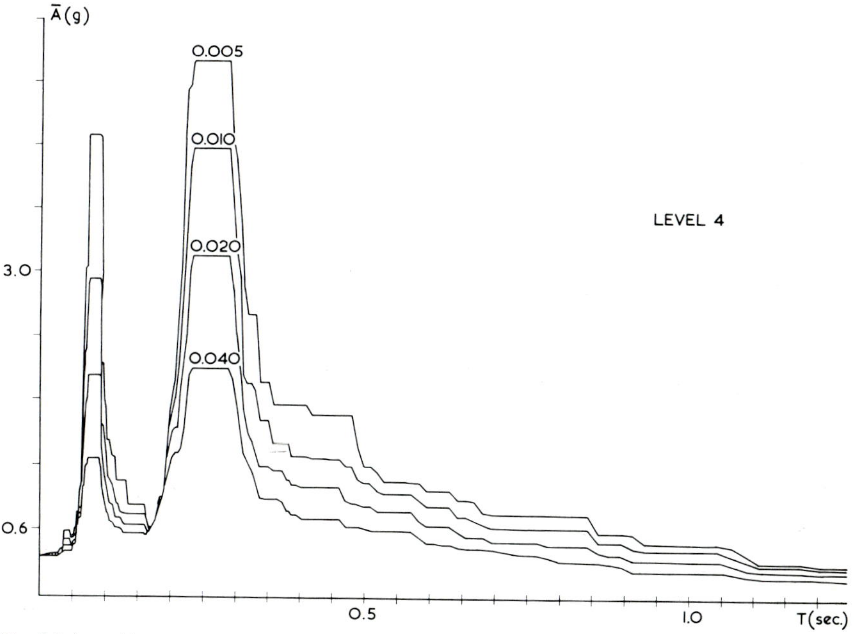
Ā(g)
0.005
0.010
0.020
0.040
3.0
0.6
LEVEL 4
0.5
1.0
T(sec.)

Fig. 8.3 (contd.)

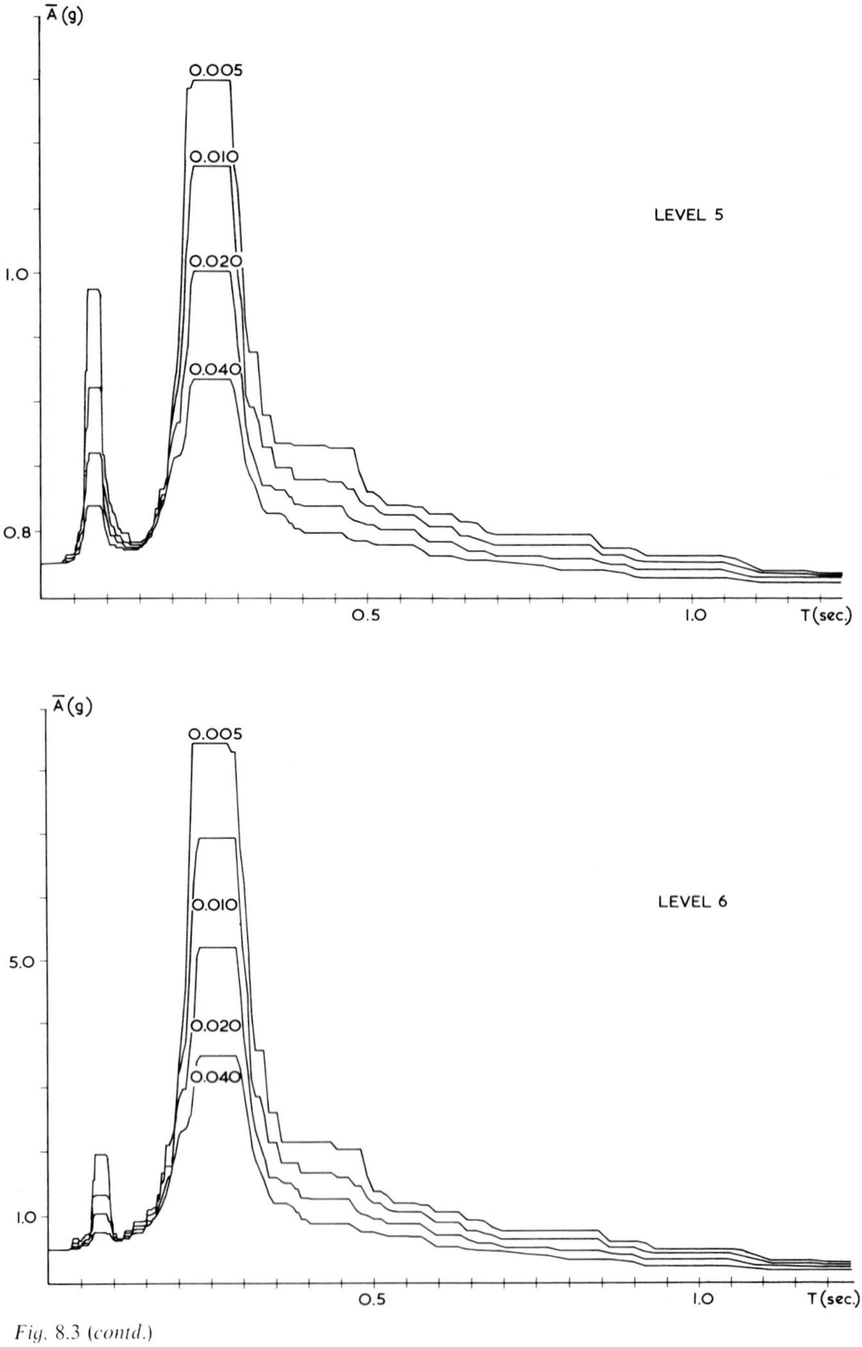

Ā (g)
0.005
0.010
0.020
0.040
LEVEL 5
1.0
0.8
0.5
1.0
T(sec.)
Ā (g)
0.005
0.010
0.020
0.040
LEVEL 6
5.0
1.0
0.5
1.0
T(sec.)

Fig. 8.3 (contd.)

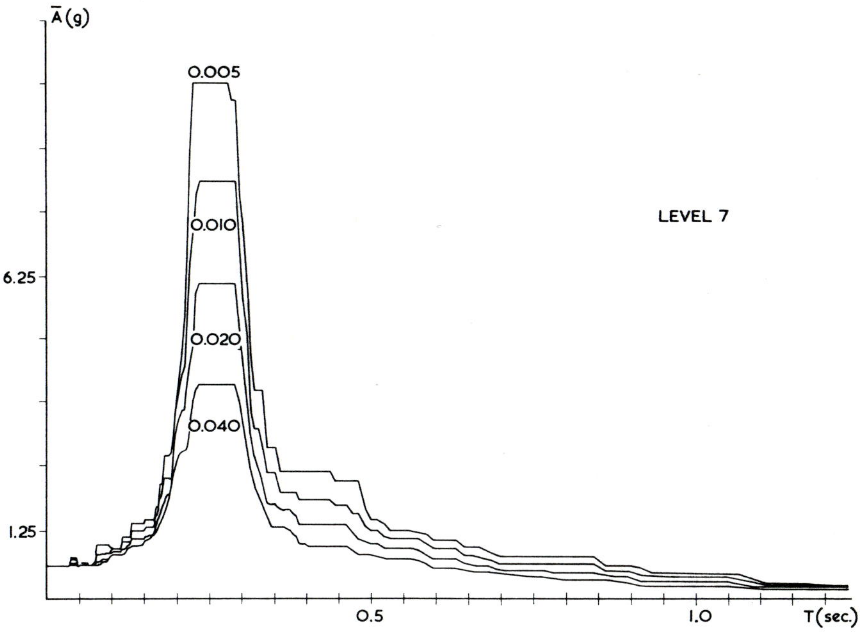

Fig. 8.3 (*contd.*)

Table 8.1

Level	$\xi_e = 0.02$	$\xi_e = 0.04$
1	3.5	2.5
2	5	3.5
3	7	5
4	9	6
5	10	6.5
6	10	7
7	11	7.5

freedom. Such is the case of a piping system, a cable tray assemblage or the internal elements of a large piece of equipment.

If the nonstructural subsystem under consideration is not important from a direct cost or from realiability of system operation or from public safety, an equivalent static analysis can be used to obtain the necessary information for design or strength requirement verification. When the element is important, a dynamic analysis should be performed using the floor response spectra as input to the subsystem. Seldom will a full dynamic analysis be required.

8.4.1 Equivalent static analysis

The subsystem should be modeled as a separate structure with fixed support conditions. The seismic coefficient to use will be the one obtained from the product of the several factors proposed in Section 8.5. This value will include the effects of the type of attachments; therefore, the nonstructural element can be modeled as fixed base condition.

Calculation of the equivalent static forces is made with the standard code formulas and the subsystem taken as fixed at ground level, but with the seismic coeficient calculated from Section 8.5. Code provisions should then be followed and the analysis procedures that apply may be those used in standard practice.

8.4.2 Dynamic analysis

A standard dynamic analysis should be performed when the nonstructural element is of importance as discussed in Section 8.2. A response spectrum solution is then acceptable. This kind of analysis is mandatory for nuclear power plants and considered good practice for facilities classified as required in the immediate post-disaster stage, i.e. hospitals.

If different supports excitation are encountered for the same structure or nonstructural element, the standard equations of dynamic have to be modified. For instance, a piping system running through different floors will be excited with a different motion at each level. A large piece of equipment restrained laterally at the top will experience there an excitation differing from its base motion. For these cases a simplified form of the modified equations of motions is[8.16,8.17].

$$M\ddot{u} + C\dot{u} + Ku = -MR\ddot{u}_g(t) \tag{8.3}$$

where $\ddot{u}_g(t)$ is a vector containing the different support accelerations and R is an influence matrix relating the support degrees of freedom to the structural ones. (See Equations 1.5 and 1.6.)

If a spectral analysis of Equation 8.3 is done, criteria for combining responses have to be applied not only to the modes but also to the excitation spectral ordinates.

8.5 GENERAL DESIGN CRITERIA

Seismic design provisions for nonstructural elements require that consideration be given to the particular element under study and to the effect its failure would have on the life safety of the public, the required continuity of the facility's function, and the overall property damage that may result as a function of public welfare. The traditional means of providing for these variables has been to establish a conservative coefficient C_p and to apply it across the board to an element regardless of its relationship to a particular location, use or occupancy.

In general, the design force is traditionally defined as:

$$F_p = C_p W_p \tag{8.4}$$

where F_p = the design force; C_p = the assigned coefficient; W_p = weight of the element under consideration. C_p usually varies from 0.10 to 2.0[8.1,8.2].

8.5.1 Occupancy–element–performance factor (*P*)

With the increasing concern since the earthquakes in 1971–73 with continuity of function for certain facilities or uses, a new factor called the "importance" factor[8.2] has been generated. The importance factor ranges from 1.0 to 1.5 based on the assigned ranking of a particular use. A variation of this approach, which more closely provides for consideration of the occupancy–element–performance concept, is that of the State of California, relating to hospitals. In this Code the "essential building" factor ranges from 1.0 to 2.5 for specific architectural nonstructural elements, systems, or components, and from 1.0 to 5.0 for specific electrical and mechanical elements, systems, and components. This approach, where a particular element, for example a suspended ceiling, is assigned a C_p value of 0.20 for nonessential buildings and a C_p value of 0.30 for essential buildings is a more rational means of providing for the relative hazard presented by the ceiling in a warehouse versus a similar ceiling in a hospital (Table 8.2).

In order to more directly relate the relative hazard of a particular element's failure to the use or occupancy involved, we must group the various occupancies together, wherein the hazard may be assumed to be comparable, and establish the several classes of building elements. A desired level of performance (*P*) of an element can then be assigned values from 0.5 to 1.5, such that the less heavily occupied uses might have the element assigned the lowest *P* value of 0.5, while in a hospital a similar or highly comparable element would be assigned the *P* value of 1.5.

The resultant equation would be:

$$F_p = C_p W_p P \tag{8.5}$$

where P = the assigned performance factor for a particular element and occupancy or use group. See Table 8.3 for factors and deflection criteria.

8.5.2 Seismic hazard exposure (SHE) groups

The term "Seismic Hazard Exposure Group" is used to denote such a grouping of uses or occupancies wherein a comparability of hazard or functional continuity is present. The simplest grouping would be two groups covering the two broad performance characteristics: essential and non-essential[8.2,8.7]. However, the non-essential group contains a broad range of uses such as warehouses for materials and high rise office buildings and hotels. It is therefore desirable to establish three groups as a minimum for

TABLE 8.2 Force factor C_p for parts or portions of buildings or other structures (portions of Table T17-23-J of State of California Title 17 – Health and Safety Code)

		Value of C_p[1]	
Parts or portions of buildings tanks, towers and equipment attached and ground supported, miscellaneous	*Direction of force*	*Essential buildings or structures*	*Non essential*[2] *buildings or structures*
All walls and partitions	Normal to flat surface	0.30[3]	0.20[3]
Cantilever parapet and other cantilever walls, except retaining walls	Normal to flat surface	1.00	1.00
When not part of a building and over 3 feet in height, masonry or concrete fences and walls	Normal to flat surface	0.30	0.30
Connections for prefabricated structural elements other than walls, with force applied at center of gravity of assembly	Any horizontal direction	0.30[7]	0.30
Exterior and interior ornamentations and appendages	Any horizontal direction	1.00[4]	1.00
Connections for exterior panels or for elements complying with Section 2314(k)5	Any horizontal direction	2.00[4]	2.00
Ceilings, including integral lighting fixtures, diffusers and attached partitions	Any horizontal direction	0.30[9]	0.20[9]
Hanging light fixtures	Any horizontal direction	1.00[4,10,12]	1.00[10,12]
Floors and roofs acting as diaphragms	Any horizontal direction	0.25[7]	0.10[7]

[1] 'C_p' shall be not less than 'F_g/W_g'. The minimum values indicated should provide considerable protection, but the use of higher C_p values is suggested for unusually important or expensive equipment not treated by a dynamic analysis, or for more conservative design, or for equipment located in the upper levels of multistory buildings. A dynamic design may be used with appropriate criteria for design and resistance.

[2] Non-essential buildings or structures are those which are not required for the complete functioning of a hospital to perform all necessary services to the public after a disaster and which are designed in accordance with Method C.

[3] See also Section 2312(b) for minimum load and deflection criteria for interior walls and partitions.

[4] Vertical 'C_p' values shall be taken as two-thirds of the horizontal 'C_p' values used.

[7] The 'W_p' shall be equal to the total load plus 25 percent of the floor live load in storage and warehouse occupancies.

[8] 'W_p' shall be the total weight plus contents.

[9] In addition to its own lateral force, suspended ceiling framing providing lateral support for walls or partitions shall be designed for the tributary reactions from all supported walls and partitions regardless of their height using the specified values for 'C_p', but not less than 5 pounds per square foot of wall. All ceiling systems shall be designed for a minimum lateral force of one pound per square foot of ceiling area. Integral lighting fixtures shall be supported separately from the ceiling system unless specifically approved otherwise.

Lightweight suspended lay-in ceiling assemblies and systems shall not be installed in rooms or spaces which exceed 300 square feet in area or in exitways unless specifically approved. When specifically shown on the approved plans, a strip of lay-in ceiling not exceeding 4 feet in width may be installed in exitways. Ceiling lay-in panels and supporting members shall not be spliced across the strip and the end of each panel shall be positively mechanically anchored to the adjacent fixed ceiling construction from below by a readily identifiable means which shall prevent the panel from falling during violent earthquake motion.

[10] Light fixtures which have passed shaking table tests approved by the Office of Architecture and Construction need not comply with this table.

[11] Emergency equipment should be located where there is the least likelihood of damage due to earthquake. Such equipment should be located at ground level and where it can be easily maintained to assure its operation during an emergency.

[12] Swinging fixtures shall have a safety cable attached to the structure and the fixture at each support capable of supporting 4 times the vertical load.

Table 8.3

*Matrix designation**	*Performance characteristic*	*P*	*Transverse bending*
S	Superior	1.5	L/240
G	Good	1.0	L/180
L	Low	0.5	L/120
N	None	–	–

* Table 8.5

assigning performance variables. These three Seismic Hazard Exposure (SHE) groups can be defined as follows[8.11]:

Group I. Buildings housing critical facilities necessary for post-disaster recovery and requiring continuoús operation during and after an earthquake.

Group II. Buildings housing dense occupancies having a high transient population, and/or sleeping conditions, or critical facilities requiring operation in the immediate post-disaster. Includes restricted movement facilities.

Group III. Low-density occupancies and generally low transient population.

Table 8.4 lists typical occupancies that would be included in each SHE group.

An example of how the element, performance, Seismic Hazard Exposure group and force factor may be related is depicted in Table 8.5 for architectural nonstructural elements and Table 8.6 for mechanical–electrical elements.

8.5.3 Seismic probability consideration

It has been the practice to design nonstructural elements using C_p values that do not provide a means for varying the design force based upon the probability of strong earthquake motion at fairly short time intervals. This approach would be too conservative for those areas of the world where medium to low earthquake motions were expected or where the time intervals would be very long between episodes. The rationale for this basis was that the cost of providing for the larger value of C_p would be negligible as regard the overall cost of the structure. For example, choice of a C_p of 0.20, 0.50, or 1.00 for parapet wall design is not sufficiently critical to warrant refinement[8.1,8.8].

Since the values of C_p will be increased by the performance factor and further modified as will be discussed later, the probability of strong earthquake motion has to be included in the design so as to avoid over-conservative and costly construction.

Factors to be used are those applicable to the structural design and are

Table 8.4 Seismic hazard exposure groupings typical occupancies or uses

Seismic hazard exposure groups

GROUP I

Buildings housing critical facilities which are necessary to post-disaster recovery and require continuous operation during and after an earthquake.

Fire facilities
Police facilities
Hospital facilities with emergency facilities
Emergency preparedness centers
Emergency communications center
Power stations and other utilities required as emergency facilities

The term "Emergency" is defined as meaning so designated by the governmental entity having jurisdiction.

GROUP II

Buildings housing high population density, occupancies having a highly transient population and/or sleeping facilities. Critical facilities requiring operation in the immediate post-disaster period. Restricted movement facilities. Large hazardous occupancies.

Public assembly for 100 or more persons
Open air stands for 2,000 or more persons
Day care centers
Schools
Colleges
Retail stores with 5,000 ft^2 floor area per floor or more than 35 ft in height
Shopping centers with covered malls, over 30,000 ft^2 gross area excluding parking
Offices over 4 stories in height or more than 10,000 ft^2 per floor
Hotels over 4 stories in height
Apartment houses over 4 stories in height
Emergency vehicle garages
Ambulatory health facilities
Wholesale stores over 4 stories in height
Factories over 4 stories in height
Printing plants over 4 stories in height
Hazardous occupancies consisting of flamable or toxic gases
Flamable or toxic liquids including storage facilities for same
Hospitals other than those in Group I

GROUP III

Buildings housing low population density occupancies and generally a less transient population.

Aircraft hangars
Factories 4 stories or less in height
Woodworking facilities
Repair garages
Service stations
Storage garages
Wholesale stores 4 stories or less in height
Printing plants 4 stories or less in height
Ice plants
Dwellings, single and two family

Table 8.4 (Continued)

Seismic hazard exposure groups
Townhouses
Retail stores less than 5,000 ft^2 per floor and 35 ft or less in height
Public assembly for less than 100 persons
Offices 4 stories or less in height or less than 10,000 ft^2 per floor
Hotels 4 stories or less in height
Apartment houses 4 stories or less in height

TABLE 8.5 Architectural force and drift performance factors

Architectural systems or components	C_p* *factor*	*Seismic hazard* I	II	*Exposure group* III
Exterior non-structural walls$_7$	0.9	S	G_2	$L_{4,8}$
Veneers	3.0	G	G_1	L_8
Roofing units	0.6	G	G_2	$N_{4,8}$
Containerized & misc. elements (free standing)	1.5	G	G	N
Stairs and shafts	1.5	S	G_3	G
Elevators and shafts	1.5	S	$L_{3,9}$	$L_{5,8}$
Vertical shafts	0.9	S	$L_{3,9}$	$L_{6,8}$
Horizontal exits	1.5	S	S	G
Public corridors includes ceilings	0.9	S	G	L_9
Private corridors	0.6	S	L_9	N
Full height separation partitions	0.9	S	G	G
Full height structural fireproofing	0.9	S	G_3	$L_{3,6,8}$
Full height other partitions	0.6	G	L_9	L_8
Partial height partitions	0.6	G	L_9	N
Ceilings — fire-rated membrane	0.9	S	G_3	G
Ceilings — non fire-rated membrane	0.6	G	G	L_8
Equipment — ceiling mounted	0.9	S	G	L_8
Equipment — wall mounted	0.9	S	G	L_8
Equipment — floor supported	0.9	S	G	L_8

[1] May be reduced one level, if landscaped 10 feet plus one foot for each floor of height. See also Note 9.
[2] May be reduced one level, if landscaped 10 feet from building face and building is only one story. See also Note 9.
[3] Must be raised one level, if building is more than three stories or 40 feet in height.
[4] Must be raised one level, if building is in urban area, system must stay attached to building.
[5] Elevator does not need to operate if building is less than 40 feet in height.
[6] Occupancy containing flammable gases, liquids and dust, must be raised one level.
[7] Connections of panelized construction to the basic building shall have a C_p of 3.0 and shall be ductile.
[8] In Seismic Hazard Zones 1 and 2, may be N.
[9] In Seismic Hazard Zone 1, may be N.
* Based on compatibility for material stresses approaching the yield level.

Table 8.6 Mechanical/electrical force level factors

Components	C_p *factor*	*Seismic hazard exposure group P factors*		
		I	*II*	*III*
Boilers, furnaces, incinerators, water heaters and other equipment using combustible or high temperature energy sources	1.5	2.0	1.0	0.5[4]
Chimneys, flues, smokestacks and vents	1.5	2.0	1.0	0.5[4]
Communications systems	1.5	2.0	1.0	0.5[4]
Conveyer systems (non-personnel)	0.75	2.0	0.0	0.0
Duct distribution systems	0.75	2.0	1.0	0.0
Electrical bus ducts and primary cable systems	1.5	2.0	1.0	0.5[4]
Electrical motor control centers, motor control devices, switchgear, transformers and substations (unit)	1.5	2.0	1.0	0.0
Electrical panelboards and dimmers	0.75	2.0	1.0	0.0
Emergency electrical systems (Code required)	1.5	2.0	2.0	2.0
Fire and smoke detection systems (Code required)	1.5	2.0	2.0	2.0
Fire suppression systems (Code required)	1.5	2.0	2.0	2.0
Life safety system components	1.5	2.0	2.0	2.0
Lighting fixtures	0.75[2]	2.0	1.0	0.5[4]
Machinery (Manufacturing or process)	0.75	2.0	1.0	0.5[4]
Piping distribution systems ($2\frac{1}{2}$ in and smaller)	0.75	2.0	0.0	0.0
Piping distribution systems (3 in and larger)	1.5	2.0	1.0	0.5[4]
Reciprocating or rotating equipment	1.5	2.0	1.0	0.5[4]
Tanks, heat exchangers and pressure vessels	1.5	2.0	1.0	0.5[4]
Utility and service interfaces	1.5	2.0[4]	1.0[4]	0.5[4]

[1] C_p values listed are for horizontal forces. Vertical C_p values shall be taken as two thirds of the horizontal.
[2] Hanging or swinging type light fixtures shall use a C_p value of 1.5 and shall have a safety cable attached to the structure and the fixture at each support point capable of supporting 4 times the vertical load.
[3] Where mechanical or electrical components are not specifically listed, the designed shall select a similarly listed component, subject to the approval of the authority having jurisdiction, and shall base the design on the performance and C_p values listed for the similar components.
[4] In Seismic Hazard Zones 1 and 2, may be 0.0

designated by the term A in the equation

$$F_p = AC_pW_pP \tag{8.6}$$

where A = the relative intensity of ground motion expected in the area for a serviceability condition design. The range of values for A may be 0.05 to 0.40[8.11]. The selected value may be different than the one used for structural design since for nonstructural elements the design is for serviceability limit states.

8.5.4 Mode of attachment of elements factor, M_e

The method of attachment of the elements to the building structural system can result in amplification of the force of the element. The magnification factor M_e depends on attachment stiffness, building period and period of the element and attachment T_a.

Using the concepts outlined in Section 8.3 a practical criteria to compute M_e is: For fixed or direct attachment of the element to the building and for seismic activated restraining devices.

$$M_e = 1$$

For elastic restraining devices, where $T_a/T < 0.6$ or $T_a/T > 1.4$,

$$M_e = 1$$

$$0.6 < T_a/T < 1.4,$$

$$M_e \doteq 2$$

$$T_a = 0.32(W_p/k)^{1/2} \tag{8.7}$$

where k = stiffness of the supporting mechanism in terms of load per unit deflection of the center of gravity (lb/in, kgf/cm). For stable resilient mountings, k = spring constant and for other resilient supports, k = slope of the load/deflection curve at the point of loading.

Theoretically the magnification factor M_e reaches very high values when T_a/T is close to 1.0. However, this, as shown in Section 8.3, is not realistic nor is it based upon observed performance of elements in actual earthquakes. Therefore, the arbitrary values of $M_e = 1.0$ and 2.0 may be used to represent the amplification characteristics of the attachments.

8.5.5 Height magnification factor, M_x

The earthquake motion affects all portions of the building to a varying degree depending upon several inter-related characteristics, including materials of construction, supporting foundation, framing system of the building, and the location of the item or element in question above the ground level. If the force applied at ground level were AC_pW_p, then as the same element was raised in the structure the coefficient would increase depending upon the overall response of the structure to the motion at the ground.

It should be noted when designing nonstructural elements that the accelerations specified for story shcars should not be used since they are values relative to the ground motion and therefore acceptable for story shears, but for nonstructural elements the total acceleration values should be used. At base level the relative acceleration is zero which will incorrectly imply the absence of earthquake forces for all nonstructural elements at such level.

Since the relative cost of nonstructural elements is included in the performance factor and the expected loss in case of underdesign is small when compared to the total cost of the structure and furthermore, it is expensive to reduce the uncertainties connected with their response, the following simple criteria for height amplification can be adopted:

$M_x = 1.0$ at the ground level or below

$M_x = 1.5$ at the roof of the building

and one may assume a straight-line variation between base and roof regardless of building height. Other design criteria establish maximum values of $M_x = 2.5$ [8.6] for multistory structures four or more stories in height. The designer should consider the overall structure response characteristics in choosing the appropriate value of M_x.

8.5.6 Design force equation: nonstructural elements

The general equation including all the variables discussed in this chapter is:

$$F_p = AC_pW_pPM_eM_x \tag{8.8}$$

where A = spectral acceleration associated with the structure's fundamental period of vibration, as a fraction of the acceleration of gravity; C_p = seismic coefficient applied to the elements (dimensionless); W_p = weight of the element; P = performance factor (dimensionless); M_e = attachment magnification factor (dimensionless), 1.0 or 2.0; M_x = height magnification factor (dimensionless), 1.0 to 1.5.

For purposes of this chapter, all values have been based upon the yield strength or equivalent of the materials involved.

For architectural elements, systems and components, M_e may be assumed to be 1.0 unless the designer desires to create an elastic attachment. In that case, the appropriate value of M_e [8.2] should be used in the design. Equation 8.8 may be reduced for architectural systems in low buildings to

$$F_p = AC_pW_pP \tag{8.9}$$

8.6 ELEMENT INTERACTION

As a result of the seismic forces acting on the base of the building, the entire structure and all its parts including nonstructural elements and furnishings are subject to deformations. These depend on the characteristics of the particular part of the building under consideration, magnititude of the seismic force, and means of connection of the particular part to the structure as a whole. These deformations include story drift, and in plane and transverse deflections.

Nonstructural elements interact with each other, for example at wall intersections and wall–ceiling intersections; with nonstructural elements of other design disciplines, such as ducts passing through walls; and with components of the structural system, as masonry nonbearing walls tightly fitted between columns. These interactions have to be considered in the design of the particular system and adequately provided for with construction details to carry out the design assumptions.

8.6.1 Interaction effects on architectural nonstructural elements

The story drift computed for the structural system may result in forces acting on nonstructural walls and partitions if these are tightly placed between structural elements. The walls will then act as shear-resisting elements and will perform as shear walls until they fail. In order to avoid such loadings on nonstructural walls, these should be separated at the top or bottom and at the sides to permit the calculated drift to occur without having the wall involved in the movement. Alternatively the walls may be offset relative to the columns, so that only the top or bottom need be separated from the structure. If on the other hand, such separations are not provided or become inoperative, the wall will be subject to these deformations and should be designed for these movements. Brittle materials, such as hollow tile, unreinforced concrete block or gypsum block, can shatter when racked by these interstory drift forces. The walls fail then to perform their primary purpose. Such is the case with a stair enclosure, an exitway wall, or an exterior wall panel.

In order to properly design nonstructural walls it is necessary to start with the interstory drift Δ_s, that the design earthquake imposes on the building. It is recommended that this drift value be multiplied by a performance factor equal to that referred to in Section 8.5.1 and Table 8.3.

The story drift primarily results in horizontal forces acting in the plane of a nonstructural wall. However, as shown in Figure 8.4, where there are cantilevered slabs or beams carrying exterior spandrels or wall panels the effect of joint rotation together with the drift Δ_s results in an offset between one spandrel and another which will place the interconnecting panel, usually glass, under bending and shear due to the differential movement of the spandrel by an amount Δ_v. This effect of the interstory drift on such systems should be taken into account in the design.

8.6.2 Interaction with the structural system

As stated in Section 8.6.1, when a stiff nonstructural wall is tightly placed in the structural framing system, it acts as a shear wall and resist the interstory drift until it fails. Similarly, other nonstructural elements, such as stairways attached directly to the columns at their platforms, can act as diaphragms and affect column response. It is important to recognize all nonstructural elements that can accidentally become force carrying members since their participation in resisting the seismic forces may change the design response with attendant failures of structural elements. Such failures have been noted in several buildings where stiffening of columns, for example by stairways or short masonry walls, has caused severe damage to columns.

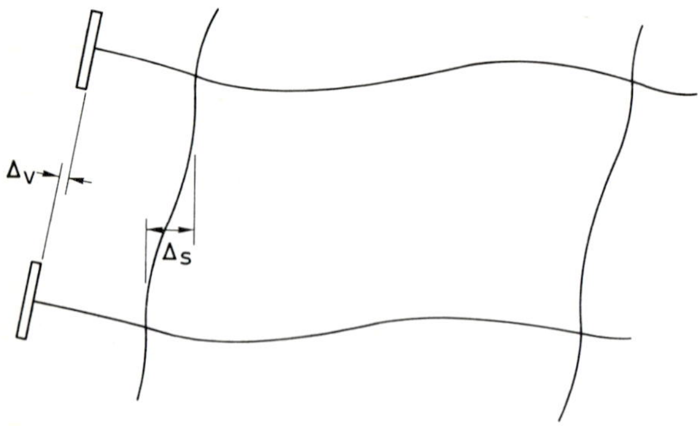

Fig. 8.4 Deflected frame

To avoid such accidental participation of nonstructural elements with the structural system requires careful attention to connection details of such elements wherever they come into contact with the structural system.

8.6.3 Provision for deflections in-plane and transverse

Design of vertical panels and walls can include provisions for separations or gaps in order to permit the computed nonstructural deformations to occur without imposition of shear on the panel. There is also need, wherever gaps are provided, to incorporate in the design, characteristics other than those directly involved in resisting earthquakes. This is the case of weather tightness for exterior panels and fire resistance and sound proofing for interior walls.

Various methods are available to provide for the exterior spaces as shown in Figure 8.5, where gap t must be equal to the modified design story drift (Δ_s × performance factor) when panels spanning single stories are used and the accumulated drift where multistory panels are involved. In addition, the degree of long-term flexibility and compressibility of the bead, rod or rope must be considered and provided for. Usually an asbestos or synthetic rope is used with a specified maximum compressibility. If this compressibility were 50% then the gap to be provided would have to be about twice the modified design story drift to prevent binding of the panels one with the other. In order to assure that the bead remains in place during installation and normal building usage, a small amount of its compressible value will be utilized during construction and to allow for cold weather widening of the joint. Therefore, the bead to be used should be designed to account for the modified design story drift divided by a percent of compressibility plus an allowance for initial construction compression and the cold weather effect. At least 1/8 in (3 mm) should be provided for as tolerance in fabrication variations.

For interior walls, the concern is to assure that the fire resistance of the assembly is not jeopardized and that sound insulation of the wall is not

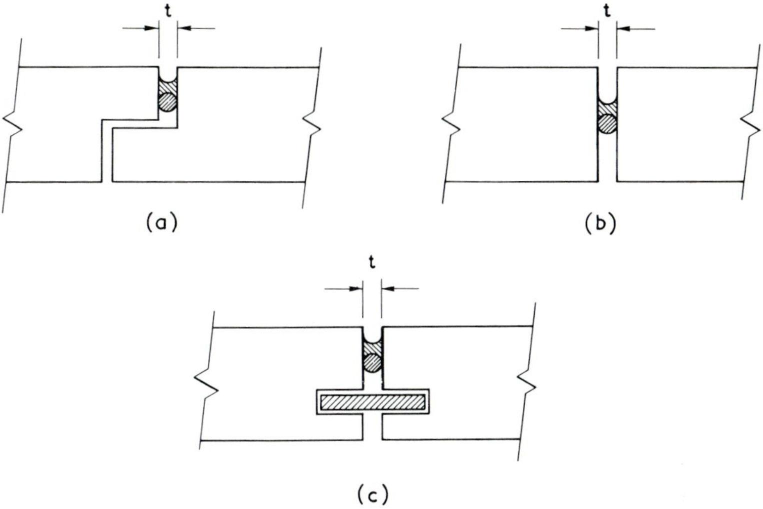

Fig. 8.5 Typical exterior panel joints; (a) shiplap type with compressible bead and sealant, (b) straight type with compressible head and sealant, (c) spline type with compressible bead and sealant

substantially reduced. In order to provide for both considerations, attention should be devoted to the details of the intersections with other elements of the building. Figure 8.6 illustrates methods of achieving the desired results.

The gap *t* in each instance in Figure 8.6 should be calculated as previously described for the exterior panels.

Similar provisions should be made for all other architectural nonstructural elements where the possibility of racking can occur that would damage or cause failure of the element.

Up to this point, we have been concerned with in-plane deflections caused by seismic drift and their effect on nonstructural architectural elements. There are also transverse deformations generated by the earthquake on every wall panel. These are the deflections due to the product of panel mass and the acceleration generated by earthquake motion at right angles to the plane of the wall as in Figure 8.7.

In order to provide for these transverse deformation effects, the panel span should be limited as set forth in Table 8.3 to avoid spalling, particularly where brittle wall components are used such as concrete, natural stone, brick or similar low-tensile strength materials. Where glass is involved, there are generally requirements already governing these installations to account for wind load and impact due to furniture or persons hitting the surface. Where such requirements do not exist, the designer is cautioned to ensure that the design avoid excessive deflection under seismic forces.

In order to assure that the design concepts for handling these deflections are fully carried out in the construction phase of the building, the required

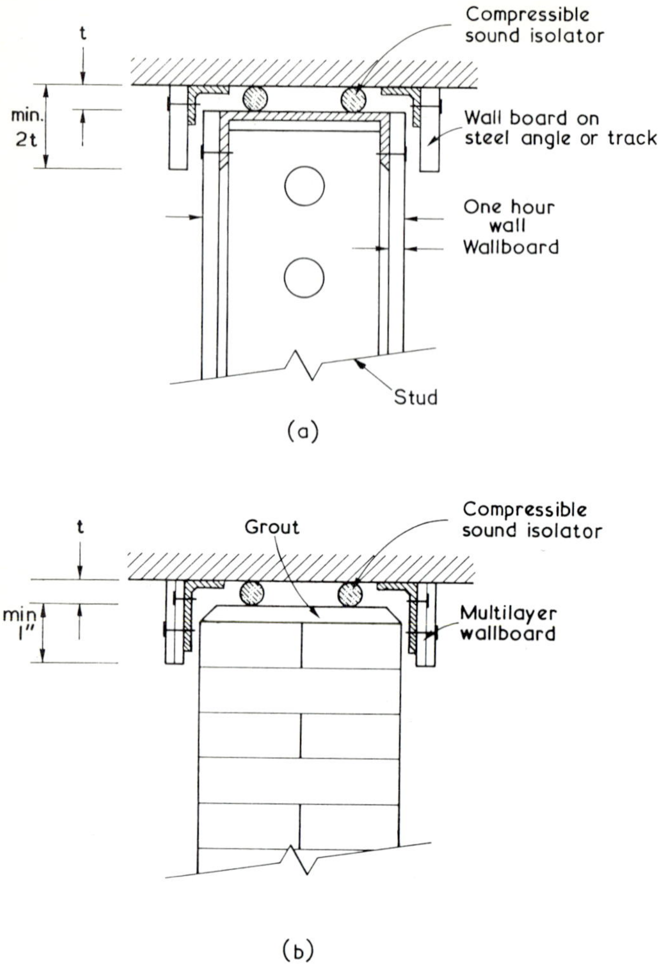

Fig. 8.6 Typical interior wall details; (a) ceiling detail for fire resistance and sound isolation, (b) masonry detail for fine resistance and sound isolation

details contemplated in the design should be shown on the drawings, with all tolerances and materials essential to the design covered in the contract documents.

8.6.4 Operational interaction of nonstructural elements

Nonstructural elements within the structure are numerous and in many instances their proper functioning is dependent upon another nonstructural element or system. For example, an exit corridor may contain two

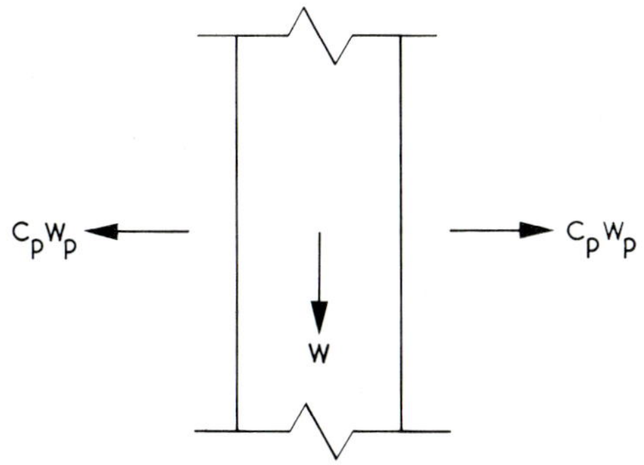

Fig. 8.7 Transverse force on wall

nonstructural walls, a suspended ceiling, emergency lighting, and exit signs. In order to perform the essential function of exiting, all these elements have to remain operative, particularly in Group I and Group II occupancies as set forth in Table 8.4.

As a general rule, where a key element adjoints a "lesser" element or one that is designed to a less restrictive performance criterion and where the failure of the lesser element would jeopardize the function of the key element, the lesser element should be designed to the same performance level as the key element. An example is a room housing the electrical switchgear of a hospital. The room partitions may be non-bearing and if not part of an exitway could be considered as low-level performance elements. But if the failure of a partition could cause failure of the switchgear and the attendant hazard to the hospital occupants, then the partitions should actually be designed to the same performance levels as the switchgear to assure its reliability.

Another area of concern involves interconnected equipment composed of several individual items tied to each other by shafts, gears, piping or other forms of operational interconnections. For the function of these elements to remain during and after an earthquake, provision must be made in the design in such a way that differences in mass of interrelated equipment, means of attachment of equipment to the structure, and the connecting devices be fully considered under dynamic loadings so that the items remain in proper spatial orientation thereby permitting their continued operation. It is not sufficient to ensure stability of each separate item of equipment on its mounts as the connecting shaft or gears may become misaligned with another piece of equipment.

8.7 CONNECTIONS, ANCHORAGES AND DETAILS

It has been noted that the methods used to attach nonstructural elements to a structure directly affect the magnitude of the forces transmitted to the element (Section 8.5.4) and the interaction that may occur due to seismic drift (Section 8.6). There are further criteria to the consider with regard the type and performance expected from the anchorage and connections incorporated in the design. These criteria include:

(1) Ductility of the connection or anchorage
(2) Connection for vertical and seismic forces
(3) Compatibility of accessory elements such as piping
(4) Uplift on anchorages

Design of connections and anchorages and their details must provide adequately for all these criteria.

8.7.1 Ductility of the connection or anchorage

Elements attached or connected to the building solely by isolated connectors, such as exterior wall panels, suspended and cantilevered elements, and other wall or ceiling supported elements, are subject to rotational forces generated by the eccentricity of the center of mass of the element and the center of support of the element. These forces are resisted by the connection or anchorage by bending or displacement within the elastic range of the attachment material. Brittle connections must clearly be avoided. See Figure 8.8.

8.7.2 Connections for combined vertical and seismic forces

With the connection or anchorage providing the sole means of attachment of some elements to the structure, such that the connection provides both the vertical load and the seismic force resistance, design should be at least conservative enough to provide for construction tolerances, unexpected loadings, greater-than-design earthquakes and the variability of workmanship. For this reason, it has been accepted practice in the United States to require a far greater design force for the attachments of exterior wall panels to the structure than for the panel itself. Thus, the panel may be designed for $C_p = 0.30$ under some conditions but the connections must be designed for $C_p = 2.0$ [8.1,8.2]. Comparable recommended criteria are contained in Table 8.5, particularly in footnote 7.

While attention has been given in the past to exterior wall panels, primarily resulting from the Anchorage earthquake damage to the J. C. Penney building, similar consideration should be given to all elements having similar characteristics. This can readily be done during the design of the connections by applying the criteria that would be used for exterior panel design. The designer should:

(1) Provide ductile connections
(2) Use a C_p for the connections 3 to 4 times larger than the C_p for the element
(3) Provide redundancy of connection and spacing of same to reduce the tendency for rotation of the center of mass

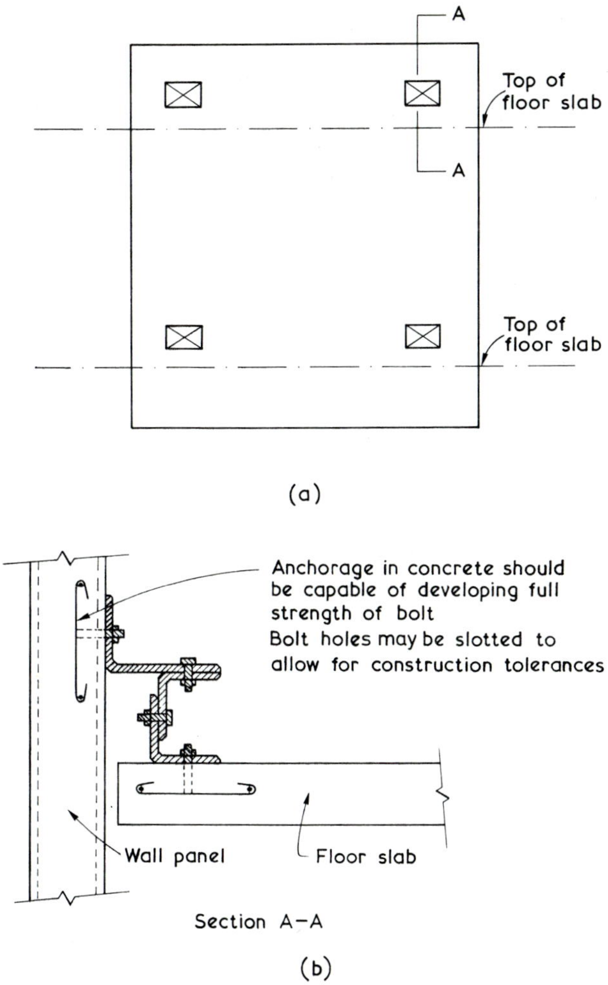

Fig. 8.8 Typical exterior panel connection; (a) typical arrangement of connection anchorages, (b) typical anchorage detail for ductility

8.7.3 Compatibility of accessory elements

In design of connections and anchorages for electrical and mechanical equipment it is necessary to also consider the accessory elements usually connected thereto. These include piping, vents, ducts, shafts, stacks, etc. Rigid or fixed mounting of a piece of equipment without either providing a flexible connection for the accessory element or comparable rigidity of attachment of the accessory can result in failure and either loss of function of the equipment

Fig. 8.9 Piping connection failure

or a creation of a secondary hazard such as fire. See Figure 8.9 (piping separation).

8.7.4 Uplift on anchorages

Earthquakes have a vertical component of movement in addition to the traditionally considered horizontal effects. Vertical motion together with the upward component due to overturn has caused considerable damage to nonstructural elements when not providing for these forces.

In order to provide for the possibility of uplift and overturn, when significant nearby earthquakes are expected, the minimum design force in the upward direction on a connection should be evaluated using not less than two-thirds of the C_p value of the element. In these cases, friction due to gravity should not be used to resist seismic motion. All connections should be by bolting, embedment, or other so-called positive means.

For rotating and oscillating equipment the possibility of the equipment natural periods coinciding with the prevailing earthquake periods should be considered and the attachments designed accordingly.

REFERENCES

8.1 'Recommended Lateral Force Requirements', Structural Engineers Association of California (1973)
8.2 'Uniform Building Code', International Conference of Building Officials (1976)

8.3 Ayres, J. M. and Sun, T., *Non Structural Damage, San Fernando Earthquake of February 9, 1971*, Vol. 1, Part B, *U.S. Department of Commerce*, National Oceanic and Atmospheric Administration, Environmental Laboratories (1973)

8.4 Marsh, R. O. and Yanev, P. I., 'Managua, Nicaragua Earthquake, December 23, 1972–Summary Report', Bechtel Power Corp. (1973)

8.5 Wright, R. N. and Kramer, S., 'Building Performance in the 1972 Managua Earthquake', *U.S. Dept. of Commerce*, National Bureau of Standards, NBS Technical Note 807 (1973)

8.6 'Seismic Design in Buildings', U.S. Depts. of Army, Navy and Air Force, TM5-809-10, Washington, D.C. (1973)

8.7 State of California Health and Safety Code, Title 17 (Hospitals) (1976)

8.8 San Francisco Building Code, Section 2314.1 (1973)

8.9 Housner, G., 'Lessons from the February 9, 1971, San Fernando Earthquake', *Proc. 40th Annual Convention Structural Engineers Association of California* (1971)

8.10 Jephcott, D. K., 'Requirements for Non-Structural Elements and Equipment – Case History of Hospitals Complex Severely Damaged in 1971 San Fernando Earthquake', *Proc. 44th Annual Convention Structural Engineers Association of California* (1975)

8.11 'Recommended Comprehensive Seismic Design Provisions for Buildings', Applied Technology Council, National Science Foundation (RANN)/National Bureau of Standards TC 3-06; NSF 78;NSB Special Publication 510

8.12 Thomson, W. T., *Vibration Theory and Applications*, Prentice-Hall, Inc. (1965)

8.13 Newmark, N. M. and Hall, W. J., 'Earthquake Resistant Design of Nuclear Power Plants', Intergovernment Conference on Assessment and Mitigation of Earthquake Risk, UNESCO (1976)

8.14 Standard Review Plan, U.S. Nuclear Regulatory Commission (1975)

8.15 Reglamento para Construcciones, Depto. del Distrito Federal, Mexico (1976)

8.16 Clough, R. W. and Penzien, J., *Dynamics of Structures*, McGraw–Hill, Inc. (1975)

8.17 Rukos, E. A., 'Response Spectrum Analysis of Multiple Supported Equipment and Components', *International Joint Petroleum Mechanical Engineering and Pressure Vessels and Piping Conference*, ASME, Mexico (1976)

8.18 Blume, J., Newmark, N. M. and Corning, L. H., *Design of Multistory Reinforced Concrete Buildings for Earthquake Motions*, Portland Cement Association (1961)

8.19 Goldberg, A., 'Seismic Design of Non-Structural Building Systems', American Society of Civil Engineers (1976)

Index

Accelerograms 2–4, 19
ACI Code 142, 143, 149, 152, 153, 156, 160–71, 184, 185, 209
AISC rules 122
Alaskan earthquake 180
American Concrete Institute 150, 155
Amplification factors 12, 31, 46
Analysis procedures 27–53
  overview 27–29
Anchor bolts 69, 136
Anchorages 286
  ductility of connection 286
  uplift on 288
Architectural nonstructural elements 281
Architectural systems, components, or elements 261, 277
Arias' scale 4
ATC recommendations 94

Base or 'overturning' moments 47
Bauschinger effect 159
Beam–column connection specimen 123, 124
Beam–column connections under cyclic loading 122
Beam–column/joints
  prestressed concrete 193
  reinforced concrete 169–73
Beam–column moment-resisting frame connection 127
Beam sidesway mechanism 164
Beams 103–9
  design 109
  instability of completely yielded 105
  lateral buckling 103
  moment reversals 190
  wide-flange 105
  see also Columns
Bearing stresses for foundations 255–56
Bending, biaxial 114
Bending moment 103, 122, 160
Bending moment diagram 107, 157
Body waves 2
Bond failure 163, 172
Braced steel frames 42
Bracing 73, 74, 132–34
Bridge piers
  foundation structure 183
  reinforced concrete 183–84
Brittle fracture 136
Brittle modes of behaviour 78

Buckling
  beams 103–7
  columns 120
  diaphragms 135
Buildings
  deformation 36
  design 37
  properties 33–36

Caracas earthquake 45
Central-difference predictor scheme 17
Centre of energy release 9
Chimney stacks 96, 136
Coefficient of variation of storey shear capacity 94
Collapse safety design conditions 83
Column flange distortion 120
Column flange thickness 121
Column sidesway mechanism 164
Column web behaviour at joint (panel web) 123
Column web buckling 120
Column web thickness 121
Columns 109–18
  bending 136
  cantilever 164
  concrete structures 167–69
  cyclic loading 116–18
  masonry 221
  prestressed concrete 192–93
  reinforced concrete 152–57
Commission on Seismic Structures 184
Compression failure 148, 153
Computer programs 28, 38
Concrete
  prestressed
    beam–column joints in 193
    building frames 184–93
    columns 192–93
    stress–strain behaviour 187
  reinforced
    beam–column joints 169–73
    columns 152–57, 167–69
    frames 57
    shear-wall buildings 41–42
    slab–column joints 173
  splitting 163
  stress-strain behaviour 144–46
Concrete structures 142–94
  axial load strength 152

Concrete structures (*contd.*)
bond and anchorage in 163–64
curvature ductility 150, 156
cyclic loading of flexural members 158–60
design loads 142
ductility 143
flexural members 165–67, 185, 190–92
flexural strength 149
moment-curvature behaviour 148
prestressed concrete building frames 184–93
prestressed concrete design philosophy 185
reinforced concrete bridge piers 183–84
reinforced concrete building frames 164–73
reinforced concrete members 148–57
reinforced concrete shear walls in buildings 173–82
rotations and deflections of members 157–58
section strength 143
strength and ductility of flexural members 185
strength and ductility of sections with flexure 148–52
strength and ductility of sections with flexure and axial load 152–57
strength of members with shear 160–63
strength design 142
ultimate curvature in member 157
Connections 119–30, 286
for combined vertical and seismic forces 286
see also Joints
Control variables 55–69
Cost-benefit analyses 70, 84
Critical damping factor 238
Cross braces 73, 74
Cross-product terms 95
Curvature diagram 157, 158
Curvature ductility 175, 182, 187
Curvature ductility factor 151, 158, 165
Cyclic loading 107, 116, 122, 158–60
Cyclic mobility 252

D'Alembert's principle 5
Damage and energy absorption 67–69
Damage probability 83
Damping, 6, 9, 15, 30, 34–36, 43, 88
Damping coefficient 241, 243
Damping ratio 9, 95, 237, 238, 266
Decision-making models 84
Deflections 45
Deformation quantities 39
Densification 249, 250, 255
Design 54–99
nature and objectives of earthquake resistant 54
optimum 92
Design force 273
and deformations 41
Design parameters 93
Design principles 70-80
Design requirements and basic principles 70
Design response spectra 29–33
Design responses 40, 45
Design spectra 86
Design values 85–89
Design variables 85
Diaphragms 76, 134–36
Displacement ductility factor 158, 164
Doppler effect 19
Drift and drift control 45, 131–32, 283
Ductile energy dissipating mechanism 165
Ductile hysteretic response 67–68
Ductile systems, failure modes in 94
Ductility 33
and damping 34–36, 88
and energy absorption 77–80
and strength 55
local and overall 59
of concrete structures 143
of connection of anchorages 286
of masonry shear walls 211–13
of members and subsystems 78–80
Ductility demands
in complex systems 64
in shear systems 65
Ductility factor 31, 34–36, 40, 43, 64, 66, 136, 158
Dynamic response of simple nonlinear systems 60–64

Earthquakes
causes of 1
characteristics of 1–26
Elastic rebound theory 1
Elastic spectra 87
Elastoplastic system 60
El Centro 184
Electrical force level factors 278
Electrical systems, components, or elements 261
Elements factor 279
Energy-absorbing devices 68
Energy absorption 211
and damage 67–69
and ductility 77–80
Energy dissipating mechanisms 214
Equivalent lateral force procedure 41–49
choice of 50–51
limitations of 49–50

Essential building factor 273
Euler buckling loads 115

Failure modes
  in ductile systems 94
  in laterally loaded reinforced concrete cantilever shear wall 174
Failure probability 83
FIP recommendations 190
Fire resistance 283
Flexible support condition 265
Flexural capacity 204
Flexural failure 149
Flexural members 158, 165–67, 185, 190–92
Flexural stiffness 203
Flexural strength 185, 187, 190
Flexural yielding 164
Floor response spectra 266–67
Force factor (C_p) 273–75, 278
Foundations 223–69
  bearing stresses 255–56
  bridge piers 183
  comparison of techniques for analysis 243–47
  complete interaction analysis 245
  deep 256
  design considerations 255–56
  direct 255
  embedded 233–34, 240, 245, 247
  impedance functions 239–41
  linear versus nonlinear analysis 247
  pile 241–43, 256
  vibration 239
  see also Site stability; Soil-structure interaction
Frame design, fully rigid 130–31
Framing systems 71–77
Free-field motions 223, 228, 229, 232, 236
Frequency range 31–33

Gaussian process 18
Geologic faults 1
Gravity loads 48–49
Ground-motion components 16–17
Ground-motion parameters 7–9
Ground motions 11–13, 19, 47
Ground-shaking 256

H rolled shapes 121
Haicheng earthquake 13
Height magnification factor 279–80
Horizontal motions 48
Housner's scale 4
Humboldt Bay nuclear plant 245, 247

Impedance functions 239–40, 242
Importance factor 273
Impulse response function 7
Inelastic systems 40
Infill panels 135–36
Instability 77, 105
Instability failure 58, 115
Instrumental scales 4
Intensity 4, 90
Interaction diagrams 181
Isolating systems 69

Jensen Filtration Plant 250
Joint core behaviour 172
Joint core hoops 171
Joint core shear force 172
Joint core shear reinforcement 171
Joints, beam-column
  prestressed concrete 193
  reinforced concrete 169–73
  see also Joints

Kanai-Tajimi power spectral density 12

L waves 2, 4
Lamellar tearing 137
Lateral displacements 132
Lateral forces 42–44
Lateral stiffness 133
Life safety 262
Linear one-dimensional filtering 13
Liquefaction 251–52
  design actions 255
  failures 249
  potential estimation 253–55
Load-deflection curves 57, 60, 63, 148, 211
Load-deflection hysteresis loops 123, 125, 126
Load-displacement curves 102
Load factors 91, 93

Magnification factors 265–67, 279
Magnitude 4
Managua 264
Masonry 195–222
  behaviour of constituent materials 198–201
  bond and anchorage 209–11
  columns 221
  compression failure 201
  construction methods 197–98
  crack formation 201
  crushing strength 201
  design codes 209
  design philosophy 203
  design shear force 206
  energy absorption 211
  energy dissipating mechanisms 214

Masonry (*contd.*)
failure forms 195
failure mechanism under uniaxial compression 199–201
flexural stiffness and building period 203
horizontal shear reinforcement 210
infilled frames 215–21
compression crushing and degradation 221
compression failure of diagonal strut 219–21
ductility potential 220
failure modes 218
influence of infill on seismic response 215
isolation of non-structural infill 215–16
reinforced tie columns 220
shear friction force 219
sliding shear failure 218
strength 218
structural infill in contact with frame 216–21
structural stiffness 216–17
tension failure mode 218
material types 195
minimum reinforcement 211
performance in earthquakes 195
reduction factors 197
reinforced 196
reinforced grouted 197, 198
reinforced hollow unit 197
reinforcement distribution for flexural strength 205–6
shear design 206–9
shear failures 206
shear walls 203–15
ductility 211
with openings 213–15
stacking bonding 198
stiffness degradation 211
stress-strain behaviour 198–99
ultimate flexural capacity 204
unreinforced 196
with shear reinforcement 208
without shear reinforcement 208
Mass and stiffness 33
Mathematical models 28
Mechanical systems, components, or elements 261
Metal bands 68
Mexico 8
Mexico City 100
Mexico City building code 77, 84, 94
Millikan Library 236, 237
Modal analysis procedure 29, 36–41, 49–51
Modal periods and shapes 37–38
Modal responses 38–39
Mode superposition method 36
Modified Mercalli scale 4, 21–23
Moment–curvature relationships 148, 158, 187, 190
Moment-panel distortion hysteresis loops 129
Moment resistant frames 101–2
buildings 41
premature failures 102
Moment rotation curves 112
MSK scale 4

Natural periods 86
New Zealand 164, 181, 184, 185
Newmark–Blume–Kapur amplification factors 12
Newmark's β method 17
Niigata earthquake 248
Nonlinear behaviour models 61
Nonlinear responses 40
Nonstructural elements 42, 261–89
architectural 281
compatibility of accessory elements 288
connections 286
deflections in-plane and transverse 282–85
design criteria 262, 272–80
design force equation 280
flexible 267–72
dynamic analysis 272
equivalent static analysis 272
interaction on structural system 282
interaction effects 281–86
operational interaction of 285–86
rigid 265–67
types of 261–62
Nonvibratory seismic effects 19–20

Occupancy-element-performance factor 273
Overturning moment 91

P waves 2, 3, 18, 224
Panel connection with beams and columns 135
Panel distortion 128
Panel joints 283
Panel stiffness and strength 135
Panel zone behaviour 127
Panel zone distortions 132
Panel zone shear stress 128
Park's stiffly stable method 17
P-delta effect 49, 156
Performance factor 278, 280, 282
Permanent loads 85
Peru 8
Pile foundations 241–43, 256
Pile-soil interactions 241
Planar models 41

Plastic deformation 103
Plastic hinges 78, 101–3, 107, 109, 158, 162, 164, 167, 168, 171, 176
Plastic moments 114
Plastic shear 127
Plate boundaries 1
Poisson process 90
Poisson's ratio 201
Pore pressure in saturated sand 250–51
Power spectral density 12–13
Power spectral density function 19
Prestressed concrete. *See* Concrete; Concrete structures
Prestressing steel 147, 185
Principal directions 16
Probability considerations 275
Probability density function 90, 91
Probability distributions 86–87, 89–90
Property damage 263

R waves 2, 4
Random-vibration theory 11–13
Recommended Comprehensive Seismic Design Provisions for Buildings 84
Reinforced concrete. *See* Concrete; Concrete structures
Reinforcing steel, stress-strain behaviour 146–47
Reliability, structural, theory of 85
Reliability analysis 89–97
Reliability function 82, 92, 93
Reliability member 93
Reliability system 93
Repair of existing structures 97
Replacement oscillators 238
Resistance-displacement curve 31
Response spectra 6, 7, 29–33, 87
Restraining devices 279
Richter magnitude 4
Rigid frames 130–31
Rolling direction 137

S waves 2, 3, 4, 18
Safety
  and uncertainty 80–84
  optimum 84
  specification 93
  structural 80–84
Safety criteria 80–97
Safety factors 85, 91
Safety levels 84, 85
Safety policies 84
San Fernando earthquake 250, 264
San Francisco 100, 101
Sand, saturation 250
Seismic effects, nonvibratory 19–20
Seismic hazard exposure groups 273–76
Seismic moment 4
Seismic-response vector 97
Seismic waves 2
  effects of topography 13
  local effects 13–16
Seismicity 90
SEOAC recommendations 45
Settlement
  earthquake-induced 250
  of dry or partially-saturated granular soils 249
SH waves 224, 245
Shear design for masonry 206–9
Shear failure 160, 162
Shear force in reinforced concrete members 160–63
Shear friction concept 177
Shear reinforcement 162
Shear walls 73, 74
  cantilever
    failure by sliding shear 177
    without openings 174–78
  coupled 178–82
  curvature ductility 175
  low-rise 178
  masonry 203–15
    ductiluty 211
    with openings 213–15
  reinforced concrete 173–82
  section detailing 177
  shear strength of multistorey 176
  squat 178
Shock-isolating devices 68
Simulation 17–19
Single-degree-of-freedom systems 5–6, 36
Site stability 247–55
  danger assessment 248–49
Slab–column joints in reinforced concrete 173
Slenderness effects 156
Slenderness limitation 176
Sliding shear failure 179
Slip-type cueves 63
Slope failures 20
Soil behaviour 15, 19
Soil compactioh 16, 19
Soil-pipe interactions 241
Soil properties 13
Soil stiffness 241
Soil-structure analysis, inertial 245
Soil–structure interaction 16, 223
  alternative formulation for deformable foundations 227–29
  definition of 223–24
  dynamic 224–34

Soil–structure interaction (*contd.*)
effects of foundation embedment 233–34
effects of foundation geometry 230–32
equations of motion 226
inertial 226, 235–41
kinematic 226, 229–34, 247
relative flexibility parameter 237
results of analysis 224
schematic representation analysis 244
techniques for analysing 224
torsion 232–33
Spandrel beams 69, 74
Spectra 6–7
for different kinds of soil 15–16
on firm ground 7–13
Spectral ordinates 9, 10, 19
Spectral response 32
Spiral columns 152
Staggered lines of defence 76
Steel structures 100–41
behaviour under strong earthquakes 100–1
bracing 132–34
brittle fracture 136
diaphragms 134–36
drift control 131–32
fully rigid frame design 130
lamellar tearing 137
moment resistant frames 101–2
structural members 102–30
Stick-slip process 1
Stiffeners 72–75, 121
Stiffness
and deformations 66
and mass 33
Stiffness degradation 57, 211
Stirrups 162–66, 177
Storey drift 281, 282
Storey forces 44–45
Storey moments 39, 44–46
Storey rotational stiffness 46
Storey shear capacity, coefficient of variation 94
Storey shears 39, 44–46
Strain compatibility and equilibrium 186
Strain rate 160
Strength reduction factor 143, 170
Strengthening of existing structures 97
Stress-strain behaviour
of concrete 144–45
of masonry 198–99
of prestressed concrete 187
of reinforcing steel 146–47
Strong column–weak girder criterion 58
Structural members 102–30
Structural steel joints 57
Superposition
of ground motion components 96
of modal contributions 95
Supplementary cross-ties 169
Surface waves 2, 3, 18

Tectonic generating mechanisms 1–2
Tension failure 148, 150
Through-thickness direction 137
Through-thickness strains 138
Tokachi-oki earthquake 232
Torsional response 45–46
Total response 39–40
Towers, 96, 136
Transfer function 12

Uncertainty and safety 80–84
Uniform Building Code 264
United States 8

Vertical component of ground motion 47
Vibration period 43
Vibration period calculation 41

Walls, interior 283
Water tanks 136
Wave trains 2
Wave velocities 2
Weld shrinkage 137
Welding of rolled shapes or plates 137

Yield point 117